Library of Congress Cataloging-in-Publication Data is available.

ISBN-13 978-1-4020-6522-4

Published by Springer,
PO Box 17, 3300 AA Dordrecht, The Netherlands

Sold and distributed in North, Central and South America
by Springer,
101 Philip Drive, Norwell, MA 02061 USA

In all other countries, sold and distributed
by Springer,
PO Box 322, 3300 AH Dordrecht, The Netherlands

Printed on acid-free paper

Printed and bound in Great Britain by MPG Books Limited, Bodmin, Cornwall.

FALK SYMPOSIUM 157

Chronic Hepatitis: Metabolic, Cholestatic, Viral and Autoimmune

Edited by

A.M. Diehl
Duke University Medical Center
Durham, NC
USA

N. Hayashi
Osaka University School of Medicine
Osaka
Japan

M.P. Manns
Medical School Hannover
Hannover
Germany

T. Sauerbruch
University Clinic Bonn
Bonn
Germany

Proceedings of the Falk Symposium 157 held in Freiburg, Germany, October 10–11 2006

Springer

Chronic Hepatitis: Metabolic, Cholestatic, Viral and Autoimmune

FALK SYMPOSIUM 157

Chronic Hepatitis: Metabolic, Cholestatic, Viral and Autoimmune

Edited by

A.M. Diehl
Duke University Medical Center
Durham, NC
USA

N. Hayashi
Osaka University School of Medicine
Osaka
Japan

M.P. Manns
Medical School Hannover
Hannover
Germany

T. Sauerbruch
University Clinic Bonn
Bonn
Germany

Proceedings of the Falk Symposium 157 held in Freiburg, Germany, October 10–11 2006

Springer

Library of Congress Cataloging-in-Publication Data is available.

ISBN-13 978-1-4020-6522-4

Published by Springer,
PO Box 17, 3300 AA Dordrecht, The Netherlands

Sold and distributed in North, Central and South America
by Springer,
101 Philip Drive, Norwell, MA 02061 USA

In all other countries, sold and distributed
by Springer,
PO Box 322, 3300 AH Dordrecht, The Netherlands

Printed on acid-free paper

Printed and bound in Great Britain by MPG Books Limited, Bodmin, Cornwall.

Contents

List of principal contributors

V Arroyo
Liver Unit
Institute of Digestive and Metabolic Diseases
Hospital Clinic
University of Barcelona
Villarroel 170
E-08036 Barcelona
Spain

HE Blum
Department of Medicine II
University Hospital Freiburg
Hugstetterstr. 55
D-79106 Freiburg
Germany

J Bruix
Hospital Clinic
University of Barcelona
BCLC Group, Liver Unit
Villarroel 170
E-08036 Barcelona
Spain

AM Diehl
Duke University Medical Center
Gastroenterology Division
Box 3256, Snyderman/GSRB-1
595 LaSalle Street
Durham, NC 27710
USA

ME Gershwin
Division of Rheumatology, Allergy and Clinical Immunology
University of California at Davis School of Medicine
GBSF Suite 6510
Davis, CA 95616
USA

P Ginès
Liver Unit
Hospital Clinic
Villarroel 170
E-08036 Barcelona
Spain

D Häussinger
Department of Gastroenterology and Hepatology
University Clinic Düsseldorf
Moorenstr. 5
D-40225 Düsseldorf
Germany

EJL Heathcote
University of Toronto
Toronto Western Hospital
Fell Pavilion 6-170
399 Bathurst Street
Toronto, Ontario M5T 2S8
Canada

N Kaplowitz
Division of Gastrointestinal and Liver Disease
Keck School of Medicine
University of Southern California
2011 Zonal Avenue, HMR 101
Los Angeles, CA 90033
USA

K Koike
Graduate School of Infection Control Sciences
Kitasato Institute for Life Sciences
Kitasato University
5-9-1 Shirokane, Minato-ku
Tokyo 108-8641
Japan

J Kondrackiene
Department of Gastroenterology
Kaunas University of Medicine
Eiveniu Street 2
50009 Kaunas
Lithuania

MK Krowka
Mayo Clinic, E-18
200 First Street SW
Rochester, MN 55905
USA

U Leuschner
Medical Centre of the Johann Wolfgang Goethe University
Stresemannallee 3
D-60596 Frankfurt-am-Main
Germany

M Nishioka
1119-20 Yashima-Nishi-Machi
Takamatsu
Kagawa 761-0113
Japan

J-M Pawlotsky
Department of Virology
Hôpital Henri Mondor
51, avenue du Marechal de Lattre de Tassigny
F-94010 Creteil
France

M Pinzani
Dipartimento di Medicina Interna
Università degli Studi di Firenze
Viale G.B. Morgagni 85
I-50134 Firenze
Italy

R Poupon
Hôpital Saint-Antoine
Service d'Hepatologie
184 rue du Faubourg St. Antoine
F-75571 Paris
France

T Sauerbruch
Medizinische Klinik I
Universitätsklinikum Bonn
Sigmund-Freud-Str. 25
D-53127 Bonn
Germany

HK Seitz
Department of Medicine and Laboratory of Alcohol Research
Liver Disease and Nutrition
Salem Medical Centre
D-69121 Heidelberg
Germany

CP Strassburg
Department of Gastroenterology, Hepatology and Endocrinology
Hannover Medical School
Carl-Neuberg-Str. 1
D-30625 Hannover
Germany

S Zeuzem
Department of Internal Medicine 1
J.W. Goethe University Hospital
Theodor-Stern-Kai 7
D-60590 Frankfurt am Main
Germany

List of chairpersons

P Berk
Mount Sinai School of Medicine
Division of Liver Diseases
Box 1039, 11th Floor, Room 23
1425 Madison Avenue
New York, NY 10029
USA

AM Diehl
Duke University Medical Center
Gastroenterology Division
Box 3256, Snyderman/GSRB-1
595 LaSalle Street
Durham, NC 27710
USA

P Ferenci
Medizinische Universität Wien
Gastroenterologie/Hepatologie
Währinger Gürtel 18-20
A-1090 Wien
Austria

G Gerken
Gastroenterologie/Hepatologie
Universitätsklinikum Essen
Hufelandstr. 55
D-45147 Essen
Germany

HJF Hodgson
Royal Free Hospital
School of Medicine
Medicine/Hepatology
Rowland Hill Street
London, NW3 2PF
UK

S Matern
Medizinische Klinik III
Universitätsklinikum Aachen
Pauwelsstr. 30
D-52074 Aachen
Germany

KH Meyer zum Büschenfelde
Trabener Str. 8
D-14193 Berlin
Germany

G Paumgartner
Innere Medizin II
Klinikum der Universität
München-Grosshadern
Marchioninistr. 15
D-81377 München
Germany

LW Powell
The University of Queensland School of Medicine
Royal Brisbane Hospital
Level 6, Ned Hanlon Building
Herston Road, Herston
Brisbane, QLD 4029
Australia

J Rodés
Hospital Clinicl y Provincial
Universidad de Barcelona
Unidad de Hepatologia
Villarroel 170
E-08036 Barcelona
Spain

W Stremmel
Innere Medizin IV
Klinikum der Universität
Im Neuenheimer Feld 410
D-69120 Heidelberg
Germany

HC Thomas
St. Mary's Hospital
Medical School
Department of Medicine
Praed Street
London, W2 1NY
UK

R Williams
University College Hospital
Medical School
Institute of Hepatology
69-75 Chenies Mews
London, WC1E 6HX
UK

JM Vierling
Baylor College of Medicine
Texas Medical Center
Baylor Liver Health
One Baylor Plaza
Houston, TX 77030
USA

Preface

The Proceedings of Falk Symposium No. 157, entitled 'Chronic Hepatitis: Metabolic, Cholestatic, Viral and Autoimmune', present an up-to-date overview on various aspects of chronic hepatitis. Chronic hepatitis is an aetiologically heterogeneous syndrome, and the various aetiologies are now clearly defined. Forty years ago, chronic hepatitis was defined on the basis of histomorphological changes under the microscope. Thanks to the introduction of molecular biology and immunology, three major hepatotropic viruses have been identified as Hepatitis B, C and D. Autoimmune hepatitis has been characterized as a disease entity on the basis of molecularly defined autoimmune markers. Recent years have seen significant progress in the understanding of toxic as well as metabolic disorders leading to chronic liver diseases. Non-alcoholic steato-hepatitis (NASH) is now the hottest topic in clinical as well as basic hepatology.

Apart from autoimmune hepatitis, cholestatic liver diseases, in particular, primary sclerosing cholangitis and primary biliary cirrhosis are also regarded as autoimmune liver diseases. Several chapters in these proceedings deal with the epidemiology and pathobiology of cholestatic autoimmune liver diseases. Research in viral hepatitis has presented us not only with the identification of all the major hepatotropic viruses but has also given us diagnostic tools as well as novel therapies that allow us, not only to diagnose, but also to treat chronic hepatitis with a viral aetiology.

Chronic hepatitis leads to end-stage liver disease with its various complications, including hepatocellular carcinoma. Therefore, a State-of-the-Art Lecture on the 'Pathogenesis of Hepatic Encephalopathy' was presented in honour of Prof. Wolfgang Gerok's 80th birthday. The State-of-the-Art chapter by Dieter Häussinger and colleagues gives an up-to-date overview of novel concepts in the pathogenesis of hepatic encephalopathy. The Sheila Sherlock Award honouring outstanding international scientists in the field of clinical hepatology was given this time to Jenny Heathcote from Toronto, Canada. Her award lecture entitled 'Past, Present and Future of Hepatology' was an absolute highlight of the Symposium; therefore, her chapter in these proceedings also highlights the clinical discipline of hepatology in its present state.

Complications of end-stage liver disease as a consequence of clinic hepatitis have been investigated for many years by various groups around the world. Significant advantages have been achieved in the understanding and treatment of type 1 and type II hepatorenal syndrome. The understanding of its pathophysiology has given us new therapeutic approaches. Hepatocellular carcinoma nowadays is the most complicated of clinical end-stage liver diseases. It is the fastest growing tumour population in the world. Very little emphasis has been given, on the contrary, to the hepatopulmonary syndrome: it may be the most neglected consequence of chronic hepatitis and liver cirrhosis. Therefore, hepatology must give more attention to this often life-threatening syndrome. The editors hope that the readers will find these proceedings of the Falk Symposium 157 a valuable source of knowledge for understanding the aetiologically heterogeneous syndrome of chronic hepatitis. Hepatology has become a fast-growing model clinical entity where 'bench to bedside' approaches have overwhelmingly shaped the everyday management of our patients.

AM Diehl, N Hayashi, MP Manns, T Sauerbruch

Section I
Metabolic and toxic hepatitis

Chair: HJF HODGSON and JM VIERLING

1
NASH: bench to bedside – lessons from animal models

A. M. DIEHL and M. F. ABDELMALEK

INTRODUCTION

Non-alcoholic fatty liver disease (NAFLD) is a spectrum of chronic liver damage, ranging from steatosis (on the most clinically benign end of the spectrum) to cirrhosis (on the opposite extreme where most liver-related morbidity and mortality occur)[1]. Non-alcoholic steatohepatitis (NASH) is a lesion of intermediate severity, characterized by overt injury and death of fatty hepatocytes. NASH may incite a sufficient fibrotic response to eventuate in cirrhosis. Hepatocellular carcinoma sometimes develops in cirrhotic livers.

It is unclear why liver disease progresses to cirrhosis in some, but not most, individuals with NAFLD. Because the ultimate severity of liver damage is generally dictated by the extent to which repair mechanisms can reconstitute healthy liver tissue following injury, treatments for NAFLD aim to reduce injury while bolstering repair. Therefore, development of successful treatments for NAFLD will be facilitated by understanding the mechanisms that mediate liver injury and repair in this disease.

MECHANISMS OF LIVER INJURY IN NAFLD

Hepatocyte steatosis and liver injury

Increased hepatocyte accumulation of triglyceride (TG) is the hallmark of NAFLD. However, the mechanisms driving TG accumulation differ among individuals, as well as within a given individual at different points in time. These differences in steatosis pathogenesis, in turn, prompt heterogeneous compensatory responses and exert different outcomes on hepatocyte viability. This concept is well illustrated by different rodent models of NAFLD. Studies of these models suggest that liver injury in NAFLD results from hepatocyte accumulation of free fatty acids (FFA), as well as various cellular stress responses that develop as fatty cells attempt to dispose of the excess lipids.

Different pathogenesis and consequences of steatosis in various animal models of NAFLD

Over the past decade multiple rodent models of NAFLD have been reported[2,3]. The most extensively studied models include forced overnutrition, genetic obesity, and ingestion of nutrient-restricted diets that limit hepatic lipid disposal.

Feeding otherwise healthy mice or rats high-calorie diets enriched with fat and/or sucrose induces hepatic steatosis, mimicking habitual overeating in humans. Hepatocytes respond to the increased load of FFA by up-regulating several mechanisms to limit exposure to potentially toxic FFA. These include increasing TG synthesis; up-regulating FA oxidation by mitochondria, microsomes and peroxisomes; and enhancing secretion of lipoproteins. In addition to increased delivery of FFA to liver, *de-novo* lipogenesis contributes to increased hepatocyte FFA exposure in rodents and humans with leptin deficiency or leptin resistance. Again, hepatocytes respond to this challenge by up-regulating TG synthesis, FA oxidation and lipoprotein export[4].

Unlike the aforementioned models in which hepatic steatosis results primarily from increased FFA 'input' into hepatocytes, methionine choline-deficient (MCD) diets promote hepatic steatosis mainly by inhibiting two mechanisms that normally dispose of FFA. Chronic consumption of such diets depletes hepatocyte carnitine pools, thereby restricting the efficiency of mitochondrial FA beta-oxidation. MCD diets also induce hepatocyte choline content and this restricts lipoprotein excretion[5]. Hence, MCD diets mimic human diseases, such as congenital enzymopathies, that limit mitochondrial beta-oxidation of FA, or abetalipoproteinaemia, which inhibits hepatocyte lipoprotein export. To compensate for decreased rates of FFA disposal, hepatocytes increase their rate of TG synthesis to store FA in a relatively inert lipid depot, while dramatically increasing non-mitochondrial mechanisms for FA oxidation via microsomal and peroxisomal pathways.

Impaired fatty acid detoxification increases toxicity from fatty acid exposure

Situations that impair FFA detoxification during increased FFA 'input' generally cause severe liver injury. This is exemplified by studies of db/db mice; db/db mice have a leptin receptor gene mutation that generates an inactive receptor. Thus, they are resistant to leptin's normal actions, and become obese and insulin-resistant[6]. Like most leptin-resistant, obese, diabetic humans, db/db mice have relatively mild NAFLD. However, they develop NASH with fibrosis when fed MCD diets that compromise lipid disposal mechanisms[7]. In db/db mice on MCD diets, treatment with antisense oligonucleotides to diacylglycerol acetyl transferase (DGAT)-2 (a rate-limiting enzyme in TG biosynthesis) reduces TG formation, removing one of their remaining 'safe' mechanisms for lipid disposal. The enhanced accumulation of FFA further exacerbates liver damage (Yamaguchi K and Diehl AM, unpublished data). Thus, increased hepatic exposure to FA is generally well tolerated when compensatory lipid disposal mechanisms can be induced, but the same lipid challenge becomes progressively hepatotoxic as lipid disposal mechanisms are lost.

Mechanisms of hepatic lipotoxicity

The specific factors that injure fatty hepatocytes are likely to be multiple, and probably vary somewhat according to the principal cause(s) of fat accumulation. In addition to direct cellular toxicity from FFA-induced lipid peroxidation, certain FA may affect global changes in liver gene expression by functioning as ligands for nuclear hormone receptors (e.g. HNF4-alpha, PPAR-alpha and PPAR-gamma)[8]. FA oxidation also generates metabolic by-products, such as reactive oxygen species (ROS) and NAD(P)H, that exert their own effects on hepatocytes by acting as signalling molecules and metabolic co-factors. ROS may also be directly cytotoxic by eliciting modifications in DNA, mitochondria, lipid membranes, and other structural components[9].

In an effort to constrain noxious consequences of ROS and NADPH accumulation, hepatocytes induce compensatory changes in mitochondrial respiration and up-regulate various antioxidant and anti-apoptotic defences. These adaptive manoeuvres, themselves, however, introduce other vulnerabilities by depleting antioxidant reserves and/or limiting the efficiency of ATP synthesis[10]. Activation of stress-related transcription factors, such as NF-κB, also increases hepatocyte production of proinflammatory cytokines, such as TNF-α and IL-6, that are NF-κB target genes[11]. Thus, even 'successful' adaptation to steatosis-imposed 'stress' increases hepatocyte vulnerability to cell death. Conversely, failure to sufficiently up-regulate survival responses results in immediate lethality.

These concepts are supported by studies of animals and humans with NAFLD (See Figures 1 and 2 for summary). In many animal models of hepatic steatosis, various antioxidant and anti-apoptotic defences have been up-regulated sufficiently to maintain hepatocyte apoptosis at relatively low (albeit increased) levels. However, these adaptations enhance vulnerability to secondary insults that overwhelm the residual capacities for antioxidant defence or ATP synthesis, such as ischaemia–reperfusion or lipopolysaccharide exposure. Hepatic injury tends to be greater in models (e.g. MCD diets) that limit antioxidant or ATP synthesis, while promoting hepatic steatosis.

Adipocytokines and liver injury in NAFLD

Adipose tissue is a rich source of several soluble factors that promote hepatic steatosis[12]; therefore it is not surprising that recent attention has focused on adiocytokines in the pathogenesis of NAFLD. Studies in several animal models of NAFLD (e.g. ob/ob mice, high-fat diet-fed mice, Agouti mice exposed to lipopolysaccharide) suggest that unbalanced production of proinflammatory cytokines (especially TNF-α) relative to anti-inflammatory factors (e.g. adiponectin) plays a pivotal role in the pathogenesis of NAFLD[11,13,14] . In such models various treatments that directly inhibit TNF-α activity (e.g. anti-TNF antibodies, genetic disruption of TNF-α or TNF receptor (TNFR)-1,

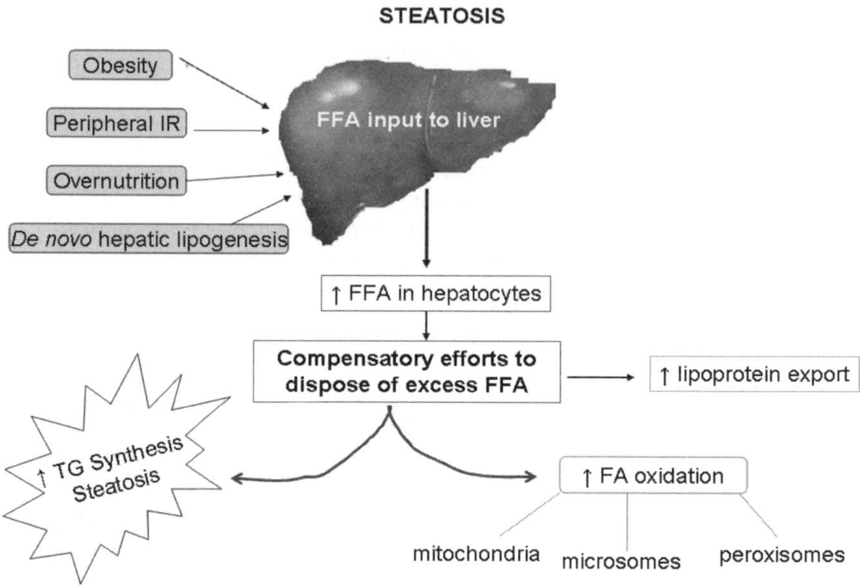

Figure 1 Pathogenesis of steatosis. Steatosis reflects the accumulation of TG in hepatocytes. TG synthesis is one of the compensatory mechanisms that hepatocytes evoke to dispose of excess FFA. Excessive accumulation of FFA within heptatocyes results from factors that increase the input of FFA into hepatocytes and/or that inhibit the disposal of hepatocyte FFA. Increased FFA input may result from various conditions, including obesity, peripheral insulin resistance, excessive dietary intake, and *de-novo* hepatic lipogenesis. Increased TG synthesis can also occur as a result of decreased disposal of hepatocyte FFA. The latter occurs when oxidation of FFA by mitochondria, microsomes and/or peroxisomes is inhibited and/or export of lipoproteins from hepatocytes is blocked

pentoxifylline, TZD) or that increase activity of adiponectin or its downstream target, AMP kinase (e.g. supplemental adiponectin, metformin), improve steatosis and steatohepatitis.

Since TNF-α and adiponectin are mutually antagonist, treatments that inhibit TNF-α enhance adiponectin activity. Conversely, interventions that boost adiponectin's actions block TNF-α's effects[12,15,16]. At the cellular level TNF-α and adiponectin exert opposing effects on hepatic lipid homeostasis. Adiponectin inhibits FA uptake and promotes FA oxidation and lipid export from hepatocytes by inducing the phosphorylation of AMP kinase, a master regulator of energy homeostasis and lipid intermediary metabolism. TNF-α inhibits adiponectin expression, as well as AMPK activation, thereby blocking adiponectin activity. Hence, when TNF-α activity predominates, hepatocytes accumulate FA and become subject to all of the outcomes discussed previously. Conversely, hepatocyte FA exposure and its consequences are limited when adiponectin activity predominates.

Adiposity generally increases serum levels of various adipose-derived factors, including both TNF-α and adiponectin[17,18], although serum levels of

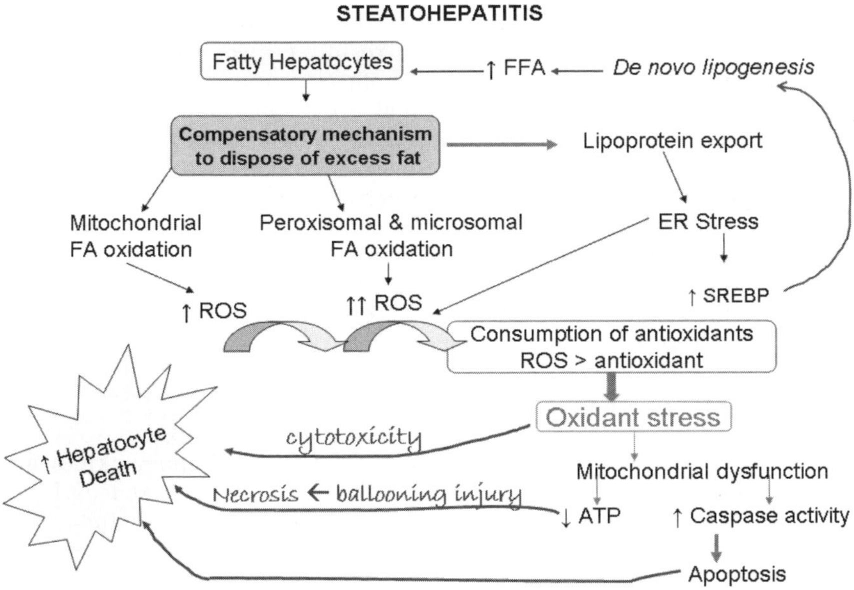

Figure 2 Pathogenesis of NASH. NASH reflects increased death of fatty hepatocytes. Various compensatory mechanisms that fatty hepatocytes mobilize in an attempt to rid themselves of excess fat eventually decrease hepatocyte viability. For example, chronic increases in mitochondrial fatty acid oxidation generate ROS that consume cellular antioxidants, eventually overwhelming antioxidant defences, leading to oxidant stress. FFA oxidation by peroxisomes and microsomes also generates ROS that place a further demand on antioxidant defences, compounding oxidant stress. Excess ROS damage various cellular structures, including mitochondria. Damaged mitochondria become dysfunctional and permeable. This reduces the efficiency of mitochondrial ATP synthesis. The resultant hepatocyte energy depletion promotes hepatocyte swelling (ballooning) and necrosis (death). Leaky mitochondria also release factors that activate apoptotic pathways, causing more hepatocyte death.. Efforts to dispose of FFA by increasing lipoprotein assembly and secretion may promote endoplasmic reticulum (ER) stress. ER stress can exacerbate ROS production, compounding all of the oxidant stress-related problems outlined above. In addition ER stress leads to the accumulation of certain transcription factors (e.g. SREBP-1c) that induce hepatocyte lipogenesis, leading to more FFA accumulation, and exacerbation of hepatocyte steatosis

TNF-α tend to be higher, and adiponectin levels tend to be lower, in individuals with more advanced liver injury (i.e. NASH) than in those with less severe liver injury (i.e. steatosis)[19]. However, neither adiocytokine is a useful biomarker for NAFLD because serum TNF-α and adiponectin levels vary considerably among individuals with any given level of liver injury. Advanced NAFLD also occurs in lipodystrophic individuals who have severely reduced adipose depots[20–22].

One explanation for these inconsistencies may be inherent differences in adipocytokine production by various adipose depots. In general, visceral adiposity (which persists in lipodystrophy and is most strongly associated with obesity-related diseases) results in greater relative exposure to proinflammatory factors[23–26]. In addition, there may be non-adipose tissue sources of adipocytokines. Such sources include resident liver cells, themselves. Adipocytic hepatic stellate cells produce both resistin and adiponectin[27] and HSC-derived resistin activates NF-κB and evokes production of cytokines, including TNF-α, IL-1β, and IL-6, by neighbouring hepatocytes (Ling L and Diehl AM, unpublished data). Variations in tissue adipocytokine production impact paracrine signalling that regulates hepatocyte intermediary metabolism and viability without causing major alterations in serum levels of adipocytokines that are derived predominantly from body fat depots. Hence, important contributions from locally produced adipocytokines help to explain the imperfect correlation between absolute serum levels of adipocytokines and the severity of liver injury in individuals with NAFLD.

Insulin resistance and NAFLD

In addition to modulating lipid homeostasis, various adipocytokines also regulate carbohydrate metabolism by influencing cellular sensitivity to insulin[28]. NAFLD is strongly associated with insulin resistance[29,30]. Moreover, the severity of NAFLD-related liver damage parallels the severity of insulin resistance, with cirrhosis and liver-related mortality being greatest in NAFLD patients with type 2 diabetes[30]. Despite the apparent robustness of this clinical correlation, it has been difficult to understand whether the association between NAFLD and type 2 diabetes reflects the negative outcomes of inhibiting insulin actions in the liver itself, or if NAFLD results from loss of insulin activity in extrahepatic tissues. If hepatic (rather than systemic) insulin resistance is, in fact, the driving force for NAFLD, this might explain why systemic insulin resistance is not a pre-requisite for hepatic steatosis[29,31]. On the other hand it does not clarify why many humans and experimental animals with NAFLD exhibit both increased rates of hepatocyte lipid synthesis (suggestive of enhanced hepatic insulin activity), and hyperinsulinaemia with excessive hepatic glucose output (typical of hepatic insulin resistance). The latter findings suggest that differential inhibition of various insulin-regulated signalling pathways within hepatocytes might be necessary for NAFLD pathogenesis[32].

The confusing role of insulin resistance in NAFLD pathogenesis/progression has not been resolved by studies of animal models of NAFLD. Indeed, such work only underscores the complexity of this relationship. Similar to many obese humans, leptin- deficient (ob/ob) and leptin- resistant (db/db) murine models of NAFLD exhibit hyperinsulinaemia, hyperglycaemia and evidence of both systemic and hepatic insulin resistance[2]. Treatment with insulin-sensitizing agents has been beneficial, albeit inconsistently. On the other hand, MCD diet-fed mice develop NASH and progressive hepatic fibrosis despite diet-induced decreases in serum insulin and glucose, suggestive of enhanced systemic sensitivity to insulin[33–35]. Nevertheless, insulin-sensitizing agents also seem to improve MCD diet-induced liver disease. Recent evidence of reduced insulin receptor- and IRS-2 tyrosine phosphorylation in the livers of MCD diet-fed mice supports the possibility that these animals have hepatic insulin resistance[36]. On the other hand, mice with hepatocyte-specific activation of AKT, a prototypical downstream target of insulin signalling, develop severe steatohepatitis, progressive fibrosis and eventual hepatocellular carcinoma, suggesting that certain insulin-initiated signals might actually promote (rather than prevent) liver damage[37,38]. At this point no unifying hypothesis has emerged to reconcile these seemingly contradictory results.

It is conceivable that part of the confusion derives from efforts to equate type 2 diabetes with extremely severe insulin resistance. The flaw in such logic is demonstrated by recent data from the Diabetes Prevention Program[38]. In that large population of insulin-resistant adults, hypo-secretion of insulin was a significantly better predictor of subsequent type 2 diabetes than the initial severity of insulin resistance. This finding led the DPP investigators to speculate that insulin resistance 'unmasked' individuals with an inherently reduced capacity for pancreatic beta cell hyperplasia. Only the latter individuals eventually became overtly diabetic. This insight raises the intriguing possibility that cirrhosis (which strongly correlates with type 2 diabetes) might result from an inadequate hepatic hyperplastic response to similar metabolic stress. The latter is a particularly attractive concept because hepatocyte hyperplasia is necessary to replace dying fatty hepatocytes, and progressive liver damage results when regeneration cannot keep pace with injury[39].

MECHANISMS OF INHIBITED LIVER REGENERATION IN NAFLD

Replicative senescence in fatty hepatocytes

Acute loss of mature hepatocytes promptly triggers a compensatory proliferative response in surviving mature hepatocytes, permitting efficient reconstitution of previously healthy livers[39]. However, in chronically fatty livers, many viable hepatocytes have sustained oxidative damage that inhibits progression through the cell cycle[40,41]. The reduced proliferative capacity of fatty hepatocytes complicates replacement of dying hepatocytes. To circumvent liver atrophy, hepatic progenitors must become engaged in the regenerative response[42]. Thus, unlike regeneration of healthy livers, repair of chronically fatty livers depends upon progenitor cell differentiation and subsequent hyperplasia, a process that is less efficient than replication of healthy, mature hepatocytes[43].

Differentiation of liver epithelial progenitors, stellate cell activation and fibrosis

Relatively little is known about the mechanisms that regulate the responses of liver progenitor populations to liver injury[39]. However, emerging evidence suggests that paracrine signalling from liver mesenchymal cells, particularly hepatic stellate cells (HSC), is important. Paradoxically, liver injury that results from hepatocyte lipid accumulation tends to reduce the lipid content of adipocytic HSC, promoting their transformation into myofibroblastic cells that promote extracellular matrix accumulation[44]. Multiple factors participate in HSC activation during liver injury, including the release of pro-fibrogenic mediators, such as platelet-derived growth factor (PDGF) and transforming growth factor (TGF) beta, from aggregating platelets and dying hepatocytes, respectively[45]. Phagocytosis of apoptotic hepatocytes also contributes to HSC activation in fatty livers[46–48].

Myofibroblastic HSC, in turn, down-regulate their production of factors, including resistin, adiponectin, IL-6, and hepatocyte growth factor (HGF), that transduce various viability signals in neighbouring hepatocytes (Diehl AM et al., unpublished data). Hence, one of the earliest consequences of HSC activation is a reduction of paracrine factors that normally maintain the survival and proliferative capacity of neighbouring mature hepatocytes. Furthermore, other factors produced by myofibroblastic HSC, such as plasminogen activator inhibitor (PAI)-1 and tissue inhibitors of metalloproteinases (TIMP), inhibit localized release of matrix-associated HGF, further reducing hepatocyte access to this mitogen[49]. Other HSC-derived factors, such as leptin and angiotensin, promote proliferation and survival of myofibroblastic HSC that generate large amounts of type 1 collagen[44,45]. Thus, the accumulation of myofibroblastic HSC, long known to signify fibrogenesis, also marks a situation in which mature hepatocytes are deprived of trophic factors that are normally produced by local adipocytic HSC. This restricts hepatocyte access to survival signals and promotes 'debridement' of the damaged liver epithelia.

Fortunately, myofibroblastic HSC produce other factors, such as stromal-derived factor (SDF)-1 and Sonic hedgehog (Shh), that promote the accumulation of hepatic epithelial progenitors in damaged livers[50,51]. Paracrine signalling between myofibroblastic HSC and immature hepatic epithelial cells enhances growth of the latter cells (Omenetti A and Diehl AM, unpublished data), thereby encouraging 're-epithelialization' of damaged livers from hepatic progenitor reserves. Ultimately, the restoration and maintenance of healthy mature liver epithelia probably requires the re-emergence of adipocytic HSC to provide a steady supply of factors that assure optimal viability of adult hepatocytes. Further research is necessary to confirm (or refute) these concepts. If validated, they would help to explain why the accumulation of myofibroblastic HSC in steatotic livers distinguishes individuals who are likely to exhibit NASH and fibrosis on a subsequent liver biopsy from those whose liver disease will not progress over time. Stated another way, the persistence of myofibroblastic HSC marks ongoing remodelling of the hepatic architecture and implies that scarring is likely because wound healing remains incomplete (see Figure 3 for summary of mechanisms that mediate progression of NASH to cirrhosis).

Much research has focused on delineating the mechanisms that modulate this process, with the intent of redirecting wound-healing responses to optimize epithelial repair while minimizing fibrosis. Interestingly, adiponectin (a product of adipocytic HSC and adipose tissue) appears to inhibit activation of adipocytic HSC to myofibroblastic cells[52]. Thus, various treatments that enhance adiponectin activity offer the promise of reducing hepatic steatosis, limiting hepatocyte death, and preventing hepatic fibrosis. Based on the earlier

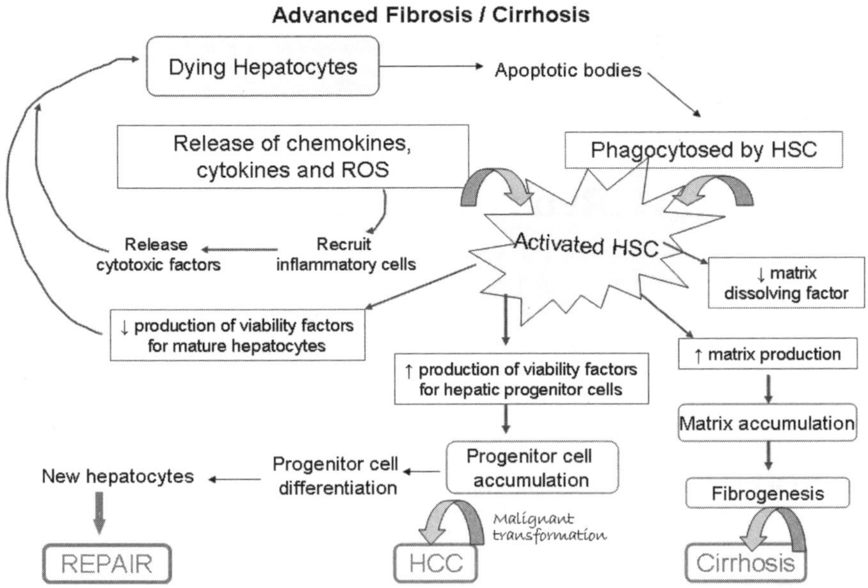

Figure 3 Pathogenesis of cirrhosis. Cirrhosis reflects unsuccessful repair of NASH. Dying hepatocytes release various factors (e.g. ROS, chemokines, cytokines) that recruit inflammatory cells into the liver. Once in the liver, these inflammatory cells release cytotoxic factors that increase hepatocyte death. Other hepatocyte-derived factors activate neighbouring hepatic stellate cells (HSC). Activated HSC produce fewer viability factors for mature hepatocytes than quiescent HSC, so this leads to more hepatocyte death, release of ROS/chemokines/cytokines, etc. Activated HSC also produce more matrix molecules and less matrix-dissolving factors. This leads to matrix accumulation (fibrogenesis), and eventually cirrhosis. Fortunately, activated HSC produce viability factors that promote the accumulation of hepatic progenitor cells within the damaged liver. Differentiation of these progenitors into mature hepatocytes permits eventual liver regeneration and reconstitution of normal hepatocyte architecture and function. However, if some of the progenitors become damaged by ROS, they may undergo malignant transformation, leading to hepatocellular cancer

discussion, therapies that inhibit HSC activating factors (i.e. TGF-β, PAI-1 and angiotensin), or that enhance HGF activity, are also attractive therapeutic candidates in NAFLD.

Progenitor accumulation and hepatocarcinogenesis

Sustained efforts to regenerate damaged livers promote hepatocarcinogenesis by exposing cells with an inherent survival advantage (i.e. progenitors) to an injurious microenvironment that might induce stochastic, transforming mutations[53,54]. Thus, the risk of cancer is reduced by removing the stimulus for progenitor accumulation (i.e. by minimizing liver injury and expediting liver repair).

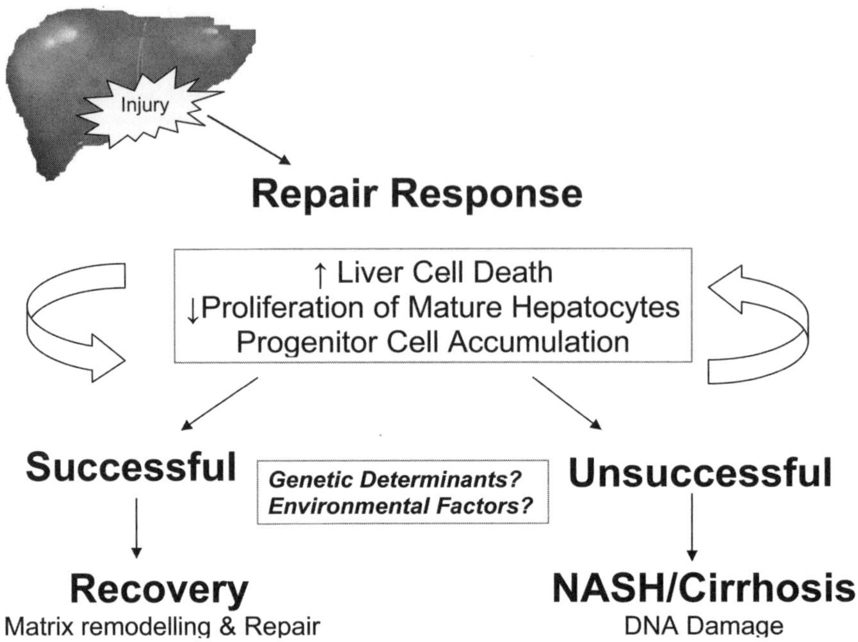

Figure 4 Spectrum of NAFLD. Risk for disease progression related to hepatic steatosis and NASH is dependent on the associate repair response(s). The cascade of events resulting in increased cell death, decreased proliferation of mature hepatocytes, and progenitor cell accumulation may result in either successful repair or recovery or injury, with risk for DNA damage depending on genetic determinants and environmental risk factors that have yet to be clearly defined

SUMMARY (SEE FIGURE 4)

The hallmark of NAFLD is triglyceride deposition within hepatocytes. Studies of experimental animals and human subjects with NAFLD suggest that liver injury in NAFLD results from hepatic accumulation of free fatty acids, as well as various cellular stress responses that develop as fatty hepatocytes attempt to dispose of the excess lipids. At least in part, inter- and intra-individual variability in the outcome of hepatic steatosis probably reflect differences in the mechanisms that drive triglyceride accumulation which, in turn, evoke heterogeneous compensatory responses that exert different effects on hepatocyte viability. Death of mature hepatocytes triggers hepatic regeneration to restore lost hepatocyte mass and function. Ultimately, the net amount of liver damage that accrues is determined by the liver's regenerative capacity. Because hepatic steatosis induces chronic oxidative and metabolic stress, it increases oxidative damage and limits energy reserves in mature hepatocytes, thereby reducing their proliferative capacity. This gradually increases the requirement for hepatic progenitors in the repair process.

Emerging evidence suggests that paracrine signalling between dying mature hepatocytes, hepatic stellate cells and epithelial progenitors is a major factor that modulates regeneration of chronically injured, fatty livers. Various adipocytokines that are produced both by adipose tissue and hepatic stellate cells regulate this process. Transient increases in myofibroblastic stellate cells probably enhance repair by mobilizing fresh crops of hepatic progenitors to replace damaged mature hepatocytes, while sustained accumulation of myofibroblastic (as opposed to adipocytic) stellate cells perpetuates hepatic injury and increases liver fibrosis. Rather than reconstituting normal liver architecture, the latter increases the risk for cirrhosis and hepatocellular carcinoma. Therefore, in order to develop treatments that will optimize recovery from NAFLD, it is important to clarify the mechanisms that regulate liver repair, as well as those that initiate liver injury.

References

1. Brunt EM, Tiniakos DG. Pathological features of NASH. Front Biosci. 2005;10:1475–84.
2. Koteish A, Diehl AM. Animal models of steatosis. Semin Liver Dis. 2001;21:89–104.
3. Farrell GC, Larter CZ. Nonalcoholic fatty liver disease: from steatosis to cirrhosis. Hepatology. 2006;43:S99–112.
4. Yen TT, Allan JA, Yu P-L, Acton MA, Pearson DV. Triacylglycerol contents and *in vivo* lipogenesis of ob/ob, db/db, and Ay/a mice. Biochim Biophys Acta. 1976;441:213–20.
5. Ghoshal AK, Ahluwalia M, Farber E. The rapid induction of liver cell death in rats fed a choline-deficient methionine-low diet. Am J Pathol. 1983;113:309–14.
6. Chen H, Charlat O, Tartaglia LA et al. Evidence that the diabetes gene encodes the leptin receptor: identification of a mutation in the leptin receptor gene in db/db mice. Cell. 1996; 84:491–5.
7. Sahai A, Malladi P, Pan X et al. Obese and diabetic db/db mice develop marked liver fibrosis in a model of nonalcoholic steatohepatitis: role of short-form leptin receptors and osteopontin. Am J Physiol Gastrointest Liver Physiol. 2004;287:G1035–43.
8. Sladek F. Desperately seeking ... something. Mol Cell. 2002;10:219–21.
9. Pessayre D, Mansouri A, Fromenty B. Nonalcoholic steatosis and steatohepatitis. V. Mitochondrial dysfunction in steatohepatitis. Am J Physiol Gastrointest Liver Physiol. 2002;282:G193–9.
10. Yang SQ, Zhu H, Li Y et al. Mitochondrial adaptations to obesity-related oxidant stress. Arch Biochem Biophys. 2000;378:259–68.
11. Feldstein AE, Werneburg NW, Canbay A et al. Free fatty acids promote hepatic lipotoxicity by stimulating TNF-alpha expression via a lysosomal pathway. Hepatology. 2004;40:185–94.
12. Chaldakov GN, Stankulov IS, Hristova M, Ghenev PI. Adipobiology of disease: adipokines and adipokine-targeted pharmacology. Curr Pharm Des. 2003;9:1023–31.
13. Xu A, Wang Y, Keshaw H, Xu LY, Lam KS, Cooper GJ. The fat-derived hormone adiponectin alleviates alcoholic and nonalcoholic fatty liver diseases in mice. J Clin Invest. 2003;112:91–100.
14. Masaki T, Chiba S, Tatsukawa H et al. Adiponectin protects LPS-induced liver injury through modulation of TNF-alpha in KK-Ay obese mice. Hepatology. 2004;40:177–84.
15. Bruun JM, Lihn AS, Verdich C et al. Regulation of adiponectin by adipose tissue-derived cytokines: *in vivo* and *in vitro* investigations in humans. Am J Physiol Endocrinol Metab. 2003;285:E527–33.
16. Stefan N, Stumvoll M. Adiponectin – its role in metabolism and beyond. Horm Metab Res. 2002;34:469–74.
17. Rajala MW, Scherer PE. Minireview: The adipocyte – at the crossroads of energy homeostasis, inflammation, and atherosclerosis. Endocrinology. 2003;144:3765–73.
18. Anderwald C, Roden M. Adipotoxicity and the insulin resistance syndrome. Pediatr Endocrinol Rev. 2004;1:310–19.

19. Hui JM, Hodge A, Farrell GC, Kench JG, Kriketos A, George J. Beyond insulin resistance in NASH: TNF-alpha or adiponectin? Hepatology. 2004;40:46–54.
20. Powell EE, Searle J, Mortimer R. Steatohepatitis associated with limb lipodystrophy. Gastroenterology. 1989;97:1022–4.
21. Shimomura I, Hammer RE, Richardson JA et al. Insulin resistance and diabetes mellitus in transgenic mice expressing nuclear SREP-1c in adipose tissue: model for congenital generalized lipodystrophy. Genes Dev. 1998;12:3182–94s.
22. Shimomura I, Hammer RE, Ikemoto S, Brown MS, Goldstein JL. Leptin reverses insulin resistance and diabetes mellitus in mice with congenital lipodystrophy. Nature. 1999;401: 73–6.
23. Atzmon G, Yang XM, Muzumdar R, Ma XH, Gabriely I, Barzilai N. Differential gene expression between visceral and subcutaneous fat depots. Horm Metab Res. 2002;34:622–8.
24. Gabriely I, Ma XH, Yang XM et al. Removal of visceral fat prevents insulin resistance and glucose intolerance of aging: an adipokine-mediated process? Diabetes. 2002;51:2951–8.
25. Gabriely I, Barzilai N. Surgical removal of visceral adipose tissue: effects on insulin action. Curr Diab Rep. 2003;3:201–6.
26. Nyholm B, Nielsen MF, Kristensen K et al. Evidence of increased visceral obesity and reduced physical fitness in healthy insulin-resistant first-degree relatives of type 2 diabetic patients. Eur J Endocrinol. 2004;150:207–14.
27. She H, Xiong S, Hazra S, Tsukamoto H. Adipogenic transcriptional regulation of hepatic stellate cells. J Biol Chem. 2005;280:4959–67.
28. Klaus S. Adipose tissue as a regulator of energy balance. Curr Drug Targets. 2004;5:241–50.
29. Marchesini G, Brizi M, Morselli-Labate AM et al. Association of nonalcoholic fatty liver disease with insulin resistance. Am J Med. 1999;107:450–5.
30. Marchesini G, Bugianesi E, Forlani G et al. Nonalcoholic fatty liver, steatohepatitis, and the metabolic syndrome. Hepatology. 2003;37:917–23.
31. Marceau P, Iron S, Hould FS et al. Liver pathology and the metabolic syndrome X in severe obesity. J Clin Endocrinol Metab. 1999;84:1513–17.
32. Matsumoto M, Seongah H, Tadahiro K, Accili D. Dual role of transcription factor FoxO1 in controlling hepatic insulin sensitivity and lipid metabolism. J Clin Invest. 2006;116: 2464–72.
33. Nanji AA. Animal models of nonalcoholic fatty liver disease and steatohepatitis. Clin Liver Dis. 2004;8:559–74, ix.
34. Rinella ME, Green RM. The methionine-choline deficient dietary model of steatohepatitis does not exhibit insulin resistance. J Hepatol. 2004;40:47–51.
35. Ng DS, Xie C, Maguire GF et al. Hypertriglyceridemia in lecithin-cholesterol acyltransferase-deficient mice is associated with hepatic overproduction of triglycerides, increased lipogenesis, and improved glucose tolerance. J Biol Chem. 2004;279:7636–42.
36. Anstee QM, Goldin RD. Mouse models in non-alcoholic fatty liver disease and steatohepatitis research. Int J Exp Pathol. 2006;87:1–16.
37. Kurlawalla-Martinez C, Stiles B, Wang Y, Devaskar SU, Kahn BB, Wu H. Insulin hypersensitivity and resistance to streptozotocin-induced diabetes in mice lacking PTEN in adipose tissue. Mol Cell Biol. 2005;25:2498–510.
38. Horie Y, Suzuki A, Kataoka E et al. Hepatocyte-specific Pten deficiency results in steatohepatitis and hepatocellular carcinomas. J Clin Invest. 2004;113:1774–83.
39. Fausto N. Liver regeneration and repair: hepatocytes, progenitor cells, and stem cells. Hepatology. 2004;39:1477–87.
40. Yang SQ, Lin HZ, Mandal AK, Huang J, Diehl AM. Disrupted signaling and inhibition of liver regeneration in obese mice with fatty livers: implications for nonalcoholic fatty liver disease pathogenesis. Hepatology. 2001;34:694–706.
41. Gao D, Wei C, Chen L, Huang J, Yang S, Diehl AM. Oxidative DNA damage and DNA repair enzyme expression are inversely related in murine models of fatty liver disease. Am J Physiol Gastrointest Liver Physiol. 2004;287:G1070–7.
42. Yang S, Koteish A, Lin H et al. Oval cells compensate for damage and replicative senescence of mature hepatocytes in mice with fatty liver disease. Hepatology. 2004;39: 403–11.

43. Roskams T, Yang SQ, Koteish A et al. Oxidative stress and oval cell accumulation in mice and humans with alcoholic and nonalcoholic fatty liver disease. Am J Pathol. 2003;163: 1301–11.
44. Bataller R, Brenner DA. Liver fibrosis. J Clin Invest. 2005;115:209–18.
45. Friedman SL. Stellate cells: a moving target in hepatic fibrogenesis. Hepatology. 2004;40: 1041–3.
46. Feldstein AE, Canbay A, Angulo P et al. Hepatocyte apoptosis and fas expression are prominent features of human nonalcoholic steatohepatitis. Gastroenterology. 2003;125: 437–43.
47. Feldstein AE, Gores GJ. Apoptosis in alcoholic and nonalcoholic steatohepatitis. Front Biosci. 2005;10:3093–9.
48. Feldstein AE, Papouchado BG, Angulo P, Sanderson S, Adams L, Gores GJ. Hepatic stellate cells and fibrosis progression in patients with nonalcoholic fatty liver disease. Clin Gastroenterol Hepatol. 2005;3:384–9.
49. Lijnen HR. Pleiotropic functions of plasminogen activator inhibitor-1. J Thromb Haemost. 2005;3:35–45.
50. Sicklick JK, Li YX, Choi SS et al. Role for hedgehog signaling in hepatic stellate cell activation and viability. Lab Invest. 2005;85:1368–80.
51. Dalakas E, Newsome PN, Harrison DJ, Plevris JN. Hematopoietic stem cell trafficking in liver injury. FASEB J. 2005;19:1225–31.
52. Marra F, Aleffi S, Bertolani C, Petrai I, Vizzutti F. Adipokines and liver fibrosis. Eur Rev Med Pharmacol Sci. 2005;9:279–84.
53. Thorgeirsson SS, Factor VM, Snyderwine EG. Transgenic mouse models in carcinogenesis research and testing. Toxicol Lett. 2000;112-113:553–5.
54. Lee JS, Heo J, Libbrecht L et al. A novel prognostic subtype of human hepatocellular carcinoma derived from hepatic progenitor cells. Nat Med. 2006;12:410–16.

2
Alcohol-induced hepatitis: pathophysiology and treatment

H. K. SEITZ and P. BECKER

INTRODUCTION

Alcohol misuse is common in the USA and in Western Europe, and is increasing in Asia. In the USA 7% of the adult population meet the definition of alcohol misuse or dependence[1]. Similar data exist for some countries in Europe, including Germany, where more than 1.5 million individuals are alcohol-dependent and approximately 3 million people have alcohol-associated organ damage[2]. Chronic alcohol consumption above a certain daily threshold results in alcoholic fatty liver (AFL) in almost 100% of cases; however, only 30% of these patients develop alcoholic fibrosis either with or without inflammation, and only 10–20% progress to liver cirrhosis. It has been speculated that genetic aspects may play a predominant role, explaining why not every alcoholic develops advanced liver disease. Although polymorphisms of genes involved in alcoholic liver disease (ALD) have been intensively studied, the results so far are disappointing[3].

Alcoholic steatohepatitis (ASH) is one feature of ALD and is characterized by a specific histomorphological appearance associated with typical clinical syndromes[4]. Severity of ASH may vary between mild, moderate and severe. Severe ASH is associated with a poor prognosis. It should be pointed out that the prerequisite for the development of ASH is AFL. The first part of this chapter will focus on pathophysiological mechanisms involved in the production of AFL, and consequently of ASH. In the second part therapeutic strategies for severe ASH will be discussed.

PATHOPHYSIOLOGY OF AFL

Various factors contribute to the occurrence of AFL; a summary is given in Figure 1. The liver is the predominant site of alcohol oxidation and alcohol is oxidized via alcohol dehydrogenase (ADH) to acetaldehyde (AA) and further by acetaldehyde dehydrogenase (ALDH) to acetate. This metabolism needs NAD+ and generates NADH. The result is an overproduction of reducing

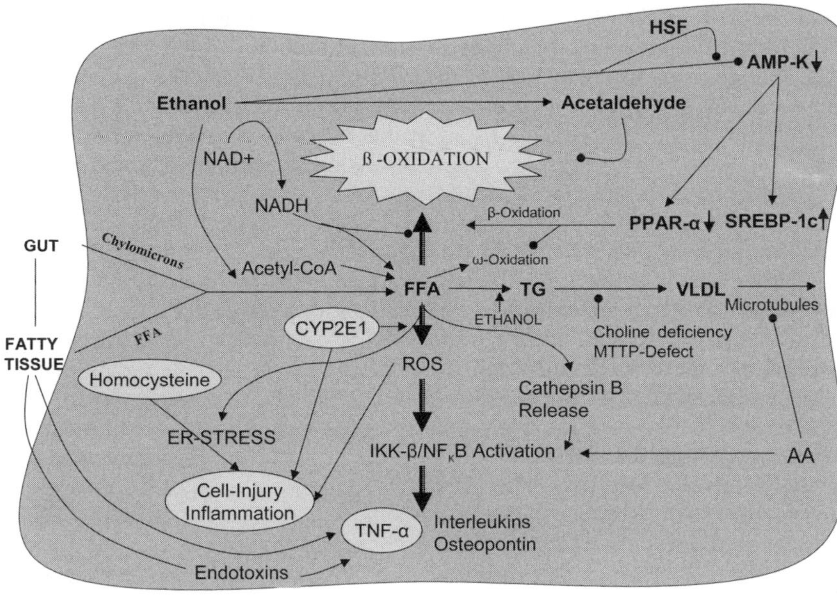

Figure 1 Pathogenesis of alcoholic fatty liver and inflammation. Ethanol is metabolized to acetaldehyde (AA) by alcohol dehydrogenase generating NADH from NAD^+. NADH stimulates free fatty acid (FFA) and triglyceride (TG) synthesis and inhibits beta-oxidation of FFA. In addition, alcohol increases the influx of chylomicrons from the gut and FFA from fat tissue into the liver. On the other hand, secretion of very low density lipoproteins (VLDL) from the liver to the blood is inhibited by AA. AA also damages mitochondria. Furthermore, alcohol inhibits adenosine monophosphate-activated kinase (AMPK) leading to a decrease in peroxisome proliferating-activated receptor (PPAR) alpha and an increase in sterol regulatory element-binding protein (SREBP) 1c, both stimulating lipogenesis and inhibiting lipolysis. The effect of ethanol and AMPK is counteracted by high saturated fats (HSF). Hepatic accumulation of FFA results in toxicity, either by the induction of cytochrome P4502E1 (CYP2E1), by activation of the IKK-beta/NF-κB pathway via a lysosomal, cathepsin B-dependent mechanism leading to an increase in TNF-α and other cytokines. TNF-α is also released from Kupffer cells after its stimulation by gut-derived endotoxins. Hepatocellular injury in ASH is primarily due to oxidative stress induced by CYP2E1 induction and to apoptosis and necrosis induced by TNF-α. ROS = reactive oxygen species; MTPP = microsomal triglyceride transport protein

equivalents in the liver and a striking change in the hepatic redox state. These reducing equivalents shift all redox partners into their reducing form and influence free fatty acid (FFA) and triglyceride metabolism significantly. NADH favours FFA– and triglyceride synthesis and inhibits mitochondrial β-oxidation of FFA. The result is an accumulation of triglycerides in the liver[4,5].

In addition, alcohol consumption leads to an increased mobilization of FFA from fat tissue and also from the gut as chylomicrons. These FFA and chylomicrons are taken up by the liver and contribute to the development of AFL.

Furthermore, alcohol results in an inhibition of lipid secretion from the liver to the blood. Thus, AA damages microtubules, leading to morphological and functional changes of these important cell organelles which are responsible for the secretion of cellular proteins and macromolecules such as albumin, transferrin and lipoproteins[4,5]. In addition, AA also damages mitochondria, interfering with the respiratory chain and the mitochondrial reoxidation of NADH to NAD^+, further stimulating AFL. Alcohol inhibits adenosine monophosphate (AMP)-activated kinase[6]. AMP-activated kinase leads to an activation of peroxisome proliferator-activated receptor alpha (PPARα), an important nuclear transfer factor for genes involved in lipolysis[7]. It has been shown that PPARα protects against alcohol-induced liver damage in mice[8]. In addition AMP-activated kinase results in a decreased activation of sterol regulatory element-binding protein 1c (SREBP1c)[7], a nuclear factor involved in lipogenesis. SREBP1c knockout mice are less likely to develop AFL[9]. Thus, alcohol may induce AFL by both mechanisms via AMP-activated kinase. It is interesting that high unsaturated fat diets may counteract the effect of ethanol on AMP-activated kinase[6,7]. Thus, in addition to chronic ethanol consumption, other factors such as overweight and obesity[10], as well as the quality of dietary fat (saturated versus polyunsaturated fatty acids) may modulate the quantity and quality of hepatic fat. Both render the liver more susceptible to a second hit, resulting in lipid peroxidation which is more pronounced with polyunsaturated fat as compared to saturated fat.

The results of a recent animal study comparing a NASH and an ASH mouse model by using microarray analysis demonstrate that the decrease of adenosine monophosphate-activated kinase (AMPK) observed in the alcohol model was not observed in the NASH model, while PPARα was found to be more decreased in NASH as compared to ASH[11].

FROM FATTY LIVER TO INFLAMMATION

Fatty liver is more sensitive towards a second hit as compared to a normal liver. This second hit may vary with respect to quantity and quality inter-individually, as in non-alcoholic fatty liver disease.

The role of reactive oxygen species (ROS)

Several enzymatic systems, including the cytochrome P4502E1 (CYP2E1)-dependent microsomal monooxygenase system, the mitochondrial respiratory chain and the cytosolic enzymes xanthine oxidase and aldehyde oxidase, have been implicated as sources of O_2^- and H_2O_2 in hepatocytes during ethanol oxidation[12]. Alcohol-mediated free radical formation may be due to enhanced electron leakage from the mitochondrial respiratory chain, along with stimulation of NADH shuttling into mitochondria, to an interaction between *N*-acetylsphingosine (from tumour necrosis factor alpha, TNF-α) and mitochondria[13], to activated phagocytes in the liver[14], to hepatic iron overload, and to nitric oxide[15]. The reaction of nitric oxide with O_2 results in the formation of peroxynitrite, which is highly reactive and impairs cell function[16]. Inducible nitric oxide synthase (iNOS) is stimulated by ethanol, and iNOS knockout mice are protected against alcohol-induced liver injury[17]. Although all these factors may contribute to the generation of ROS following alcohol ingestion, animal experiments have convincingly demonstrated the important role of CYP2E1 in the production of ROS and ALD.

Chronic alcohol consumption leads to a 10–20-fold induction of CYP2E1, which metabolizes ethanol to acetaldehyde. This cytochrome is also involved in the activation of various co-carcinogens to their ultimate carcinogens, including nitrosamines, aflatoxins, and polycyclic hydrocarbons. The events in this setting are reviewed elsewhere[18].

It has been shown that the concentration of CYP2E1 in the liver correlates with the generation of hydroxy-ethyl radicals (HER) and thus with lipid peroxidation[19]. HER bind to proteins and form neoantigens[20]. Antibodies against HER-derived epitopes are detectable in the sera of patients with ALD[21,22]. Induction of CYP2E1 resulted in ALD in animals, and inhibition of CYP2E1 by chlomethiazole was associated with an improvement in the damage[23]. It has been concluded that this is mainly due to the stimulation and inhibition of free radical formation, respectively. The role of CYP2E1 induction and cell injury has been studied in detail in the liver. For example, oxidized DNA products have been found to be lower in CYP2E1 knockout mice compared to wild-type mice[24,25], whereas more pronounced hepatic damage was observed in transgenic mice overexpressing CYP2E1[26]. Although NADPH oxidase knockout mice do not produce a radical detectable by spin trapping, oxidative DNA damage does occur in these animals. Moreover, since CYP2E1 knockout mice reveal less DNA damage, at least two sources of oxidants may have additive effects on each other.

In humans the extent of CYP2E1 induction is individually determined, but may be significant following the ingestion of 40 g alcohol per day, corresponding to 400 ml of 12.5 vol.% wine, for 1 week[27]. ROS produced by CYP2E1 led to lipid peroxidation, with the generation of malondialdehyde and 4-HNE[28]. 4-HNE not only leads to a mutation of the p53 gene at codon 249, which makes cells more resistant to apoptosis and shows some growth advantages[29], but also causes the production of exocyclic DNA etheno adducts, which can be measured in urine and determined immunohistologically in the liver[30]. It has been shown that these adducts are present in AFL, but are more pronounced in advanced ALD such as cirrhosis. Some evidence suggests that the degree of CYP2E1 induction in cell cultures can be correlated with the extent of etheno DNA adducts. With respect to the effect of ethanol on the generation of mitochondrial ROS, readers are referred to a recent review article[31].

It should be pointed out that hepatic FFA may also induce CYP2E1 and may lead to ROS. Thus fatty liver enhances the production of ROS and is especially sensitive to these species, since increased lipid peroxidation may occur and increased lipid peroxidation products such as 4-HNE and malondialdehyde are generated, which may bind to protein, resulting in protein adducts (with change in function and antibody production) and which may also bind to DNA as described above.

Role of TNF-α and other cytokines

In fatty liver, FFA can directly activate the IKK-β/NF-κB pathway in hepatocytes via a lysosomal, cathepsin B-dependent mechanism[32]. Hereby a translocation of Bax to lysosomes with lysosomal destabilization and release of cysteine protease (cathepsin B) is required, resulting in an activation of NF-κB via IKK-β and a consecutive increase in TNF-α, and other interleukins including osteopontin[33]. In addition, acetaldehyde also leads to an increase in NF-κB[34]. The degree of inflammation and fibrosis correlates with NF-κB expression, which in turn coincided with apoptosis albeit Bcl-2 and uncoupling protein 2 (UCP-2)[35].

Most of TNF-α may, however, be secreted by Kupffer cells due to endotoxin stimulation. Chronic alcohol consumption may lead to intestinal bacterial overgrowth and to mucosal injury. As a result lipopolysaccharides and bacterial products are entering the portal vein and these endotoxins activate hepatic Kupffer cells via CD14/toll 4 receptor complex, leading to intracellular signal transduction, with the release of proinflammatory cytokines, particularly TNF-α, eicosanoids, ROS and nitric oxide[33,34,36,37]. These cytokines induce inflammation and fibrogenesis. TNF-α results after binding to its receptor in cell apoptosis. It should be pointed out that Fas-induced apoptosis is significantly enhanced in the presence of CYP2E1[38].

As ALD can be abolished by the use of a CYP2E1 inhibitor, TNF-α antibodies may also prevent ASH in animals[37]. Furthermore the application of antibiotics leading to a decrease in gut endotoxins, or the knockout of Kupffer cells by gadolinium chloride also improve hepatic histology in the Tsukamoto–French alcohol model[39]. Subsequently, when TNF-α receptor knockout mice were studied, ASH could also be prevented[40,41]. All this supports the central role of TNF-α in the pathophysiology of ASH. Figure 1 shows a simplified scheme of the pathogenesis of ASH.

Besides other factors, interleukin 6 (IL-6) is significantly elevated in alcoholic patients with advanced liver disease[42]. It should be pointed out that IL-6 inhibits HOGG1, an important repair enzyme for 8-oxoguanosine adducts. IL-6 is also anti-apoptotic through up-regulation of the anti-apoptotic gene mcl-1[34].

More recently intrahepatic gene expression was studied in ASH as compared to AFL and normal livers[43]. Gene expression was analysed by DNA microarray on RNA isolated from patients' livers. The most important results confirmed by RT-PCR was an up-regulation of IL-8, osteopontin and TNFRSF14, a member of the TNF-α gene family. Osteopontin is probably the most interesting protein which has also been found to be elevated in experimental alcohol-induced liver injury[44], and may deserve further attention.

Liver cirrhosis may develop from ASH; however, it is not clear whether ASH with enormous inflammation-driven oxidative stress and hyperproliferation is an important prerequisite for cancer development some decades later in the state of cirrhosis. Therefore it would be important to know whether cirrhotic patients who went through ASH have an increased risk for hepatocellular cancer compared to those without ASH.

Effect of ethanol on the ubiquitin–proteasome pathway

Another mechanism by which alcohol may contribute to the occurrence of ASH is inhibition of the ubiquitin–proteasome pathway. The ubiquitin–proteasome pathway regulates protein digestion within the cell. Many liver cell functions are regulated by this mechanism of protein degradation; these include cell cycle checkpoints and activation of transcription factors such as nuclear factor NF-κB, and hypoxia-inducible factor 1 α (HIF1α). The loss of proteasomes or the inhibition of the ubiquitin–proteasome pathway can lead to hepatocellular injury including proliferation and apoptosis, and hepatic inclusions of aggregated cytokeratins[45]. Liver cell gene expressions, dependent on transcription factor activation by the proteasome, could impede the inflammatory response of the liver and the response to hypoxic injury. Ethanol stabilizes proteins normally degraded by the proteasomes. This may include among others CYP2E1, NF-κB, and HIF1α[45]. The inhibition does not occur for 2 weeks after ethanol feeding; this suggests that an oxidized product of CYP2E1 generated by ethanol CYP2E1 would be a likely mechanism of inhibition. Indeed, transgenic mice in which the gene of CYP2E1 had been knocked out, failed to develop proteasome activity inhibition even after 1 month of ethanol feeding[46]. One candidate product is 4-HNE, an end-product of lipid peroxidation which results from HER generation by CYP2E1 during ethanol oxidation. Thus, oxidative stress leading to 4-HNE may further potentiate oxidative stress by inhibiting degradation of CYP2E1.

Another possible mechanism that could explain the loss of proteasome activity after alcohol feeding is phosphorylation of a regulatory subunit of the proteasome[47]. For more detailed information please refer to a recent review article[45].

The occurrence of Mallory bodies (MB) may be linked to polyubiquinated proteins in the liver, especially cytokeratins. Hyperphosphorylation of cytokeratin 8 has been found in MB, and it is believed that hyperphosphorylation induces MB formation[48]. The turnover of cytokeratins is signalled by phosphorylation followed by ubiquitination. Proteolysis of the cytokeratins then occurs by the proteasome. In addition MB also stained positive with antibodies against heat-shock protein[49], as well as p62 (Z1P)[50]. In summary, proteasome inhibitors such as ethanol induce MB formation by leading to an inhibition of proteolysis and an accumulation of protein in the cell. Since cytokeratins are major components of MB they disappear from the cell (ghost cell formation). MB cells lose their bile canaliculi and cytokeratin in the plasma associated with an enlargement of the cell. The functional consequences are yet unclear. Lack of cytokeratins may make cells more vulnerable to injury or physical stress, may disturb the progression of the cell cycle and may develop apoptosis in response to FAS.

THERAPY OF ACUTE ALCOHOLIC STEATOHEPATITIS

On the basis of the pathophysiology of ASH various therapeutic strategies have been developed:

1. Abstinence
2. Nutritional therapy
3. Drugs (glucocorticosteroids, pentoxifyllin, antioxidants, TNF-α antibodies)
4. Extracorporal liver support
5. Liver transplantation

It has already been pointed out that ASH may vary from mild to severe, and that in most cases alcohol abstinence is mandatory to cure ASH. In more severe cases prognosis worsens and therefore prognostic scores have been developed to identify those patients with poor prognosis. The Maddrey index or discriminant function (DF) is probably the most-used score with bilirubin and prothrombin time as independent risk factors[51] (Table 1). A DF of more than 32 identifies patients with a 1-month mortality of 50%. In addition, the Orrego index has also been used and is a combined clinical and laboratory index (CCLI)[52]. CCLI correlates well with 1-year mortality. Furthermore, the Model for Endstage Liver Disease (MELD), considers kidney function in addition to other variables[53] (Table 1). A MELD index over 21 gives a 3-month mortality of 20%. Finally, the Glasgow Alcoholic Hepatitis Score (GAHS) seems even better, since a recent comparison with DF showed a significantly better prediction of 3-month outcome with 75% as compared to 53%[54]. In general, ASH with ascites and encephalopathy has a 90-day mortality of approximately 75%. If either ascites or encephalopathy is present, mortality is approximately 25%[55].

Table 1 Prognosis scores in alcoholic hepatitis

Maddrey score: Discriminant Function (DF)

$4.6 \times$ (PTT patient − PTT control) + bilirubin (mg%) $\geqslant 32$ = 1-month mortality 50%

Model for Endstage Liver Disease (MELD)

$9.57 \times \log_e$ (creatinine mg%) + $3.78 \times \log_e$ (bilirubin mg%) + $11.2 \times \log_e$ (INR) + 6.4 $\geqslant 21$ = 3-month mortality 20%

Glasgow Alcoholic Hepatitis Score (GAHS)

	Score 1	Score 2	Score 3
Age	<50	>50	–
White cell count (10^9/L)	<15	>15	–
Urea (mmol/L)	<5	>5	–
Prothrombin time	<1.5	1.5–2.0	>2.0
Bilirubin (mg%)	<7.25	7.25–14.5	>14.5

Nutritional therapy

Oral and/or enteral nutrition

As assessed in a large United States trial from Veterans Affairs Hospitals, the prevalence of malnutrition reaches 100% in patients with severe ASH, and a significant correlation of malnutrition with short- and long-term survival was established[56]. So far five studies have been published that investigated the efficacy of oral or enteral nutritional support in patients with ASH[57–61]. The design, sample size, duration and nutritional therapies administered differed substantially; hence it is not surprising that the outcomes were variable. In the first four studies[57–60], nitrogen balance and/or serum albumin levels were improved along with nutritional support, while mortality was unaffected. This may have been due to the small sample size in all studies, which could have caused a type II error masking a beneficial effect of nutritional therapy and supporting an ongoing need of trials. In this respect the study of Cabré et al. merits attention[61]. In this randomized multicentre trial 71 patients with severe alcoholic hepatitis were enrolled to receive either 40 mg/day prednisolone and a diet containing 2000 kcal/day, or total enteral nutrition via a nasoduodenal tube also providing 2000 kcal/day (n = 35), for 4 weeks. The severity of ASH was confirmed using the DF. Although the differences in both the short-term mortality and the 1-year probability of survival between the two groups were not statistically different, several important issues became apparent.

1. There was a significantly better outcome during the 1-year follow-up after 4 weeks of treatment for the patients treated with total enteral nutrition (p = 0.04, intention-to-treat analysis), with only two patients dying (8%), whereas 10 patients (37%) who were treated with prednisolone alone died during the 1-year follow-up period, mostly due to infections within the first 6 weeks of follow-up.

2. Furthermore, there was a high number of drop-outs in the total enteral nutrition group, resulting from the lack of patient acceptance of the nasogastric tube rather than from serious adverse events. From these data it may be possible that a combination of total enteral nutrition and steroids may yield even better clinical results than either therapy alone.

Two further studies with a reasonable number of patients showed no treatment benefit for patient survival, but surrogate parameters of nutritional status, such as serum albumin, retinol-binding protein and nitrogen balance, improved[57,60]. The studies by Galambos et al.[57] and Soberon et al.[60] included only 16 and 14 patients, respectively, which appears to be an insufficient sample size to draw firm conclusions. In two multicentre trials of patients from United States Veterans Administration Hospitals, Mendenhall et al. attempted to enhance the efficacy of enteral nutrition by adding oxandrolone, a synthetic anabolic steroid[62,63]. In the first study oxandrolone therapy was compared against prednisolone, with nutritional support not being the primary intervention in a cohort of 57 patients with ASH[62]. In the second study 273 patients with alcoholic hepatitis on active treatment received a standard hospital diet in addition to a nutritional formula preparation containing 1600 kcal/day (Hepatic Aid), thereby providing a supplement of 60 g/day of proteins during a period of 30 days while patients were in the hospital, together with oxandrolone at 80 mg/day[63]. After discharge, patients received 1200 kcal/day of Hepatic Aid, containing 45 g/day of proteins in addition to 40 mg/day oxandrolone, for the following 60 days as outpatients. In control patients oxandrolone was replaced by placebo tablets accompanied by a low-calorie, low-protein food supplement providing 6.8 g/day of proteins or 264 kcal/day while in the hospital, and 5.1 g/day or 198 kcal/day as outpatients. Because of the large sample size, subgroups were analysed, such as those with either severe or moderate malnutrition at inclusion into the study. Thereby, it was demonstrated that patients with moderate malnutrition experienced a significant improvement through active treatment, as estimated on the basis of DF, and the level of malnutrition defined as the degree of protein energy malnutrition (p = 0.03 and 0.05, respectively). The most important finding, however, was the relationship of therapy to survival, which showed that, at 30 days of treatment, 9.4% of patients on active treatment had died, while this figure reached 20.9% in the placebo group. After 6 months this difference was even more pronounced, showing that 20.3% of actively treated patients had died, while mortality in the placebo group was 37.3% (p = 0.037). Another important result was that the treatment benefit could be observed only with successful nutritional therapy and not with oxandrolone medication. Patients from both studies were recently subjected to a meta-analysis, in which the authors concluded that the efficacy of oxandrolone treatment resulted from sufficient nutritional therapy rather than from the administration of oxandrolone[64].

Parenteral nutrition

The data available do not allow us to distinguish the effect of parenteral nutrition on ASH from that on cirrhosis separately. Parenteral nutrition in patients with ASH and cirrhosis was performed in 244 patients with different study protocols[65–70]. Only randomized controlled trials will be discussed here. The treatment duration ranged from 21 to 30 days. Unfortunately, none of the studies could demonstrate an improved survival, although the first study by Nasrallah and Galambos provoked this expectation[65]. All trials showed an improvement of visceral protein as assessed by serum albumin levels, and three studies reported a positive nitrogen balance in the treatment groups. Surprisingly, no long-term studies exist which address intermediate or long-term effects on survival in patients with alcoholic hepatitis or cirrhosis, and the most recent trial which investigated the treatment benefit of parenteral nutrition in patients undergoing hepatic resection for hepatocellular carcinoma dates back to 1994[71]. However, the proportion of patients in the study population with ALD, i.e. ASH and cirrhosis, was not reported. To our knowledge no study on total parenteral nutrition in exclusively alcoholic cirrhotics has been performed.

In patients with ASH only one trial investigated whether beneficial effects on nutritional status may be achieved by the combination of intravenous nutritional therapy and oral oxandrolone[72,73]. Thirty-nine patients with clinical criteria of alcoholic hepatitis, but without confirmatory liver biopsy, were allocated to four groups receiving: (1) nutritionally adequate standard diet; (2) standard diet and oxandrolone 80 mg/day; (3) standard diet plus 2 L of 3.5% branched-chain amino acid enriched solution/day (Aminsyn II, Abbott); or (4) standard diet and intravenous branched-chain amino acid plus 80 mg oxandrolone/day for 21 days. While the first paper of the study described improvements of liver function tests such as galactose elimination capacity, antipyrine clearance, as well as liver and spleen sizes, following parenteral nutrition, regardless of oxandrolone administration, a second analysis of the same study demonstrated a significant improvement of nutritional status, nitrogen balance and visceral proteins in the groups receiving parenteral nutritional therapy with and without oxandrolone. Oxandrolone administration again failed to result in further improvement of measured parameters with the exception of serum pre-albumin and transferrin, which were highest in the patients treated with parenteral nutrition plus oxandrolone. Tolerability of nutritional therapy and oxandrolone was apparently good, since all patients completed the trial. However, in addition to the fact that the study groups were relatively small, the significance of liver function tests to reflect the patients' prognosis is questionable, and long-term survival was not reported. Due to the lack of clear-cut evidence, oxandrolone, either in combination with nutritional therapy or alone, cannot be advised for patients with ASH.

Drugs

Antioxidants

Since ROS have been implicated in the pathogenesis of ASH, various antioxidants have been used in the treatment of ASH with negative results. More recently two studies have used vitamin E and a cocktail of various antioxidants (vitamin C, vitamin E, selenium, methionine, allopurinol, desferrioxamine, *N*-acetylcystein, b-carotin) to patients with various severities of ASH[74,75]. Neither study could show a beneficial effect on hepatic function or mortality.

Glucocorticosteroids

Therapy with steroids inhibits Kupffer cell and neutrophilic granulocyte function, decreases proinflammatory cytokines such TNF-α, TGF-β, IL-1 and IL-8, and also decreases the immune response towards acetaldehyde protein adducts. A great number of studies on the use of steroids in ASH has been published[51,62,76–87]. Various meta-analyses have been performed, coming to different results. While the meta-analysis from the 1990s shows a therapeutic benefit[88], the meta-analysis of 1995 did not show such a benefit[89]. A final meta-analysis from 2002, having integrated the new studies again, concluded a beneficial effect of steroids in a certain subgroup of patients with ASH[90].

Analysing the more recent placebo-controlled double-blind prospective carefully designed studies, it can be concluded that, under certain conditions, steroids may be of benefit to a subgroup of patients with severe ASH. These are patients with a DF of more than 32, with encephalopathy and jaundice and who do not have gastrointestinal haemorrhages or infections. Early change in bilirubin following 7 days of steroid treatment is a simple and powerful criterion for helping to identify patients who are responders to steroid treatment. Approximately 75% of patients with a DF greater than 32 are responders to steroids. At 6 months survival of patients with an early change in bilirubin levels is around 80%[91]. Under 30 mg prednisolone given for 4 weeks, these patients may have a significantly better 1-month and 1-year survival; this benefit is lost after 2 years. An important side-effect of steroid treatment is sepsis and infections, which have been taken carefully into consideration.

Pentoxifylline

One study in patients with ASH has been performed with pentoxifylline, which is a non-selective phosphodiesterase inhibitor[92]. The clinical indication for that drug is peripheral arterial occlusive disease. Pentoxifylline inhibits synthesis and secretion of TNF-α, decreases the activation of neutrophilic granulocytes and lymphocyte proliferation, and seems to have antifibrotic effects in animal experiments. The study has been performed in 101 patients with severe ASH (DF 32 and more) with three times 400 mg pentoxifylline over 1 month in a placebo-controlled double-blind fashion. Short-term mortality was significantly reduced in the pentoxifylline group, primarily due a decrease in hepatorenal syndrome. In addition, TNF-α levels were found to be significantly reduced. The effect was greater compared to the effect expected by the use of glucocorticosteroids. No significant side-effects have been noted.

TNF-α antibodies

Another approach to eliminate TNF-α has been performed by the use of a chimeric monoclonal TNF-α antibody. Two studies with 20 and 12 patients (DF over 32 and biopsy-proven ASH) have been performed. In the first study, by Spahr et al.[93], the antibody (infliximab) was given to 20 patients with ASH in a dose of 5 mg/kg body weight in addition to 40 mg prednisolone. No survival difference was observed; however, there was a decrease in serum concentrations of IL-6 and IL-8, and of the DF. In the second study, by Tilg et al.[94], infliximab was given at the same dose in addition to norfloxacin to ASH patients with ascites. The 15-month survival was 83%, but two patients died due to septicaemia. More recently Naveau et al.[95] performed a study with 36 ASH patients with a DF over 32. Infliximab in a dose of 10 mg/kg body weight was given three times on weeks 0, 2, and 4. In addition, 40 mg prednisolone were administered over 28 days. This regimen was compared to 40 mg prednisolone with placebo. The authors recorded a mortality of 39% in the verum group, and 18% in the placebo group, which was not statistically significant. The mortality due to infections was 22% versus 11% and to severe infections 83% versus 28%, which was highly significant ($p = 0.002$). No difference in DF was found at any time.

The dose of infliximab used in this study is in our opinion far too high, since with this dose a significant immune suppression occurs. In addition, a combination of steroids plus TNF-α antibodies should not be performed.

Preliminary data have been published with etanercept, which is a soluble TNF-α receptor Fc fusion protein. A study was performed in 13 patients with a DF of more than 15 and/or hepatic encephalopathy for 2 weeks[96]; 32 mg/kg body weight of etanercept was given as loading dose, followed by 25 mg subcutaneously at days 4, 8 and 12. Thirty-day survival was 92% and 90-day survival was 33%. Side-effects were reported in 23%.

Extracorporeal liver support

Extracorporeal liver support by the use of molecular absorbent recirculating system (MARS) has been performed in eight patients with severe ASH and hepatic encephalopathy[97]. Five of these patients had hepatorenal syndrome I, two had hepatorenal syndrome II. The overall 90-day mortality was 50%, whereas the predicted 90-day mortality with the MELD score was 27%. Of the five patients with hepatorenal syndrome I, two died within the 30 days, one within 90 days, one had an orthotopic liver transplantation, and one was alive at day 90. Of the two patients with hepatorenal syndrome II, one died and one was alive at day 90. One patient had no hepatorenal syndrome and was still alive at the end of the study.

Liver transplantation

Orthoptic liver transplantation in ASH has been performed, and the outcome has been compared to patients with alcoholic cirrhosis alone. This Spanish study showed that there was no difference in 5-year survival between 36 patients with alcoholic cirrhosis plus ASH and 32 patients with alcoholic cirrhosis alone following liver transplantation[98]. There was also no difference in 5-year survival between those patients who had DF over 32 and those below 32. Ten per cent of patients returned to alcohol consumption after transplantation. Despite these data a 6-month abstinence from alcohol is generally required in most liver transplantation centres before liver transplantation will be performed.

References

1. Grant B, Harfort T, Dawson D, Chou P, DuFour M, Pickering R. Prevalence of DSM-IV alcohol abuse and dependence. United States: Alcohol Health Res World. 1992;18:243–8.
2. Jahrbuch Sucht, Deutsche Hauptstelle für Suchtfragen, editor. Geesthacht, Germany: Neuland Verlagsgesellschaft mbh, 2003.
3. Stickel F, Österreicher CH. The role of genetic polymorphisms in alcoholic liver disease. Alcohol Alcohol. 2006;41:209–24.
4. Seitz HK. Alkoholische Lebererkrankungen. In: Riemann JF, Fischbach W, Galle PR, Mössner J, editors. Gastroenterologie in Klinik und Praxis. Stuttgart: Thieme Verlag. 2007 (In press).

5. Seitz HK, Stickel F. ASH and NASH. Fuchs M, editor. Rationale Diagnostik und Therapie von chronischen Hepatitiden. Wissenschaftliche Verlagsgesellschaft Stuttgart. 2007 (In press).
6. You M, Considine RV, Leone TC, Kelly DP, Crabb DW. Role of adiponectin in the protective action of dietary saturated fat against alcoholic fatty liver in mice. Hepatology. 2005:42:568–77.
7. Anania FA. Adiponectin and alcoholic fatty liver: is it, after all, about what you eat? Hepatology. 2005;42:530–2.
8. Nakajima T, Kamijo Y, Tanaka N et al. Peroxisome proliferator-activated receptor α protects against alcohol-induced liver damage. Hepatology. 2004;40:972–80.
9. Ji C, Chan C, Kaplowitz N. Central role of SREBP lipogenic pathways in alcohol-induced hepatic steatosis. Gastroenterology. 2006;130:A596.
10. Bellentani S, Tiribelli C. The spectrum of liver disease in the general population: lesson from the Dionysos study. J Hepatol. 2001; 35:531–7.
11. Tsukamoto H, Ding QG, She H. The comparison of experimental NASH with ASH pathogenesis. Alcohol Clin Exp Res. 2006;30:98A.
12. Albano E. Free radicals and alcohol-induced liver injury. In: Sherman CDIN, Preedy VR, Watson RR, editors. Ethanol and the Liver. London: Taylor & Francis, 2002:153–90.
13. Garcia-Ruiz C, Colell A, Paris R, Fernandez-Checa JC. Direct interaction of GD3 ganglioside with mitochondria generates reactive oxygen species followed by mitochondrial permeability transition, cytochrome c release and caspase activation. FASEB J. 2000;14:847–50.
14. Bautista AP. Neutrophilic infiltration in alcoholic hepatitis. Alcohol. 2002;27:17–21.
15. Chamulitrat W, Spitzer JJ. Nitric oxide and liver injury in alcohol fed rats after lipoplysaccharide administration. Alcohol Clin Exp Res. 1996;20:1065–70.
16. Beckmann JS, Koppenol WH. Nitric oxide, superoxide and peroxynitrite: the good, the bad and the ugly. Am J Physiol. 1996;271:C1424–37.
17. Arteel GE. Oxidant and antioxidant in alcohol-induced liver disease. Gastroenterology. 2003;124:778–90.
18. Seitz HK, Osswald B. Effect of ethanol on procarcinogen activation. In: Watson RR, editor. Alcohol and Cancer. Boca Raton, FL: CRC Press, 1992:55–72.
19. Dupont I, Lucas D, Clot P, Ménez C, Albano E. Cytochrome P4502E1 inducibility and hydroxyethyl radical formation among alcoholics. J Hepatol. 1998;28:564–71.
20. Albano E, Clot P, Morimoto M, Tomasi A, Dianzani MU. Evidence for the covalent binding of hydroxyethyl radicals to rat liver microsomal proteins. Alcohol Alcohol. 1993;28:453–9.
21. Clot P, Bellomo G, Tabone M, Arico S, Albano E. Detection of antibodies against proteins modified by hydroxyethyl free radicals in patients with alcoholic cirrhosis. Gastroenterology. 1995;108:207–10.
22. Clot P, Albano E, Elliasson E et al. Cytochrome P4502E1 hydroxyethyl radical adducts as the major antigenic determinant for autoantibody formation among alcoholics. Gastroenterology. 1996;111:206–16.
23. Gouillon Z, Lucas D, Li J et al. Inhibition of ethanol-induced liver disease in the intragastric feeding rat model by chlormethiazole. Proc Soc Biol Med. 2000;224:302–8.
24. Bardag-Gorce F, Yuan QX, Li J et al. The effect of ethanol induced cytochrome P4502E1 on the inhibition of proteasome activity by alcohol. Biochem Biophys Res Commun. 2000; 279:23–9.
25. Bradford BU, Kona H, Isayama F et al. Cytochrome P4502E, but not nicotinamide adenine dinucleotide phosphate oxidase is required for ethanol-induced oxidative DNA damage in rodent liver. Hepatology. 2005;41:336–44.
26. Morgan K, French SW, Morgan TR. Production of a cytochrome P4502E1 transgenic mouse and initial evaluation of alcoholic liver damage. Hepatology. 2002;36:122–34.
27. Oneta CM, Lieber CS, Li JJ et al. Dynamics of cytochrome P4502E1 activity in man: induction by ethanol and disappearance during withdrawal phase. J Hepatol. 2002;36:47–52.
28. Aleynik SI, Leo MA, Aleynik MK, Lieber CS. Increased circulating products of lipid peroxidation in patients with alcoholic liver disease. Alcohol Clin Exp Res. 1998;22:192–6.

29. Hu W, Feng Z, Eveleigh J et al. The major lipid peroxidation product, trans-4-hydroxy-2-nonenal, preferentially forms DNA adducts at codon 249 of human p53 gene, a unique mutational hot spot in hepatocellular carcinoma. Carcinogenesis. 2002;23:1781–9.
30. Frank A, Seitz HK, Bartsch H, Frank N, Nair J. Immunohistochemical detection of 1,N^6-ethenodeoxyadenosine in nuclei of human liver affected by diseases predisposing to hepatocarcinogenesis. Carcinogenesis. 2004;25:1027–31.
31. Hoek JB, Cahill A, Pastorino JG. Alcohol and mitochondria: a dysfunctional relationship. Gastroenterology. 2002;122:2049–63.
32. Feldstein AE, Werneburg NW, Canbay A et al. Free fatty acids promote hepatic lipotoxicity by stimulating TNF-α expression via a lysosomal pathway. Hepatology. 2004;40:185–94.
33. Day CP. From fat to inflammation. Gastroenterology. 2006;130:207–10.
34. Seitz HK, Stickel F. Risk factors and mechanisms of hepatocarcinogenesis with special emphasis on alcohol and oxidative stress. Biol Chem. 2006;387:349–60.
35. Ribeiro PS, Cortez-Pinto H, Solá S et al. Hepatocyte apoptosis, expression of death receptors, and activation of NF-κB in the liver of non-alcoholic and alcoholic steatohepatitis patients. Am J Gastroenterol. 2004;99:1708–17.
36. Jarvelainen HA, Fang C, Ingelman-Sundberg M, Lindros KO. Effect of chronic coadministration of endotoxin and ethanol on rat liver pathology and proinflammatory and antiinflammatory cytokines. Hepatology. 1999;29:1503–10.
37. Hines IN, Wheeler MD. Recent advances in alcoholic liver disease. III. Role of the innate immune response in alcoholic hepatitis. Am J Physiol. 2004;287:G310–14.
38. Wang X, Lu Y, Cederbaum AI. Induction of cytochrome P4502E1 increases hepatotoxicity caused by Fas agonistic Jo2 antibody in mice. Hepatology. 2005;42:400–10.
39. Thurman RG. Alcoholic liver injury involves activation of Kupffer cells by endotoxins. Am J Physiol. 1998;275:G605–11.
40. Uesugi T, Froh M, Arteel GE, Bradford, BU, Thurman RG. Toll-like receptor 4 is involved in the mechanism of early alcohol induced liver injury in mice. Hepatology. 2001;34:101–8.
41. Yin M, Bradford BU, Wheeler MD et al. Reduced early alcohol induced liver injury in CD 14 deficient mice. J Immunol. 2001;166:4737–42.
42. Urbaschek R, McCuskey RS, Rudi V et al. Endotoxin, endotoxin-neutralizing capacity sCD14, sICAM-1, and cytokines in patients with various degrees of alcoholic liver disease. Alcohol Clin Exp Res. 2001;25:261–8.
43. Seth D, Gorrell MD, Cordoba S, McCaughan GW, Haber PS. Intrahepatic gene expression in human alcoholic hepatitis. J Hepatol. 2006;45:306–20.
44. Apte UM, Banerjee A, McRee R, Wellberg E, Ramaiah SK. Role of osteopontin in hepatic neutrophil infiltration during alcoholic steatohepatitis. Toxicol Appl Pharmacol. 2005;207: 25–38.
45. French SW, Bardag-Gorce F. Ubiquitin–proteasome pathway in the pathogenesis of liver disease. In: Dufour JF, Clavien PA, editors. Signalling Pathways in Liver Disease. Berlin: Springer, 2005:377–87.
46. Bardag-Gorce F, Yuan QX, Li J et al. The effect of ethanol-induced cytochrome P4502E1 on the inhibition of proteasome activity by alcohol. Biochem Biohpys Res Commun. 2000; 279:23–9.
47. Kawahara H, Cadrin M, French SW. Ethanol-induced phosphorylation of cytokeratin in cultured hepatocytes. Life Sci. 1990;47:859–63.
48. Cadrin M, Anderson M, Aasheim LH et al. Modifications in cytokeratin and actin in cultured liver cells derived from griseofulvin-fed mice. Lab Invest. 1995;72:453–60.
49. Riley NE, Li J, McPhaul LW et al. Heat shock proteins are present in Mallory bodies (cytokeratin aggresomes) in human liver biopsy specimens. Exp Mol Pathol. 2003;74:168–72.
50. Stumptner C, Fuchsbichler M, Lehner M et al. Sequence of events in the assembly of Mallory body components in mouse liver: clues to the pathogenesis and significance of Mallory body formation. J Hepatol. 2001;34:665–75.
51. Maddrey WC, Boitnott JK, Bedine MS, Weber FL, Mezey E, White RI. Corticosteroid therapy of alcoholic hepatitis. Gastroenterology. 1978;75:193–9.
52. Orrego H, Israel Y, Blake YE, Medline A. Assessment of prognostic factors in alcoholic liver disease: toward a global quantitative expression of severity. Hepatology. 1983;3:896–905.

53. Wiesner R, Edwards E, Freeman R et al. Model for end-stage liver disease (MELD) and allocation of donor livers. Gastroenterology. 2003;124:91–6.
54. Forrest EH, Evans CDJ, Stewart S et al. Analysis of factors predictive of mortality in alcoholic hepatitis and derivation and validation of the Glasgow alcoholic hepatitis score. Gut. 2005;54:1174–9.
55. Dunn W, Jamil LH, Brown LS et al. MELD accurately predicts mortality in patients with alcoholic hepatitis. Hepatology. 2005;41:353–8.
56. Mendenhall CL, Anderson S, Weesner RE et al. Protein–calorie malnutrition associated with alcoholic hepatitis. Veterans Administration Cooperative Study Group on Alcoholic Hepatitis. Am J Med. 1984;76:211–22.
57. Galambos JT, Hersh T, Fulenwider JT et al. Hyperalimentation in alcoholic hepatitis. Am J Gastroenterol. 1979;72:535–41.
58. Mendenhall CL, Bongiovanni G, Goldberg S et al. VA cooperative study on alcoholic hepatitis. III: changes in protein–calorie malnutrition associated with 30 days of hospitalization with and without enteral nutritional therapy. J Parenter Enteral Nutr. 1985;9:590–6.
59. Calvey H, Davis M, Williams R. Controlled trial of nutritional supplementation, with and without branched chain amino acids enrichment, in treatment of acute alcoholic hepatitis. J Hepatol. 1985;1:141–51.
60. Soberon S, Pauley MP, Duplantier R et al. Metabolic effects of enteral formula feeding in alcoholic hepatitis. Hepatology. 1987;7:1204–9.
61. Cabré E, Rodriguez-Iglesias P, Caballeria J et al. Short-term and long-term outcome of severe alcohol-induced hepatitis. Hepatology. 2000;32:36–42.
62. Mendenhall CL, Anderson S, Garcia-Pont P et al. Short-term and long-term survival in patients with alcoholic hepatitis treated with oxandrolone and prednisolone. N Engl J Med. 1984;311:1464–70.
63. Mendenhall CL, Moritz TE, Roselle GA et al. A study of oral nutritional support with oxandrolone in malnourished patients with alcoholic hepatitis: results of a Department of Veterans Affairs Cooperative Study. Hepatology. 1993;17:564–76.
64. Mendenhall CL, Roselle GA, Gartide P et al. Relationship of protein calorie malnutrition to alcoholic liver disease: a re-examination of data from two Veterans Administration Cooperative Studies. Alc Clin Exp Res. 1995;19:635–41.
65. Nasrallah JM, Galambos JT. Amino acid therapy of alcoholic hepatitis. Lancet. 1980; II:1276–7.
66. Diehl AM, Boitnott JK, Herlong HF et al. Effect of parenteral amino acid supplementation in alcoholic hepatitis. Hepatology. 1985;5:57–63.
67. Naveau S, Pelletier G, Poynard T et al. A randomized clinical trial of supplementary parenteral nutrition in jaundiced alcoholic cirrhotic patients. Hepatology. 1986;6:270-4.
68. Achord JL. A prospective randomized clinical trial of peripheral amino acid–glucose supplementation in acute alcoholic hepatitis. Am J Gastroenterol. 1987;82:871–5.
69. Simon J, Galambos JT. A randomized controlled study of peripheral parenteral nutrition in moderate and severe alcoholic hepatitis. J Hepatol. 1988;7:200–7.
70. Mezey E, Caballeria J, Mitchell MC et al. Effect of parenteral amino acid supplementation on short-term and long-term outcomes in sever alcoholic hepatitis: a randomized controlled trial. Hepatology. 1991;14:1090–6.
71. Fan ST, Lo CM, Lai ECS et al. Perioperative nutritional support in patients undergoing hepatectomy for hepatocellular carcinoma N Engl J Med. 1994;331:1547–52.
72. Bonkovsky H, Fiellin DA, Smith GS et al. A randomized, controlled trial of treatment of alcoholic hepatitis with parenteral nutrition and oxandrolone. I: Short-term effects on liver function. Am J Gastroenterol. 1991;86:1200–8.
73. Bonkovsky H, Singh RH, Jafri ICH et al. A randomized, controlled trial of treatment of alcoholic hepatitis with parenteral nutrition and oxandrolone. II: Short-term effects on nitrogen metabolism, metabolic balance, and nutrition. Am J Gastroenterol. 1991;86:1209–18.
74. Phillips M, Curtis H, Portmann B, Donaldson N, Bomford A, O'Grady J. Antioxidants versus corticosteroids in the treatment of severe alcoholic hepatitis – a randomized clinical trial. J Hepatol. 2006;44:784–90.
75. Mezey E, Potter JJ. A randomized placebo controlled trial of vitamin E for alcoholic hepatitis. J Hepatol. 2004;40:40–60.

76. Ramond MJ, Poynard T, Rueff B et al. A randomized trial of prednisolone in patients with severe alcoholic hepatitis. N Engl J Med. 1992;326:507–12.
77. Carithers Jr RL, Herlong HF, Diehl AM et al. Methylprednisolone therapy in patients with severe alcoholic hepatitis: a randomized multicenter trial. Ann Intern Med. 1989;110:685–90.
78. Helman RA, Temko MH, Nye SW, Fallon HJ. Natural history and evaluation of prednisolone therapy. Ann Intern Med. 1971;74:311–21.
79. Porter HP, Simon FR, Pope CE, Volwiler W, Fenster F. Corticosteroid therapy in severe alcoholic hepatitis. N Engl J Med. 1971;284:1350–5.
80. Campra JL, Hamlin EM, Kirshbaum RJ, Olivier M, Redeker AG, Reynolds TB. Prednisone therapy of acute alcoholic hepatitis. Ann Intern Med. 1973;79:625–31.
81. Blitzer BL, Mutchnick MG, Joshi PH, Phillips MM, Fessel JM, Conn HO. Adrenocorticosteroid therapy in alcoholic hepatitis: a prospective, double-blind randomized study. Am J Dig Dis. 1077;22:477–84.
82. Lesesne HR, Bozymski EM, Fallon HJ. Treatment of alcoholic hepatitis with encephalopathy. Comparison of prednisolone with caloric supplements. Gastroenterology. 1987;74:169–73.
83. Shumaker JB, Resnick RH, Galambos JT, Makopour H, Iber FL. A controlled trial of 6-methylprednisolone in acute alcoholic hepatitis. Am J Gastroenterol. 1978;69:443–9.
84. Theodossi A, Eddleston ALWF, Williams R. Controlled trial of methylprednisolone therapy in severe acute alcoholic hepatitis. Gut. 1982;23:75–9.
85. Depew W, Boyer T, Omata M, Redeker A, Reynolds T. Double-blind controlled trial of prednisolone therapy in patients with severe acute alcoholic hepatitis and spontaneous encephalopathy. Gastroenterology. 1980;78:524–9.
86. Bories P, Guedj JY, Mirouze D, Yousfi A, Michel H. Traitement de l'hépatite alcoolique aigue par la prednisolone. Presse Med. 1987;16:769–72.
87. Mathurin P, Duchatelle V, Ramond MJ et al. Survival and prognostic factors in patients with severe biopsy-proven alcoholic hepatitis treated by prednisolone: randomized trial, new cohort, and simulation. Gastroenterology. 1996;110:1847–53.
88. Imperiale TF, McCullough AJ. Do corticosteroids reduce mortality from alcoholic hepatitis? Ann Intern Med. 1990;113:299–307.
89. Christensen E, Gluud C. Glucocorticoids are ineffective in alcoholic hepatitis: a metaanalysis adjusting for confounding variables. Gut. 1995;37:113–18.
90. Mathurin P, Mendenhall CL, Carithers Jr RL et al. Corticosteroids improve short-term survival in patients with severe alcoholic hepatitis (AH): individual data analysis of the last three randomized placebo controlled double-blind trials of corticosteroids in severe AH. J Hepatol. 2002;36:480–7.
91. Mathurin P. Corticosteroids for alcoholic hepatitis – what's next? J Hepatol. 2005;43:526–33.
92. Akriviadis E, Botla R, Briggs W, Han S, Reynolds T, Shakil O. Pentoxifylline improves short-term survival in severe acute alcoholic hepatitis: a double-blind, placebo-controlled trial. Gastroenterology. 2000;119:1637–48.
93. Spahr L, Rubbia-Brandt L, Frossard J. Combination of steroids with infliximab or placebo in severe alcoholic hepatitis: a randomized controlled pilot study. J Hepatol. 2002;37:448–56.
94. Tilg H, Jalan R, Kaser A et al. Anti tumor necrosis factor-alpha monoclonal antibody therapy in severe alcoholic hepatitis. J Hepatol. 2003;38:419–25.
95. Naveau S, Chollet-Martin S, Dharancy S et al. A double-blind randomized controlled trial of infliximab associated with prednisolone in acute alcoholic hepatitis. Hepatology. 2004; 39:A5.
96. Menon KVN, Stadheim L, Kamath PS et al. A pilot study of the safety and tolerability of etanercept in patients with alcoholic hepatitis. Am J Gastroenterol. 2004;99:255–61.
97. Jalan R, Sen S, Steiner C, Kapoor D, Alisa A, Williams R. Extracorporeal liver support with molecular adsorbents recirculating system in patients with severe acute alcoholic hepatitis. J Hepatol. 2003;38:24–31.
98. Tomé S, Martinez-Rey C, González-Quintela A et al. Influence of superimposed alcoholic hepatitis on the outcome of liver transplantation for end-stage alcoholic liver disease. J Hepatol. 2002;36:793–8.

3
Drug-induced hepatitis

N. KAPLOWITZ

INTRODUCTION

Drug-induced liver injury (DILI) has emerged in the past decade as an increasingly important medical problem[1]. Among the issues which make DILI so prominent are: (1) It is the leading cause of acute liver failure in the USA and Europe; (2) it is an important cause of jaundice in adults; (3) it can mimic any form of acute or chronic liver disease; (4) it is the leading cause of failure in drug development and warnings and withdrawals postmarketing. In this chapter I will concisely address some of the key clinical and pathophysiological points about DILI.

CLINICAL–PATHOLOGICAL PRESENTATION: THE SIGNATURE

DILI can induce virtually any form of acute or chronic liver disease. Table 1 lists the major presentations. The vast majority of DILI cases present as acute hepatitis (ALT/ULN $\div$ alk ptase/ULN $\geqslant 5$) or cholestatic/mixed injury. The distinction is important since acute icteric DILI hepatitis is potentially life-threatening. This is referred to as Hy's law after the late Hyman Zimmerman who noted a mortality $\geqslant 10\%$ in this situation[2] and has been validated in several recent reports[3,4]. DILI hepatitis typically resolves over weeks when drug is discontinued. On other hand, cholestatic reactions tend to resolve slowly over months when drug is discontinued. The fate of acute DILI after drug cessation is to progress to cirrhosis in a small percentage. Chronicity is more likely with cholestatic reactions and can lead to ductopenia and biliary cirrhosis on rare occasions following any cause of acute drug-induced cholestasis[5]. A common misconception is that each drug has a very narrow signature. Indeed, a typical signature is seen with some drugs (e.g. isoniazid hepatitis) but many drugs cover a broad spectrum from hepatitis to cholestasis (e.g. amoxicillin-clavulanate).

Aside from the common presentation as acute hepatitis and/or cholestasis, less common modes of injury are seen in special cases, such as microvesicular steatosis, resembling Reyes' syndrome, caused by mitochondrial toxicity of nucleoside reverse transcriptase inhibitors used for HIV and veno-occlusive

Table 1 The spectrum of clinical pathological signatures of drug-induced liver injury

Active injury
Hepatocellular (hepatitis-like)
acetaminophen, isoniazid, troglitazone
Cholestatic/mixed
erythromycin, sulindac, amoxicillin-clavulanate
Microvesicular steatosis (Reyes-like)
nucleosides, valproate
Veno-occlusive disease
chemotherapy–cyclophosphamide, azathioprine, dacarbazine, busulphan
Chronic injury
Chronic hepatitis (mostly autoimmune-like)
methyldopa, nitrofurantoin, minocycline, diclofenac, germander
Ductopenic cholestatic (PBC-like)
chlorpromazine, etc.
Steatohepatitis (NASH-like)
amiodarone, methotrexate, tamoxifen
Nodular regenerative hyperplasia
azathioprine, 6-MP, 6-TG, androgens

disease mainly observed in the setting of cytoreductive preparation for bone marrow transplantation. Chronic hepatitis is seen mostly resembling autoimmune hepatitis and exhibits a characteristically long latency (e.g. nitrofurantoin). Other chronic patterns include non-alcoholic steatohepatitis (NASH) and nodular regenerative hyperplasia (NRH). Drug-induced NASH is difficult to distinguish from underlying NASH but at least tends to exacerbate it (e.g. tamoxifen). The clinical significance of thiopurine-induced NRH remains uncertain – however, concern for development of portal hypertension is real.

IDIOSYNCRATIC DILI: ALLERGIC VERSUS NON-ALLERGIC

DILI can be viewed as predictable or unpredictable (idiosyncratic). Predictable DILI is dose-related and frequently occurs if exposure to a potentially toxic dose has occurred. Suicidal overdose of acetaminophen is one of the few examples. Unpredictable DILI is the type usually encountered in practice. It may or may not be dose-related but is most importantly characterized by a low incidence of overt DILI (0.01–1.0%) superimposed on a higher incidence of asymptomatic ALT abnormalities (1.0–10%). It is idiosyncratic because the occurrence of DILI reflects the unique susceptibility to an adverse reaction of a small subgroup of the population, presumably based on genetics and environmental factors (concomitant medicine, diet, and illnesses). Idiosyncrasy is characterized as allergic or non-allergic. The former is relatively easy to identify due to the concomitant appearance of systemic manifestations of allergy, such as fever, rash, eosinophilia, adenopathy, other organ involvement, characteristic initial latency of a few weeks and rapid positive rechallenge. Liver biopsy typically shows non-specific hepatitis but may show eosinophilic infiltrate and/or granuloma suggestive of allergy. Less commonly, some cases of allergic idiosyncrasy closely resemble autoimmune hepatitis (AIH). As noted above, some cases exhibit a very long latency, either having subclinical ongoing low-grade hepatitis or presenting with a rapid late progression to severe overt hepatitis. These patients have less prominent systemic allergic features but have antinuclear antibody (ANA), smooth muscle antibody (SMA), hypergammaglobulinaemia, and plasma cell hepatitis with variable fibrosis. The key distinction from idiopathic AIH is that the disease resolves when the drug is withdrawn, so in the clinical setting, with concomitant use of a drug known to induce this picture, it is prudent to withhold steroids if possible and assess response to cessation of the drug. An unresolved question is whether this AIH-like reaction represents true drug-induced allergy or drug-induced unmasking of underlying quiescent AIH, especially when the activity of the hepatitis does not subside after drug cessation.

The assignment of non-allergic idiosyncratic DILI is somewhat more problematic. The conclusion that it is not allergic is based upon the absence of systemic or autoimmune features of allergy, the fact that rechallenge is often not associated with recurrence of the problem, and the latency is variable, in some cases being short and in others being long. As a working approach, this distinction between non-allergic and allergic idiosyncrasy seems reasonable, although adaptive immune responses restricted to the liver cannot be completely excluded.

Adaptation to mild injury is commonly observed with idiosyncratic hepatotoxins so that continued used of the drug leads to return to normal liver tests. However, some patients do not improve, and they progress to overt and sometimes fatal injury. Whether this progression is due to a failure to adapt to the first hit or the addition of a pathophysiological second hit to the liver in response to the first hit in certain susceptible individuals is presently unknown, and the focus of research in this field.

OVERVIEW OF PATHOGENESIS

Current concepts of pathogenesis of DILI have centred on upstream factors and downstream participants (Figure 1). Upstream, the hepatocyte is the key target and most drug hepatotoxicity involves the participation of reactive or toxic metabolites which induce chemical consequences which cause cellular stress and/or death, which tends to be relatively mild. The reactive metabolite may serve as a hapten, eliciting an adaptive immune response, but this usually also requires the co-stimulation provided by the low-grade hepatocyte injury or the participation of other danger signals (e.g. cytokines generated because of infections)[6]. Indeed the clinically overt allergic hepatotoxic reactions with drugs almost always occur on a background of more frequent mild asymptomatic drug-induced injury (e.g. allopurinol, diclofenac, propylthiouracil, and halothane). This is also the case with apparent non-allergic idiosyncratic DILI (e.g. isoniazid, troglitazone, flutamide, bosentan,

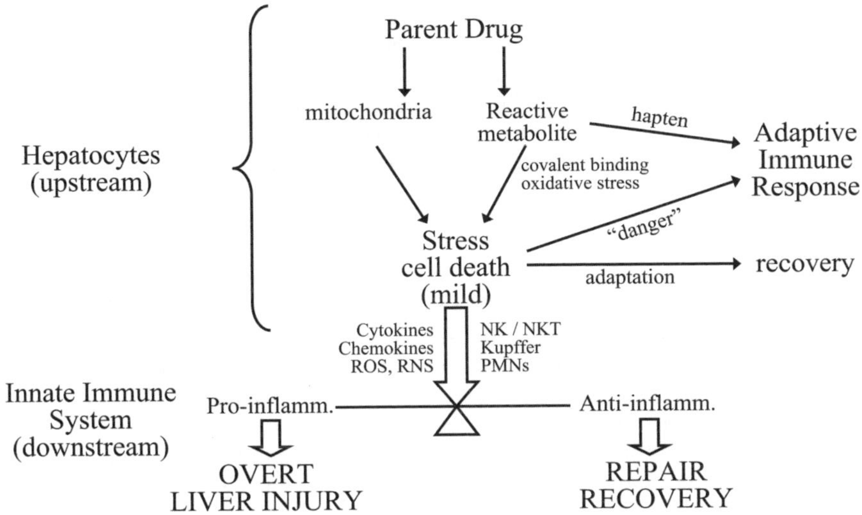

Figure 1 Pathogenesis of drug-induced liver injury (DILI)

ximelagatran). The cellular stress and mild injury may lead to an adaptive response (i.e. recovery) which could be due to changes in exposure/ detoxification of the toxic metabolite or the counter-regulatory protective responses to stress or cell death pathways. Alternatively, in some individuals the mild injury may activate the innate immune system to promote or inhibit inflammation, depending on the genetic and environmental status of the individual. This concept is illustrated below, with experimental acetaminophen hepatotoxicity.

The key role of reactive metabolite exposure as the most upstream initiator which is necessary but not sufficient to induce DILI is underscored by the wide variety of drugs[7] which have been withdrawn or not approved in the USA, which have been documented to be converted to reactive metabolites in test

Table 2 Importance of reactive metabolites in liver injury (adapted from ref. 7)

Withdrawn	*Warning*		*Not approved in the USA*	
Benoxaprofen	Acetaminophen	Ketoconazole*	Alpidem	Glafenine
Flutamide*	Carbamazepine	Methyldopa	Amineptine	Ibufenac
Iproniazid	Clozapine	Pyrazinamide	Amodiaquine	Isoxicam
Nefazodone	Dacarbazine*	Rifampin	Cincophen	Niperotidine
Troglitazone	Dantrolene*	Tacrine	Dihydralazine	Perhexiline
	Diclofenac	Tamoxifen	Dilevalol	Pirprofen
	Disulfiram	Terbinafine	Ebrotidine	Tilbroquinol
	Felbamate*	Ticlopidine		Ximelegatran
	Halothane	Tolcapone*		
	Isoniazid*	Valproic Acid*		
	Leflunomide	Zileutin		

*Black box.

systems (Table 2). Many pharmaceutical companies screen for reactive metabolites (covalent binding, GSH conjugates, in-silico structural alerts) while screening for therapeutic targets, and attempt to minimize financial risk by excluding such compounds from future development. Although this is probably a sound economic strategy, the occurrence of reactive metabolites does not necessarily predict adverse events in humans.

DIAGNOSIS

For the practising physician the key to diagnosis is suspicion, and in general diagnosis can be viewed as guilt by association[8]. Thus, if a patient develops hepatitis-like, cholestatic, or especially mixed injury within a period of weeks to months after starting the continuous use of a medication, the condition disappears over weeks (hepatitic) to months (cholestatic) after cessation of the drug and, most importantly, if no other plausible alternative cause can be found, one should have a strong suspicion of DILI. Distinctions such as possible, probable, definite, unlikely or unrelated are less important than simply possible/probable versus unlikely/unrelated. Two other features are quite helpful: (1) positive rechallenge – this is rarely justified but happens unintentionally at times, and is more characteristic of allergic DILI; (2) the track record – knowledge of case reports, series and regulatory labelling are helpful in characterizing an expected frequency, possible risk factors, and the clinical signature to be expected. There are several scoring scales[9,10] in use for DILI, but they are in need of optimization and are not necessarily of value to the clinician other than as a reminder of the type of issues to consider: latency, response to dechallenge, response to rechallenge, presence of known risk factors, exclusion of other causes, track record, and presence of systemic or allergic features. Finally, the typical work-up to rule out other causes includes hepatitis A, B, C, CMV and EBV virology and serology, and autoimmune markers, imaging of the biliary tract, cultures to exclude infection, and noting hypotension, hypoxia, or congestive failure.

ACETAMINOPHEN HEPATOTOXICITY

Clinical issues

Acetaminophen toxicity is the leading cause of acute liver failure (ALF) in the USA, accounting for >40% of cases. Interestingly about half of these cases are not single suicidal overdoses but rather are unintentional overdoses in a therapeutic attempt[11]. The median dose and duration of the latter cases was 7.5 g/day for 3 days, and a small number of cases appeared not to exceed the upper recommended limit (4 g/day), although accuracy of history in patients with ALF can be questioned. Nevertheless, these data underscore the somewhat narrow safety window of acetaminophen. This issue is further underscored by the finding that nearly 25% of cases of 'idiopathic' ALF had circulating serum adducts of the toxic metabolite of acetaminophen (NAPQI) covalently bound to protein[12]. These adducts would not be expected to form until nearly all the GSH had been preferentially depleted, at least in the hepatocytes releasing the protein adducts, presumably during necrosis. Recently, acetaminophen (NAPQI)–protein adducts were identified in 12.5% of cases of ALF attributed to acute hepatitis A or B[13]; this raises the question of whether it is dangerous to ingest acetaminophen while experiencing symptoms of acute viral hepatitis; could acetaminophen toxicity be more likely in viral hepatitis and then even be additive or potentiative to viral hepatitis in worsening injury, leading to acute liver failure? Finally, a recent study in healthy volunteers given 1g acetaminophen every 6 h for up to 14 days showed a high incidence of serum ALT abnormalities (sometimes very high) beginning after the first 3 days of drug[14]. This scenario, although implying that mild ALT increases are not of concern in individual patients, does underscore once again the limited safety range of maximal doses of acetaminophen. Although serious overt liver injury in patients taking $\leqslant$4 g/day has rarely been described, and mainly in alcoholics, concomitant isoniazid-treated patients, and during poor nutritional intake (starvation), the finding of the recent healthy volunteer study reminds us that, although in the absence of risk factors mentioned above the risk of serious injury with recommended doses is very low, it becomes very worrisome when patients ingest a little more than the maximum recommended dose, e.g. 6 g/day[15]. This is easy to achieve inadvertently since hundreds of combination medications contain acetaminophen, including cold remedies and opiates. In any case, since it is likely that far fewer individuals develop unintentional acetaminophen hepatotoxicity than the number who exceed the recommended maximum daily doses, and since only a subgroup of healthy subjects develops ALT abnormalities at therapeutic doses, it is reasonable to hypothesize that there is an idiosyncratic basis for acetaminophen toxicity in the setting where maximum or supra-maximum doses are consumed for therapeutic intent.

Experimental model of DILI

Acetaminophen (APAP) is the most extensively studied experimental hepatotoxin. Although arguably a possible occasional idiosyncratic toxin, it had been hoped that understanding the genetic and environmental influences on the susceptibility of animal species, especially mice, to APAP would provide insights or clues for hypothesis testing in human idiosyncratic DILI. Thus, by manipulating hepatocellular and inflammatory cell responses to APAP exposure, it has been possible to modulate the susceptibility to injury[16]. Conceptually this follows the scheme in Figure 1.

First, it is important to consider upstream events that occur in hepatocytes, beginning with initiation of the process of APAP toxicity, namely metabolism of APAP and exposure to the reactive metabolite, NAPQI. NAPQI is preferentially detoxified by binding to GSH. When the cellular GSH is nearly fully consumed, especially including mitochondria GSH, NAPQI covalently binds to protein thiols. Binding to protein thiols is a biomarker of the state of profound GSH depletion, and correlates with the appearance of toxicity. At the same time profound mitochondrial GSH depletion unleashes enhanced

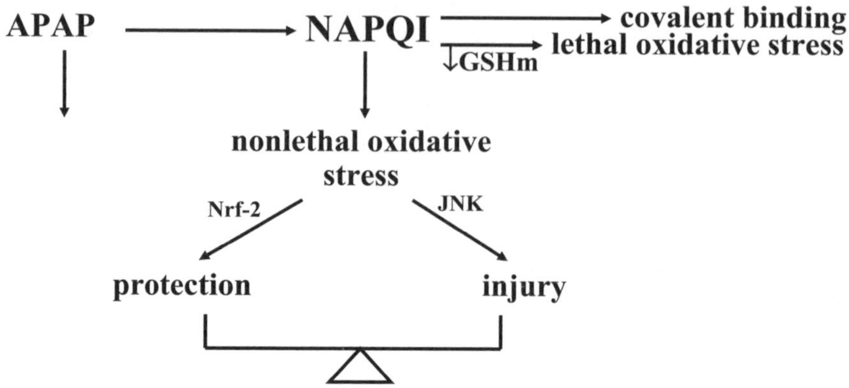

Figure 2 Hepatocellular upstream mechanisms of acetaminophen (APAP) toxicity

exposure to reactive oxygen species[17]. Although it is conceivable that massive ensuing covalent binding and oxidative stress lead to cell lysis, mounting evidence suggests that progression to injury (e.g. cell necrosis) is determined by the balance of cellular signal transduction and transcription events that are triggered by the oxidative stress (Figure 2). Thus, redox perturbations associated with APAP activate the nuclear translocation of Nrf-2, a transcription factor which activates the antioxidant response element in the promoters of antioxidant genes leading to transcriptional up-regulation of gene expression of protective proteins involved in GSH synthesis and phase 2 detoxification[18]. Therefore, the exposure to NAPQI is modulated by the rapid and dynamic effects of APAP metabolism on gene expression. At the same time oxidative stress in response to APAP leads to sustained activation of c-jun-N-terminal kinase (JNK)[19]. JNK activation leads to toxicity since small molecule JNK inhibitors and antisense oligonucleotide-mediated silencing of JNK 1 and 2 are remarkably effective in protecting against APAP liver injury, even when given several hours after APAP when peak covalent binding and GSH depletion have already occurred[19]. The mitochondrion is likely to be the target of JNK, although this could be because JNK promotes Bax translocation to mitochondria, or because of the observed direct translocation of JNK itself to mitochondria; both are blocked by JNK inhibitor treatment. The target of direct JNK activity in mitochondria remains to be identified. It is hypothesized that the necrotic cell death is initiated by multiple hits to the mitochondria; covalent binding and GSH depletion in mitochondria may sensitize to JNK actions (direct or indirect via Bax) which leads to collapse of mitochondrial function. Necrosis is observed rather than apoptosis, perhaps because oxidative stress and profound ATP depletion inhibit progression of the apoptosis cascade. A key conclusion from these studies is that the stress of toxic metabolites is not necessarily directly lethal, but can activate or recruit the participation of built-in death programmes, a type of programmed necrosis or assisted suicide.

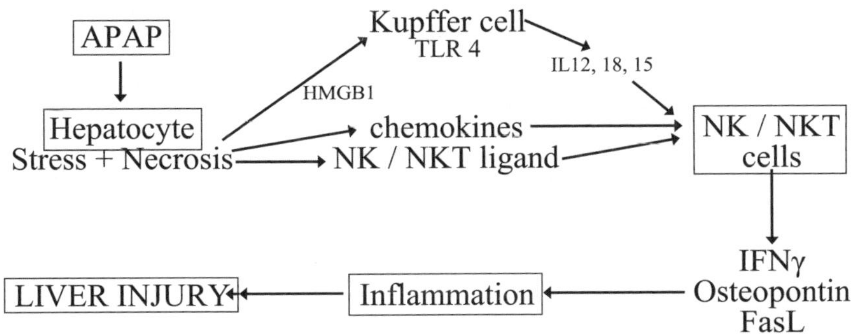

Figure 3 Working model of participation of the innate immune system and the pivotal role of NK/NKT cells in acetaminophen (APAP)-induced liver injury

The upstream events in the hepatocyte in response to APAP with cell stress and cell death then trigger downstream participation of the innate immune system in modulating the extent of organ damage[16]. By genetic manipulations it has been possible to increase or decrease the susceptibility of mice to APAP (creating a kind of mouse idiosyncrasy) without effects on the upstream events (covalent binding, GSH depletion). Thus, for example, IL-10, and IL-6 null mice are much more susceptible to APAP injury[20,21], whereas IFN-γ[22,23] or osteopontin[24] null mice are more resistant. Depletion of NK/NKT cells protects, and deficiency of Fas and FasL also protects[23,25]. The role of neutrophils remains controversial. Currently it appears that the NK/NKT cells play a pivotal role as they are the major source of increased expression of IFN-γ, osteopontin and FasL in APAP treated mice – these cytokines, chemokines and death ligands can increase hepatic inflammation as well as perhaps directly targeting hepatocytes leading to aggravation of necrosis in APAP-induced stressed and vulnerable hepatocytes. What triggers the recruitment and activation of NK/NKT cells is uncertain, but probably it is initiated by release of substances from stressed or necrotic hepatocytes, as illustrated in Figure 3. Candidates included: (1) increased or decreased hepatocyte expression of NK/NKT cell ligands which bind to respective activating or inhibiting receptors on NK/NKT cells; (2) direct production of chemokines by stressed hepatocytes or (3) passive release of 'cytokines' such as high-mobility group box 1 (HMGB1), a chromatin-associated protein known to be released from necrotic hepatocytes, which binds to toll-like receptor 4 on Kupffer cells activating the production of proinflammatory cytokines (e.g. IL-12, IL-18) which promote recruitment and activation of NK/NKT cells[26].

References

1. Kaplowitz N. Idiosyncratic drug hepatotoxicity. Nature Rev Drug Discovery. 2005;4:489–99.
2. Kaplowitz N. Rules and laws of drug hepatotoxicity. Pharmacoepidemiol Drug Safety. 2006;15:1–3.

3. Andrade RJ, Lucena MI, Fernandez MC et al. Drug-induced liver injury: an analysis of 461 incidences submitted to the Spanish registry over a 10-year period. Gastroenterology. 2005; 129:512–21.
4. Bjornsoon E, Olsson R. Outcome and prognostic markers in severe drug-induced liver disease. Hepatology. 2005;42:481–9.
5. Andrade R, Lucena M, Kaplowitz N et al. Outcome of acute idiosyncratic drug induced liver injury: long-term follow-up in a hepatotoxicity registry. Hepatology. 2006;44:1581–8.
6. Liu Z-X, Kaplowitz N. Immune-mediated drug-induced liver disease. Clin Liver Dis. 2002; 6:467–86.
7. Walgren J, Mitchell M, Thompson D. Role of metabolism in drug-induced idiosyncratic hepatotoxicity. Crit Rev Toxicol 2005;35:325–61.
8. Kaplowitz N. Causality assessment versus guilt by association in drug hepatotoxicity. Editorial. Hepatology. 2001;33:308–10.
9. Maria V, Victorino R. Diagnostic value of specific T cell reactivity to drugs in 95 cases of drug induced liver injury. Gut. 1997;41:534–40.
10. Lucena M, Camargo R, Andrade R et al. Comparison of two clinical scales for causality assessment in hepatotoxicity. Hepatology. 2001;33:123–30.
11. Larson AM, Polson J, Fontana RJ et al. Acute liver failure study group: acetaminophen-induced acute liver failure: results of a United States multicenter, prospective study. Hepatology. 2005;42:1364–72.
12. Davern T, James L, Hinson J et al. Measurement of serum acetaminophen–protein adducts in patients with acute liver failure. Gastroenterology. 2006;130:687–94.
13. Polson J, James L, Davern T et al. Acetaminophen as a co-factor in acute liver failure due to viral hepatitis determined by measurement of acetaminophen-protein adducts. Gastroenterology. 2006;130:A772 (abstract).
14. Watkins P, Kaplowitz N, Slattery J et al. Aminotransferase elevations in healthy adults receiving 4 grams of acetaminophen daily: a randomized controlled trial. J Am Med Assoc. 2006;296:87–93.
15. Kaplowitz N. Acetaminophen hepatotoxicity: what we know, what we don't know and where do we go? Hepatology. 2004;40:23–6.
16. Liu Z-X, Kaplowitz N. Role of innate immunity in acetaminophen toxicity. Expert Opin Drug Metab Toxicol. 2006;2:1–11.
17. Nelson SD, Bruschi SA. Mechanism of acetaminophen-induced liver disease. In: Kaplowitz N, Deleve LD, editors. Drug-Induced Liver Disease. New York: Dekker, 2003:287–325.
18. Goldring C, Kitteringham NR, Elsby R et al. Activation of hepatic Nrf2 *in vivo* by acetaminophen in CD-1 mice. Hepatology. 2004;39:1267–76.
19. Gunawan B, Liu ZX, Han D et al. c-Jun-N-terminal kinase plays a major role in murine acetaminophen hepatotoxicity. Gastroenterology. 2006;131:165–78.
20. Bourdi M, Masubuchi Y, Reilly TP et al. Protection against acetaminophen-induced liver injury and lethality by interleukin 10: role of inducible nitric oxide synthase: Hepatology. 2002;35:289–98.
21. Masubuchi Y, Bourdi M, Reilly T et al. Role of interleukin-6 in hepatic heat shock protein expression and protection against acetaminophen-induced liver disease. Biochem Biophys Res Commun. 2003;304:207–12.
22. Ishida Y, Kondo T, Ohshima T et al. A pivotal involvement of IFN-γ in the pathogenesis of acetaminophen-induced acute liver injury. FASEB J. 2002;16:1227–36.
23. Liu ZX, Govindarajan S, Kaplowitz N. Innate immune system plays a critical role in determining the progression and severity of acetaminophen hepatotoxicity. Gastroenterology. 2004;127:1760–74.
24. Welch KD, Reilly TP, Bourdi M et al. Genomic identification of potential risk factors during acetaminophen-induced liver disease in susceptible and resistant strains of mice. Chem Res Toxicol. 2006;19:223–33.
25. Zhang H, Cook J, Nickel J et al. Reduction of liver Fas expression by an antisense oligonucleotide protects mice from fulminant hepatitis. Nat Biotechnol. 2000;18:862–7.
26. Scaffidi P, Misteli T, Bianchi ME. Release of chromatin protein HMGB1 by necrotic cells triggers inflammation. Nature. 2002;418:191–5.

Section II
Cholestatic hepatitis

Chair: AM DIEHL and G GERKEN

4
Liver disease in pregnancy: diagnosis and treatment

J. KONDRACKIENE and L. KUPCINSKAS

INTRODUCTION

Pregnancy is associated with many normal physiological changes, which are classically associated with chronic liver disease and usually disappear after delivery: spider angiomas, palmar erythema, elevated alkaline phosphatase due to placental production, increased plasma volume by approximately 50%, hypoalbuminaemia. An increase in serum aminotransferase, bilirubin, or fasting total bile acid concentrations during pregnancy is pathological and should prompt further evaluation[1]. Prothrombin time is unchanged, and serum fibrinogen increases in late pregnancy. Abnormalities in liver tests occur in 3% of pregnancies with causes ranging from self-limiting to rapidly fatal[2]. The types and presentation of hepatobiliary disease during pregnancy are varied. There are pregnancy-related liver disorders, diseases occurring coincidentally, and preexisting chronic liver disease. Unlike diseases such as viral hepatitis, which may occur at any time, liver diseases unique to pregnancy occur at a characteristic time. Hyperemesis gravidarum begins in the first trimester, intrahepatic cholestasis of pregnancy in the second or third trimester, preeclampsia, HELLP syndrome and acute fatty liver of pregnancy usually in the third trimester[3].

HYPEREMESIS GRAVIDARUM

Hyperemesis gravidarum (HG) is characterized by intractable nausea and vomiting, usually resolving by week 20 of gestation. Some degree of nausea with or without vomiting occurs in 50–90% of all pregnancies[4]. HG may be considered the severe end of the spectrum of symptoms; however, there is no clear demarcation between common pregnancy-related morning sickness and the infrequent pathological form[5]. The true incidence of HG is not well documented and may vary from as little as three cases per 1000 pregnancies to as many as 1 in 100 (see ref. 6). Pregnancy outcome is favourable.

Hyperemesis has been attributed to a number of factors, but the aetiology remains uncertain. Psychological, hormonal factors may be involved in the pathogenesis of HG[7]. Several other theories to explain HG have been suggested, including abnormal gastric motility, specific nutrient deficiencies (e. g. zinc), alterations in lipid levels, changes in the autonomic nervous system, or genetic factors[8]. None is consistently associated with or highly predictive of the disease.

HG has its onset in early pregnancy and resolves by 16–18 weeks of gestation, but in some patients may persist until delivery. Symptoms range from mild to severe. Patients may present with weight loss exceeding 5% of their body weight, dehydration, ketosis and electrolyte derangements such as hypokalaemia and metabolic alkalosis. Rarely HG may lead to serious complications, including Wernicke encephalopathy (from deficiency of vitamin B_1), sequelae of malnutrition (immunosuppression, poor wound healing), and spontaneous oesophageal rupture[9,10]. Mild hyperthyroidism may be associated with HG, perhaps due to high serum concentrations of human chorionic gonadotropin which may have more thyroid-stimulating activity in pregnant women[11].

Abnormal liver enzyme values occur in approximately 50% of patients who are hospitalized with HG. Alanine aminotransferase (ALT) and aspartate aminotransferase (AST) are mildly elevated, and rarely as high as 1000 U/L. Hyperbilirubinaemia can also occur, but rarely exceeds 4 mg/dl[12,13]. Serum amylase and lipase may increase, but their origin is from salivary glands. The degree of these abnormalities in liver tests correlates with the vomiting. Abnormal liver biochemical tests resolve promptly upon resolution of the vomiting. Liver biopsy shows necrosis with cell dropout, steatosis, centrilobular vacuolization, and rare bile plugs[13].

HG is generally a diagnosis of exclusion. Other conditions unrelated to pregnancy that can cause persistent nausea and vomiting include gastrointestinal disorders (gastroenteritis, hepatitis, pancreatitis, cholelithiasis), genitourinary tract disorders, metabolic (e.g. diabetes, porphyria) and neurological disease (e.g. migraine, tumour, vestibular lesions), drug toxicity, and psychological problems.

Treatment is symptomatic, though hospitalization is occasionally necessary for rehydration and nutritional support. Many patients respond to rehydration and a short period of gut rest, followed by reintroduction of a diet rich in carbohydrates and low in fat. Further therapy in selected patients may include antiemetics, and enteral and parenteral nutrition[14]. No medication is currently approved by the United States Food and Drug Administration for treatment of HG. Thus, a reasonable approach is to begin therapy with agents that have minimal maternal side-effects and, if these are ineffective, substitute other drugs in a step-wise progression. Pyridoxine (vitamin B_6) (10–25 mg orally t.i. d.) may help women with mild to moderate nausea and minimal vomiting. Antiemetics, including promethazine (12.5–25 mg q. 4 h p.o., i.m. or p.r.), metoclopramide (5–10 mg q. 8 h i.v. or p.o.), ondansetron (8 mg q. 12 h i.m. or p.o.), prochlorperazine (5–10 mg q. 3 to 4 h i.m. or p.o. or 25 mg b.i.d. p.r.), or a combination of droperidol and diphenhydramine, may benefit selected patients[15]. The safety of antihistamines was affirmed in a meta-analysis that examined the association between antihistamine use and major malformations[16]. The role of corticosteroids for hyperemesis is unsettled. There may be a slightly increased risk of oral clefts when the drugs are administered before 10 weeks of gestation[17]. Other options are P6 acupuncture or acupressure wristbands and powdered ginger (1 g daily). A Cochrane-type review did not find P6 acupuncture or acupressor to be clearly effective[14]. Although ginger has been shown to be more effective than placebo, and equivalent to vitamin B_6 (pyridoxine), its safety in pregnancy has not been firmly established[18]. Hypnosis has been reported to be helpful in selected patients[19].

INTRAHEPATIC CHOLESTASIS OF PREGNANCY

Intrahepatic cholestasis of pregnancy (ICP) is the most common pregnancy-related liver disorder. In many areas of the world ICP is a rare disease occurring at a rate of 1 in 1000–10 000 pregnancies[20]. However, the incidence of ICP is markedly higher in Sweden and other Scandinavian countries (up to 2% of deliveries in 1950–60) and even more in Chile (up to 14% of deliveries in 1960–70)[21]. In these countries the incidence of ICP has decreased more recently, whereas it has increased in other parts of Europe, the USA, Asia, Australia and some Latin American countries. The latter phenomenon might be explained at least in part by a greater awareness of the disease[20].

ICP is characterized by pruritus starting in the second or third trimester of pregnancy, and disappearing after delivery. It is often generalized but predominates on the palms and the soles of the feet, and is worse at night. Pruritus may precede laboratory abnormalities[22]. Physical examination is non-specific, but may show excoriations due to scratching. Jaundice occurs in 10–20%, typically within 4 weeks of the onset of itching.

Serum total bile acid concentrations increase in ICP, and may be the first or only laboratory abnormality. Serum cholic acid increases more than chenodeoxycholic acid, resulting in a marked elevation of the cholic/chenodeoxycholic acid ratio compared to pregnant women without ICP[22,23]. The ratio of glycine/taurine conjugates is decreased. Other laboratory findings include elevations in the serum concentrations of alkaline phosphatase (up to four times normal), 5′-nucleotidase, and total and direct bilirubin concentrations. Total bilirubin levels rarely exceed 6 mg/dl. Surprisingly, the serum concentration of gammaglutamyltranspeptidase is normal or modestly elevated. Serum aminotransferases are also elevated, and may reach values greater than 1000 U/L, making distinction from viral hepatitis important[24]. Prothrombin time is usually normal. When present, prolonged prothrombin times reflect vitamin K deficiency due to cholestasis or to the use of bile acid sequestrants (such as cholestyramine) rather than liver dysfunction. Liver biopsy is rarely necessary for the diagnosis. When performed, histopathology is characterized by cholestasis without inflammation. Bile plugs in hepatocytes and canaliculi predominate in zone 3. The portal tracts are unaffected[25].

The cause of ICP is still under discussion. The pathogenesis can be related to abnormalities in the metabolism and disposition of sex hormones and/or bile acids, determined by a genetic predisposition and environmental factors[26,27]. There is increasing evidence that genetically determined dysfunction in the canalicular ABC transporters bile salt export pump (BSEP, ABCB11) and multidrug resistance protein 3 (MDR3, ABCB4) might be risk factors for development of ICP[28,29]. Hormonal factors may trigger the transient decompensation of the heterozygous state for a MDR3 gene defect during pregnancy, leading to ICP[30–32].

At present the optimal treatment of ICP is still under debate. Treatment for ICP focuses on reducing symptoms and preventing maternal and fetal complications. Several drugs have been studied, and most focus on relieving symptoms. Anthistamines, benzodiazepines, phenobarbital, dexamethasone, epomediol, *S*-adenosyl-L-methionine, and cholestyramine have been used without clear evidence of efficacy. Cholestyramine has been used for reducing pruritus. Observational studies suggest that cholestyramine may be associated with improved maternal morbidity without a documented improvement in fetal outcome[33–35]. Cholestyramine may worsen the malabsorption of fat-soluble vitamins, especially vitamin K. A case report of severe fetal intracranial haemorrhage during treatment with cholestyramine for ICP has raised the possibility that severe maternal vitamin K deficiency may lead to fetal vitamin K deficiency and coagulopathy[35]. Currently, the most promising treatment is the synthetic bile acid ursodeoxycholic acid (UDCA). Improvement in maternal and fetal morbidity was suggested in eight clinical trials and several observational studies, although these studies were small and in some respects inconsistent[33,34]. The largest trial included 84 symptomatic patients who were randomly assigned to UDCA 8–10 mg/kg body weight per day, or cholestyramine 8 g/day for 14 days. The results of the study showed a significant improvement of pruritus severity, aminotransferase activities and serum bile acid concentrations and a more favourable outcome of pregnancy and absence of adverse events after treatment with a moderate dose of UDCA. In contrast, cholestyramine alleviated pruritus only mildly and caused side-effects (nausea and vomiting)[36].

Maternal prognosis is benign. Affected women generally have no hepatic sequelae. Cholestasis recurs during subsequent pregnancies in 60–70%. The administration of oral contraceptives to women with a history of ICP rarely results in recurrent cholestasis. In contrast, ICP may have serious consequences for the fetus. Clinical studies show that the disease is a strong risk factor for premature deliveries in 19–60%[37], stillbirths in 1–2%[38,39] and fetal distress in 22–33%[40]. The cause is unknown. There is no ideal method for fetal surveillance in ICP. Some obstetricians recommend non-stress tests or biophysical assessment[41]; however, no test reliably predicts the risk of fetal demise. Measurement of the serum total bile acid concentration has also been suggested for fetal assessment in ICP[42]. In a study from Sweden the probability of fetal complications (defined as preterm delivery, asphyxial events, meconium staining of amniotic fluid, placenta, and membranes) was directly related to concentration of bile acid levels even after controlling for other risk factors[42]. Fetal complications were not observed until bile acid levels were 40 μmol/L[42]. The best approach is early delivery, the timing of which should be guided by the patient's symptoms (mostly pruritus), the gestational age, and whether the cervix is favourable[41]. In most patients delivery should be accomplished by 38 weeks. However, when cholestasis is severe (such as when patients are jaundiced), delivery should be considered at 36 weeks gestation if lung maturity is achieved, or as soon thereafter as fetal lung maturity is established[38].

PREECLAMPSIA

Hypertensive disorders complicate 12–22% of pregnancies. Preeclampsia occurs in approximately 3–14% of all pregnancies worldwide[43]. Preeclampsia refers to the new onset of hypertension and proteinuria after 20 weeks of gestation in a previously normotensive woman.

The aetiology of this condition is unknown, but it seems that uteroplacental ischaemia plays a major role. The placental ischaemia causes activation of the endothelium. All of the clinical features of preeclampsia can be explained as maternal responses to generalized endothelial dysfunction[44,45]. Disturbed endothelial control of vascular tone causes hypertension, increased vascular permeability results in oedema and proteinuria, and abnormal endothelial expression of procoagulants leads to coagulopathy. These changes also cause ischaemia of target organs, such as the brain, liver, kidney, and placenta. Fibrin deposition, periportal haemorrhage, ischaemic lesions, and microvesicular fat deposition are histological findings observed in the livers of preeclamptic women[46]. The clinical manifestations of liver involvement are right upper quadrant or epigastric pain, elevated liver enzymes and, in severe cases, subcapsular haemorrhage or hepatic rupture. HELLP syndrome develops in 10–20% of women with severe preeclampsia.

HELLP SYNDROME

The haemolysis (H), elevated liver enzymes (EL), and low platelets (LP) syndrome occurs in approximately 1 of 1000 pregnancies. The majority of cases are diagnosed between 28 and 36 weeks of gestation[47]. The disease presents prior to delivery in 70%, postpartum in 30%, usually within 48 h, but occasionally as long as 7 days after delivery. The HELLP syndrome probably represents a severe form of preeclampsia, but this relationship remains controversial. As many as 15–20% of patients do not have antecedent hypertension or proteinuria, leading some experts to believe that HELLP is a separate disorder from preeclampsia[48].

The clinical presentation of HELLP is varied. Although some patients have no symptoms, most complain of abdominal pain and tenderness[47]. The pain is located in the midepigastrium, right upper quadrant, or below the sternum. Many patients also have nausea, vomiting, and malaise. Jaundice is evident in less than 5%. Bleeding related to thrombocytopenia is an uncommon presentation. Hypertension and proteinuria are present in approximately 85% of cases, pulmonary oedema (6%), ascites (8%), and acute renal failure, usually occurring in the setting of disseminated intravascular coagulation (DIC) (20%)[47].

The diagnosis of HELLP syndrome is based on the presence of the laboratory abnormalities comprising its name: microangiopathic haemolytic anaemia with characteristic schistocytes on blood smear, an elevated indirect bilirubin and a low serum haptoglobin concentration (25 mg/dl), serum LDH >600 U/L or total bilirubin >1.2 mg/dl, AST >70 U/L, platelet count <100 000 cells/μl[47]. Women who do not meet all of the above laboratory abnormalities are considered to have partial HELLP syndrome. A liver biopsy is rarely necessary to establish the diagnosis, and may be hazardous to perform because of the coagulopathy that may be present. It typically shows periportal haemorrhage and fibrin deposition[49]. There is little correlation between the histological findings and clinical presentation.

The differential diagnosis includes acute fatty liver of pregnancy or, because of similar symptoms or laboratory findings, gastroenteritis, hepatitis, appendicitis, gallbladder disease, idiopathic thrombocytopenic purpura, haemolytic–uraemic syndrome, or thrombotic thrombocytopenic purpura[50]. Marked elevations in serum aminotransferases (usually 1000–2000 IU/L or higher) are not typical of HELLP; when they occur they may indicate hepatic infarction or subcapsular haematoma rather than viral hepatitis. HELLP may be difficult to distinguish clinically from acute fatty liver of pregnancy. Prolongation of the prothrombin and activated partial thromboplastin times, low glucose concentration, and elevated creatinine concentration are more common in women with acute fatty liver than HELLP[51].

Patients with HELLP may develop a haematoma beneath Glisson's capsule[47]. Patients who develop a haematoma typically have abdominal pain and many have thrombocytopenia (with platelet count that may be below 20 000 μl), shoulder pain, nausea, and vomiting. If hepatic rupture occurs, swelling of the abdomen from haemoperitoneum and shock rapidly ensue. The aminotransferases are usually modestly elevated, but values of 4000–5000 U/L can occasionally be seen. Body imaging using CT or MRI is more dependable than ultrasonography in detecting these lesions[52]. The management of a contained haematoma is to support the patient with volume replacement and blood transfusion, as needed, with consideration of percutaneous embolization of the hepatic arteries[53]. If rupture has occurred, patients are best managed by a team experienced in liver trauma surgery[54]. Operative management includes packing, drainage, hepatic artery ligation, and/or resection of affected areas of the liver. When the haemorrhage could not be contained, liver transplantation has been reported[55]. Patients who survive have no hepatic sequelae; however, recurrent episodes in subsequent pregnancies have been reported, suggesting that there may be a specific predisposition to this condition[56].

The cornerstone of therapy is delivery[57]. Delivery is indicated for pregnancies 34 weeks of gestation, non-reassuring tests of fetal status (e.g. biophysical profile, fetal heart rate testing), or presence of severe maternal disease: multiorgan dysfunction, disseminated intravascular coagulation (DIC), liver infarction or haemorrhage, renal failure, or abruptio placenta[57]. Many patients will have a normal vaginal delivery following the induction of labour. However, cesarean delivery can be considered in very preterm gestations (under 30–32 weeks) when the cervix is unfavourable. In pregnancies <34 weeks of gestation in whom the mother and fetus are stable, corticosteroids to enhance fetal lung maturation should be given. Magnesium sulphate is often given to prevent convulsions (6 g i.v. followed by 2 g/h as a continuous infusion). Severe hypertension should be controlled (i.e. blood pressure less than 160/105 to 110 mmHg) with labetalol, hydralazine, nifedipine or, in severe cases, with sodium nitroprusside[58]. Corticosteroids (e.g. dexamethasone 10 mg i.v. every 12 h) can be considered until remission is noted[59]. The outcome for mothers with HELLP is generally good. With treatment, maternal mortality is about 1%. Maternal complications and gestational age at delivery are strongly associated with fetal prognosis[60]. Fetal complications may include prematurity (70%). The overall perinatal mortality is 7–20%, primarily related to extreme prematurity in association with abruptio placenta or growth restriction[57]. However, surviving babies do not have an increased risk of liver disease or thrombocytopenia, and have an outcome similar to babies of a similar gestational age.

ACUTE FATTY LIVER OF PREGNANCY

Acute fatty liver of pregnancy (AFLP) was first described in 1940 and was initially thought to be exceedingly rare and universally fatal[61]. The incidence is estimated to be 1 in 7000 to 1 in 16 000. AFLP occurs in the second half of pregnancy, usually close to term; however, some patients may present and be diagnosed after delivery[62].

Recently, an association between AFLP and one of the inherited defects in mitochondrial beta-oxidation of fatty acids, long-chain 3-hydroxyacyl CoA dehydrogenase deficiency (LCHAD), was suggested[63]. This enzyme is one of the four enzymes which break down long-chain fatty acids in the liver. The accumulation of long-chain 3-hydroxyacyl metabolites produced by the fetus or placenta is toxic to the liver and may be the cause of the liver disease. The most common mutation is E474Q. Testing for the known genetic variants of this LCHAD is available and should be performed in affected women, their infants, and fathers. Infants with LCHAD are at risk to develop fatal non-ketotic hypoglycaemia, imitating Reye's syndrome or defects in urea cycle function[63]. In addition, some forms of LCHAD deficiency are associated with neonatal dilated cardiomyopathy or progressive neuromyopathy[51]. These babies should be managed with foods with medium-chain fatty acids, low-fat diet and should be followed up for sequelae.

The spectrum of clinical involvement is broad, ranging from asymptomatic elevations in aminotransferases to fulminant hepatic failure. AFLP may present with malaise, fatique, anorexia, nausea, vomiting, headache, abdominal pain, or jaundice. Within 1–2 weeks of onset of symptoms the disease may rapidly worsen. Patients then develop acute liver failure, including hepatic encephalopathy, ascites, oedema, or renal insufficiency. Over one-half of patients have preeclampsia[64]. Extrahepatic complications include infection, intra-abdominal bleeding, transient polyuria and polydipsia due to central diabetes insipidus, acute renal failure, pancreatitis, or severe hypoglycaemia. Women with AFLP have abnormal liver tests, with the serum aminotransferase elevations ranging from modest values up to 1000 IU/L. The white blood cell count may be higher than is usually seen in pregnancy. The platelet count is normal unless the patient has progressed to disseminated intravascular coagulation (DIC), which is associated with marked reduction in antithrombin III[65]. Severely affected patients also have elevations in serum ammonia and hypoglycaemia. In cases of diagnostic uncertainty liver biopsy is indicated. Using special stains (oil red O on frozen section, or electron microscopy), the typical microvesicular fatty infiltration, pericentral pallor with lobular disarray and vacuolization of the centrizonal hepatocytes can be seen[66]. The differential diagnosis includes acute viral hepatitis, especially hepatitis E in endemic areas, or herpes simplex, drug-induced hepatitis, HELLP syndrome, or biliary tract disorders.

The treatment of AFLP is delivery after maternal stabilization. Prothrombin time usually starts to normalize shortly thereafter although, in severe cases, there may be many more days of illness requiring maximal supportive management in an intensive-care unit, including mechanical ventilation because of coma, dialysis for acute renal failure, parenteral nutrition because of associated pancreatitis, or even surgery to treat bleeding from a preceding cesarean section. The presence of DIC does not appear to influence the clinical outcome[65]. Maternal stabilization requires glucose infusion and reversal of coagulopathy (e.g. administration of fresh-frozen plasma, cryoprecipitate, packed red blood cells, and, rarely, platelets), as needed. Lethargic patients should have their blood ammonia concentration evaluated and receive lactulose therapy if it is elevated. Occasional patients with AFLP have undergone liver transplantation, but this should not be needed with early diagnosis and prompt delivery[67].

Most severely ill patients recover and have no sequelae of the liver disease itself[62]. The maternal mortality rate was very high before 1970 (92%). The maternal prognosis is currently greatly improved. With early delivery and advances in supportive management maternal mortality is now 10–12% and fetal mortality 9–20%[46,68]. Recurrence in subsequent pregnancies is uncommon. Affected women should be warned of this possibility and should be tested, along with their newborn infants who may be affected, for LCHAD[69].

References

1. Bacq Y, Zarka O, Bréchot JF et al. Liver function tests in normal pregnancy: a prospective study of 103 pregnant women and 103 matched controls. Hepatology. 1996;23:1030–4.
2. Ch'ng CL, Morgan M, Hainsworth I, Kingham JG. Prospective study of liver dysfunction in pregnancy in Southwest Wales. Gut. 2002;51:876–80.
3. Riely CA et al. Liver disease in the pregnant patient. Am J Gastroenterol. 1999;94:1728–32.
4. Association of Professors of Gynecology and Obstetrics. Nausea and vomiting of pregnancy. Washington, DC: APGO, 2001.
5. Goodwin TM. Hyperemesis gravidarum. Clin Obstet Gynecol. 1998;41:597.
6. Kallen B. Hyperemesis during pregnancy and delivery outcome: a registry study. Eur J Obstet Gynecol Reprod Biol. 1987;26:291–302.
7. Lagiou P, Tamimi R, Mucci LA et al. Nausea and vomiting in pregnancy in relation to prolactin, estrogens, and progesterone: a prospective study. Obstet Gynecol. 2003;101:639–44.
8. Goodwin TM. Nausea and vomiting of pregnancy: an obstetric syndrome. Am J Obstet Gynecol. 2002;186:S184–9.

9. Spruill SC, Kuller JA. Hyperemesis gravidarum complicated by Wernicke's encephalopathy. Obstet Gynecol. 2002;99:875–7.
10. Paauw JD, Bierling S, Cook CR, Davis AT. Hyperemesis gravidarum and fetal outcome. J Parent Ent Nutr. 2005;29:93–6.
11. Goodwin TM, Montoro M, Mestman JH et al. The role of chorionic gonadotropin in transient hyperthyroidism of hyperemesis gravidarum. J Clin Endocrinol Metab. 1992;75: 1333–7.
12. Abell T, Riely C. Hyperemesis gravidarum. Gastroenterol Clin N Am. 1992;21:835–49.
13. Larrey D, Rueff B, Feldmann G et al. Recurrent jaundice caused by recurrent hyperemesis gravidarum. Gut. 1984;25:1414–15.
14. Jewell D, Young G. Interventions for nausea and vomiting in early pregnancy. Cochrane Database Syst Rev. 2003:CD000145.
15. Magee LA, Mazzotta P, Koren G. Evidence-based view of safety and effectiveness of pharmacologic therapy for nausea and vomiting of pregnancy (NVP). Am J Obstet Gynecol. 2002;186:S256–61.
16. Seto, A, Einarson, T, Koren, G. Pregnancy outcome following first trimester exposure to antihistamines: meta-analysis. Am J Perinatol. 1997;14:119–24.
17. Shepard TH, Brent RL, Friedman JM et al. Update on new developments in the study of human teratogens. Teratology. 2002;65:153–61.
18. Smith C, Crowther C, Willson K et al. A randomized controlled trial of ginger to treat nausea and vomiting in pregnancy. Obstet Gynecol. 2004;103:639–45.
19. Simon EP, Schwartz J. Medical hypnosis for hyperemesis gravidarum. Birth. 1999;26:248–54.
20. Reyes H, Sjovall J. Bile acids and progesterone metabolites in intrahepatic cholestasis of pregnancy. Ann Med. 2000;32:94–106.
21. Reyes H. Review: Intrahepatic cholestasis: a puzzling disorder of pregnancy. Gastoenterol Hepatol. 1997;12:211–16.
22. Lammert F, Marschall HU, Glantz A, Matern S. Intrahepatic cholestasis of pregnancy: molecular pathogenesis, diagnosis and management. J Hepatol. 2000;33:1012–21.
23. Brites D, Rodrigues CM, Oiveira N et al. Correction of maternal serum bile acid profile during ursodeoxycholic acid therapy in cholestasis of pregnancy. J Hepatol. 1998;28:91–8.
24. Milkievicz P, Elias E, Williamson C, Weaver J. Obstetric cholestasis. Br Med J. 2002;324: 123–4.
25. Rolfes D, Ishak K. Liver disease in pregnancy. Histopathology. 1986;10:555–70.
26. Meng LJ, Reyes H, Palma J, Hernandez J, Ribalta J, Sjovall J. Progesterone metabolism in normal human pregnancy and in patients with intrahepatic cholestasis of pregnancy. In: Reyes HB, Leuschner U, Arias IM, editors. Pregnancy, Sex Hormones and the Liver. Dordrecht: Kluwer, 1996:91–100.
27. Jacquemin E. Role of multidrug resistance 3 deficiency in pediatric and adult liver disease: one gene for three diseases. Semin Liver Dis. 2001;21:551–62.
28. Jacquemin E, De Vree JM, Cresteil D et al. The wide spectrum of multidrug resistance 3 deficiency: from neonatal cholestasis to cirrhosis of adulthood. Gastroenterology. 2001; 120:1448–58.
29. Jacquemin E, Cresteil D, Manouvrier S, Boute O, Hadchouel M. Heterozygous non-sense mutation of the MDR3 gene in familial intrahepatic cholestasis of pregnancy. Lancet. 1999; 353:210–11.
30. Pauli-Magnus C, Lang T, Meier Y et al. Sequence analysis of bile salt export pump (ABCB11) and multidrug resistance p-glycoprotein 3 (ABCB4, MDR3) in patients with intrahepatic cholestasis of pregnancy. Pharmacogenetics. 2004;14:91–102.
31. Pauli-Magnus C, Meier P. Pharmacogenetics of hepatocellular transporters. Pharmacogenetics. 2003;13:189–98.
32. Savander M, Ropponen A, Avela K et al. Genetic evidence of heterogeneity in intrahepatic cholestasis of pregnancy. Gut. 2003;52:1025–9.
33. Jenkins JK, Boothby LA. Treatment of itching associated with intrahepatic cholestasis of pregnancy. Ann Pharmacother. 2002;36:1462–5.
34. Burrows RF, Clavisi O, Burrows E. Interventions for treating cholestasis in pregnancy. Cochrane Database Syst Rev. 2001;CD000493.

35. Sadler LC, Lane M, North R. Severe fetal intracranial haemorrhage during treatment with cholestyramine for intrahepatic cholestasis of pregnancy. Br J Obstet Gynaecol. 1995;102: 169–70.
36. Kondrackiene J, Beuers U, Kupcinskas L. Efficacy and safety of ursodeoxycholic acid versus cholestyramine in intrahepatic cholestasis of pregnancy. Gastroenterology. 2005; 129:894–901.
37. Bacq Y, Sape, Bréchot MC, Pierre F, Fignon A, Dubois F. Intrahepatic cholestasis of pregnancy: a French prospective study. Hepatology. 1997;26:358–64.
38. Rioseco AJ, Ivankovic MB, Manzur A et al. Intrahepatic cholestasis of pregnancy: a retrospective case–control study of perinatal outcome. Am J Obstet Gynecol. 1994;170:890–5.
39. Alsulyman OM, Ouzounian JG, Ames-Castro M, Goodwin TM. Intrahepatic cholestasis of pregnancy: perinatal outcome associated with expectant management. Am J Obstet Gynecol. 1996;175:957–60.
40. Fisk NM, Bye WB, Storey GN. Maternal features of obstetric cholestasis: 20 years experience at King George V Hospital. Aust NZ J Obstet Gynaecol. 1988;28:172–6.
41. Heinonen, S, Kirkinen, P. Pregnancy outcome with intrahepatic cholestasis. Obstet Gynecol. 1999;94:189–93.
42. Glantz A, Marschall HU, Mattsson L. Intrahepatic cholestasis of pregnancy: relationships between bile acid levels and fetal complication rates. Hepatology. 2004;40:467–74.
43. Sibai BM, Gordon T, Thom E et al. Risk factors for preeclampsia in healthy nulliparous women: a prospective multicenter study. National Institute of Child Health and Human Development Network of Maternal–Fetal Medicine Units. Am J Obstet Gynecol. 1995;172: 642–8.
44. Redman CW, Sacks GP, Sargent IL. Preeclampsia: an excessive maternal inflammatory response to pregnancy. Am J Obstet Gynecol. 1999;180:499–506.
45. Roberts JM, Taylor RN, Goldfien A. Clinical and biochemical evidence of endothelial cell dysfunction in the pregnancy syndrome preeclampsia. Am J Hypertens. 1991;4:700–8.
46. Minakami H, Oka N, Sato T et al. Preeclampsia: a microvesicular fat disease of the liver? Am J Obstet Gynecol. 1988;159:1043–7.
47. Sibai BM, Ramadan MK, Usta I et al. Maternal morbidity and mortality in 442 pregnancies with hemolysis, elevated liver enzymes, and low platelets (HELLP syndrome). Am J Obstet Gynecol. 1993;169:1000–6.
48. Sibai BM. The HELLP syndrome (hemolysis, elevated liver enzymes, and low platelets): much ado about nothing? Am J Obstet Gynecol. 1990;162:311–16.
49. Barton J, Riely C, Adamec T et al. Hepatic histopathologic condition does not correlate with laboratory abnormalities in HELLP syndrome (hemolysis, elevated liver enzymes, and low platelet count). Am J Obstet Gynecol. 1992;167:1538–43.
50. Weiner CP. Thrombotic microangiopathy in pregnancy and the postpartum period. Semin Hematol. 1987;24:119–29.
51. Ibdah JA, Bennett MJ, Rinaldo P et al. A fetal fatty-acid oxidation disorder as a cause of liver disease in pregnant women. N Engl J Med. 1999;340:1723–31.
52. Barton, JR, Sibai BM. Hepatic imaging in HELLP syndrome (hemolysis, elevated liver enzymes, and low platelet count). Am J Obstet Gynecol. 1996;174:1820–5.
53. Rinehart BK, Terrone DA, Magann EF et al. Preeclampsia-associated hepatic hemorrhage and rupture: mode of management related to maternal and perinatal outcome. Obstet Gynecol Surv. 1999;54:196–202.
54. Stevenson J, Graham D. Hepatic hemorrhage and the HELLP syndrome: a surgeon's perspective. Am Surg. 1995;61:756–60.
55. Hunter S, Martin M, Benda J, Zlatnik FJ. Liver transplant after massive spontaneous hepatic rupture in pregnancy complicated by preeclampsia. Obstet Gynecol. 1995;85:819–22.
56. Wust MD, Bolte AC, de Vries JI et al. Pregnancy outcome after previous pregnancy complicated by hepatic rupture. Hypertens Pregnancy. 2004;23:29–35.
57. Sibai BM. Diagnosis, controversies, and management of the syndrome of hemolysis, elevated liver enzymes, and low platelet count. Obstet Gynecol. 2004;103:981–91.
58. Visser W, Wallenburg HC. Temporising management of severe pre-eclampsia with and without the HELLP syndrome. Br J Obstet Gynaecol. 1995;102:111–17.

59. Martin JR Jr, Perry KG Jr, Blake PG et al. Better maternal outcomes are achieved with dexamethasone therapy for postpartum HELLP (hemolysis, elevated liver enzymes, and thrombocytopenia) syndrome. Am J Obstet Gynecol. 1997;177:1011–17.
60. Abramovici D, Friedman SA, Mercer BM et al. Neonatal outcome in severe preeclampsia at 24 to 36 weeks' gestation: does the HELLP (hemolysis, elevated liver enzymes, and low platelet count) syndrome matter? Am J Obstet Gynecol. 1999;180:221–5.
61. Sheehan H. The pathology of acute yellow atrophy and delayed chloroform poisoning. J Obstet Gynaecol. 1940;47:49.
62. Castro MA, Fassett MJ, Reynolds TB et al. Reversible peripartum liver failure: a new perspective on the diagnosis, treatment, and cause of acute fatty liver of pregnancy, based on 28 consecutive cases. Am J Obstet Gynecol. 1999;181:389–95.
63. Treem WR, Rinaldo P, Hale DE et al. Acute fatty liver of pregnancy and long-chain 3-hydroxyacyl-coenzyme A dehydrogenase deficiency. Hepatology. 1994;19:339–45.
64. Riely C, Latham P, Romero R, Duffy T. Acute fatty liver of pregnancy: a reassessment based on observations in nine patients. Ann Intern Med. 1987;106:703–6.
65. Castro MA, Goodwin TM, Shaw KJ et al. Disseminated intravascular coagulation and antithrombin III depression in acute fatty liver of pregnancy. Am J Obstet Gynecol. 1996; 174:211–16.
66. Rolfes D, Ishak K. Acute fatty liver of pregnancy: a clinicopathologic study of 35 cases. Hepatology. 1985;5:1149–58.
67. Ockner S, Brunt E, Cohn S, Krul E. Fulminant hepatic failure caused by acute fatty liver of pregnancy treated by orthotopic liver transplantation. Hepatology. 1990;11:59–64.
68. Pereira SP, O'Donohue J, Wendon J, Williams R. Maternal and perinatal outcome in severe pregnancy-related liver disease. Hepatology. 1997;26:1258–62.
69. Treem WR, Shoup ME, Hale DE et al. Acute fatty liver of pregnancy, hemolysis, elevated liver enzymes, and low platelets syndrome, and long chain 3-hydroxyacyl-coenzyme A dehydrogenase deficiency. Am J Gastroenterol. 1996;91:2293–300.

5
Geoepidemiology of primary biliary cirrhosis

C. SELMI, P. INVERNIZZI, M. ZUIN, P. M. BATTEZZATI and M. E. GERSHWIN

INTRODUCTION

Primary biliary cirrhosis (PBC) is a chronic cholestatic liver disease of unknown aetiology that is characterized by progressive destruction of the intrahepatic bile ducts[1]. It eventually leads to liver cirrhosis and failure over a period of time that can vary widely among patients[2]. The only treatments currently accepted are lifelong therapy with ursodeoxycholic acid (UDCA) and liver transplantation[1]. The former has been shown to delay disease progression when used at appropriate doses[3] while the latter is often burdened by disease recurrence but overall characterized by high survival rates[4]. It is currently accepted that PBC pathogenesis is multifactorial, with genetic and environmental factors interplaying to determine disease onset and progression. Epidemiological data are a powerful tool to determine new potential candidates for the pathogenesis and aetiology of diseases. In the case of PBC, epidemiological data can be divided into three main categories with different implications: i.e. familial, twin, or clustered cases; geographical patterns; and risk factors.

PBC IN FAMILY MEMBERS, TWINS, AND CLUSTERS

Family members of patients with PBC present an increased risk of developing the disease, and cases occurring within a family have been termed 'familial PBC'[5]. Reported rates for familial cases of PBC are illustrated in Table 1. Data from the United Kingdom report PBC frequencies among family members spanning between 1% and 2.4%[6,7]. A geoepidemiological pattern seems to apply also to prevalence rates of familial PBC. Moreover, similar to what is suggested for incidence rates in the general population[8], familial PBC frequencies reported thereafter have also been increasing, possibly due to the increasing awareness of physicians and to more sensitive diagnostic tools. Dr

Table 1 Prevalence rates of familial PBC reported from different geographical areas

Country	*Year*	*Patients (n)*	*Familial prevalence (%)*	*Reference*
USA	1994	396	4.3	9
UK	1983	117	1	6
UK	1990	347	2.4	7
UK	1995	736	1.33	5
UK	1999	157	6.4	14
Sweden	1990	111	4.5	10
Italy	1997	156	3.8	11
Japan	1999	156	5.1	12
Iceland	2004	14	7.1	15
USA	2005	1032	5.9	16

Bach and colleagues observed a 4.3% familial prevalence following a retrospective analysis of clinical records and a prospective mail survey prevalence that ascertained 396 PBC cases from the United States[9]. Occurrence of familial PBC was observed in 4.5% of cases in Sweden[10], 3.8% in Italy[11], and 5.1% in Japan[12]. In 2000 a population-based study in Olmsted County, Minnesota[13] identified 46 cases of PBC and reported a prevalence among family members higher than in the general population. The familial component of PBC onset was also estimated based on the sibling relative risk (i. e. the risk for a sibling of an affected individual to be diagnosed with the disease); such a relative risk in PBC is 10.5[14]. In the same series from the UK, 6.4% of patients with PBC had one family member also affected, with such prevalence being 1% when only first-degree relatives were studied[14]. It is of note that a study of the genetically unique population of Iceland has indicated that patients with PBC from such a genetic background did not present similar antimitochondrial patterns, while only one familial case of PBC could be identified among 85 first-degree relatives[15]. Finally, our recent epidemiological study on PBC risk factors[16] indicated a prevalence of familial cases of 5.9% among 1032 patients with PBC; accordingly, a positive familial history of PBC was also the strongest risk factor for the disease.

Disease concordance rate in monozygotic twin pairs can be defined as the proportion of affected pairs concordant for the disease. Its determination allows estimation of the weight of genetic factors in determining the susceptibility to complex diseases. Moreover, the comparison of such a rate with observed concordance in dizygotic twin sets, who share only environmental factors, provides an estimate of the role of such factors. In autoimmune diseases other than PBC concordance rates in monozygotic twins vary widely between 12% for rheumatoid arthritis[17] and 0.83 for coeliac disease[18], being on average below 50% in most cases. Data on PBC concordance in twins have, for a long time, been limited to two reports, with one describing a concordant[19] and one a discordant[20] pair of twins, although in both cases monozygosity was not genetically ascertained. Following a world-wide effort for data and sample collection we have recently reported our findings in a group of eight monozygotic and eight dizygotic twin pairs. Concordance rates for PBC in the two types of twins were 63% and zero, respectively, with 3 out of 8 monozygotic twin sets being discordant for disease occurrence[21].

In recent years several reports have described the case of specific locations presenting a higher prevalence of PBC cases compared to accepted rates. Although we cannot rule out that higher rates might in fact better represent PBC prevalence in the absence of population-based data, or be due to serendipity, these anedoctal reports are interesting and may stress the importance of the environment (and possibly genetics in some cases) in determining PBC onset. Among such reports our group has described the case of a small town in Alaska where seven cases of PBC could be identified, or the case of a husband and wife (who had grown up in the same part of a major US city) both being diagnosed with PBC[22]. More recently a Canadian study from Vancouver, BC, has demonstrated that 25% of patients referred for liver transplantation for PBC in that area were First Nations people, who constitute only 4% of the general population[23]. Finally, a work from Columbia University, NY, suggested that PBC cases referred for liver transplantation in New York City might be more common in the vicinity of the toxic waste sites (Superfunds)[24]. Data from NYC therefore seem to indicate that factors derived from such sites might be responsible for PBC onset, or rather progression, thus explaining the need for liver transplantation.

OPEN ISSUES IN PBC EPIDEMIOLOGY

Among the most intriguing observations in PBC is the comparison of epidemiological data obtained from different geographical areas and from anedoctal reports of clusters in specific locations. These have led to the concept of the geoepidemiology of PBC, with obvious implications for the proposed role of environmental factors (mostly chemicals and infectious agents) in the pathogenesis of the disease. Roughly, data seem to indicate that PBC is more commonly diagnosed in northern European areas such as the UK and Scandinavian countries. Some methodological issues, however, warrant further discussion to allow a correct interpretation of data. First, we need to consider that most of the data of PBC incidence and prevalence are descriptive. Some studies have potential flaws, including insufficiently precise or non-uniform case definition, with non-uniform criteria for case ascertainment often used. Population-based studies for PBC are challenging since there is not a non-invasive screening marker with 100% sensitivity. Second, cases are often drawn from a poorly defined population, thus raising the issue of the correct denominator to be used for calculating prevalence and incidence rates. Third, the difference between prevalence and incidence is not always clear, due to limited accuracy in the definition of time periods. Fourth, and most importantly, conclusions about trends over time of incidence and comparisons of prevalence across countries must consider the potential bias of growing disease awareness among health-care providers as well as of health-care quality standards. For example, the presence in a specific area of a tertiary referral centre might in some cases lead to the identification of a larger number of patients with PBC. Similarly, we also note that the prevalence of neither PBC nor serum antimitochondrial antibody (AMA) have been investigated in non-Westernized countries that in fact include the major part of the world population. Lastly, the quality of health care might in part explain observed differences, as in the case of African Americans in the US who seem to manifest the disease less often than other ethnicities.

GEOEPIDEMIOLOGY OF PBC

Reported prevalence and incidence rates of PBC are highest in the UK[25,26], the Scandinavian region[10], and the northern United States[13]. However, we emphasize again that this assumption might be the result of more rigorous epidemiological studies from these countries compared with other areas, rather than reflecting a true higher prevalence of cases. This section will discuss available data from various geographical regions (illustrated in Table 2).

United Kingdom

Triger and colleagues reported in 1980 the first rigorous epidemiological English study on PBC[26]. PBC cases were ascertained using physician surveys and follow-up of subjects with positive serum AMA; 34 PBC cases were identified, thus indicating a prevalence of 54/million and an incidence of 5.8/million per year. The authors also suggested for the first time a connection between PBC and environmental factors possibly contained in a residential water reservoir. PBC prevalence among people receiving water from this reservoir was approximately 10 times higher than that observed in the remaining population. The water supply was also analysed, but no significant toxins were found. These observations have recently received further attention following a similar report from Ala and colleagues from New York City[27].

The Newcastle-upon-Tyne area appears to present the highest PBC prevalence and incidence. In 1983 Hamlyn and colleagues[6] reported that PBC had higher incidence and prevalence rates than those observed by Triger et al. in Sheffield, and we note that authors used more strict case-finding criteria, later accepted internationally[28]. According to the study, PBC yearly incidence was 10/million, and prevalence varied between 37 and 144/million in different areas. Further, prevalence among older women (i.e. over 65) was 200/million. The same group of investigators conducted a similar study in 1990 on a larger area in northern England[7] using the same case-finding criteria as the Sheffield study. The observed yearly incidence of PBC was 19/million, while prevalence rates ranged between 129 and 154/million, thus confirming previous data. In 1997 stricter case definition criteria were introduced again, and the incidence rate was estimated as 22/million per year, with the prevalence rate being 240/million[25]. In this case, and in the study proposing an increasing incidence over the 1990s for PBC in the UK[8], the potential causative effect of increased awareness cannot be ruled out. Two studies were also performed in Scotland. A study performed in Dundee[29] included 21 PBC cases and reported an incidence rate of 10.6/million per year with a prevalence of 40.2/million. A larger population from Glasgow was studied in 1987. The results of the study, so far presented only in an abstract[30], indicated PBC incidence to be similar to that described by Hislop and colleagues, while prevalence was higher (70–92/million). We note, however, that both studies used case-finding criteria less strict than the English studies previously discussed and, in some cases, these were not consistent.

Sweden

PBC prevalence and incidence rates in Sweden are similar to those reported for the UK. Thirty-three PBC cases were ascertained in Malmö[31] from 1973 to 1982 and indicated that PBC yearly incidence varied between 4 and 24/million. Interestingly, the authors also reported that PBC prevalence appeared to be increasing from 28/million to 92/million each year. Another group from Orebro reported an incidence rate of 14/million and a prevalence rate of 128/million using the same case-finding criteria[32]. We should note, however, that the series of patients identified by both groups was small. Conversely, in 1990 Danielsson and colleagues described over 100 PBC cases in Umea diagnosed between 1973 and 1982 that resulted in the highest PBC prevalence rates[10]. However, the use of a compatible liver histology as the only diagnostic criterion for defining cases might have influenced the observation of PBC incidence estimated in 13.3/million per year with a prevalence of 151/million. It is of note also that, in contrast to English reports, cases ascertained in this study were more commonly found in rural rather than urban areas.

Other European countries

In 1995 researchers from Estonia used surveys for health-care professionals and described 69 cases of PBC[33] using positive serum AMA and either a compatible liver histology or elevated liver enzyme levels as diagnostic criteria. The reported PBC incidence (2.27/million) and prevalence (26.9/million) rates were significantly lower than in England or Sweden. Data from a 1995 multinational epidemiological study involving 10 Western European countries should be carefully evaluated since they resulted from non-uniform case-definition criteria across countries[34]. PBC prevalence presented a significant variability among countries, ranging, for example, from 25 to 64/million in England. We note that these data were in most cases different from published data while countries that had not previously investigated, such as West Germany, France, Ireland, and Spain, presented PBC point prevalence rates below 20/million.

Non-European countries

Data on PBC epidemiology are not available in large parts of the world such as Africa or Asia, with the possible exception of Japan, where serum AMA prevalence in the general population has been recently inferred[35]. Anedoctal case series or personal communications have been reported from these countries, thus making data semiquantitative and obviously poorly reliable.

Table 2 Data on prevalence, incidence, and sex ratio in PBC from epidemiological studies

Year	*Location*	*No. of cases*	*Case finding*	*Diagnosis criteria*	*Annual incidence (per million)*	*Prevalence (per million)*	*Gender ratio (MF)*
1980	Sheffield, UK	34	PS, lab reports	AMA+ and LFT or liver histology	5.8	54	1:16
1980	Dundee, UK	21	Liver histology	AMA+ and liver histology	10.6	40.2	1:9.5
1983	Newcastle, UK	117	Hospital registers, lab reports, death certificates	AMA+, LFT, and liver histology	10	37–144	1:14
1984	Malmö, Sweden	33	PS, lab reports, death certificates	AMA+, LFT, and liver histology	4–24	28–92	1:3
1984	Western Europe	569	PS	Non uniform	4	23 (5–75)	1:10
1985	Orebro, Sweden	18	Lab reports	AMA+, LFT, and liver histology	14	128	1:3.5
1987	Glasgow, UK	373	Lab reports	AMA+, liver histology	11–15	70–93	–
1990	Umea, Sweden	111	PS, hospital registers, lab reports	Liver histology	13.3	151	1:6
1990	Ontario, Canada	225	PS	AMA+, liver histology	3.26	22.4	1:13
1990	Northern England	347	PS, hospital admission data, lab reports	AMA+ and LFT or liver histology	19	129–154	1:9

Table 2 (Continued)

Year	*Location*	*No. of cases*	*Case finding*	*Diagnosis criteria*	*Annual incidence (per million)*	*Prevalence (per million)*	*Gender ratio (MF)*
1995	Victoria, Australia	84	PS, hospital records	AMA+, LFT, and liver histology	–	19.1	1:11
1995	Estonia	69	PS, hospital admission data, lab reports	AMA+, LFT and liver histology	2.27	26.9	1:22
1997	Newcastle, UK	160	PS, hospital admission data, lab reports, death certificates	AMA+, LFT, and liver histology	14–32	240	1:10
2000	Olmsted County, MN (USA)	46	Hospital records	LFT, and AMA+ or liver histology	27	402	1:8
2005	Japan	5805	Ministry of Health database	Diagnosis in hospital records	–	–	1:8
2005	USA	1032	Tertiary referral centres	Two out of: AMA+, LFT, or liver histology	–	–	1:12

Abbreviations used: PS, physician survey; AMA, anti-mitochondrial antibodies; LFT, liver function tests

Two studies have been conducted in Australia. First, Watson and colleagues identified 84 cases of PBC from the region of Victoria using rigorous case-finding methods that included all three major diagnostic criteria[36]. Reported PBC prevalence was 19.1/million, among the lowest rates ever reported. More recently, a study from the same region reported different data[37]. Authors ascertained 249 cases of PBC that resulted in a cumulative prevalence of 151/million, thus almost 10-fold higher than that reported in 1995. More interestingly, prevalence rates were significantly higher for British, Italian, and Greek immigrants, reaching 208 cases/million. These observations suggest that native Victorians might carry a genetic background protective against PBC onset.

A study from Ontario, Canada, performed in 1987[38] ascertained 225 PBC cases identified with positive serum AMA and compatible liver histology and reported a low cumulative annual incidence (3.26/million) and a prevalence rate of 22.4/million. In 2000 a large study performed on the population of Olmsted County, Minnesota[13] estimated the age- and sex-adjusted PBC prevalence to be 27/million, among the highest ever reported, based on data from a computerized index of diagnoses for inpatients and outpatients available in that State. We note that case-finding criteria for this study included elevated liver enzyme levels and a positive AMA or a compatible liver histology.

RISK FACTORS

The study of risk factors in PBC might indeed fit within the models proposed by geoepidemiology. Some of the observed factors may be associated with specific geographical areas as well as direct basic science in searching for the cause of the disease.

Beside genetics, a number of factors have been proposed to play a role in the aetiology of PBC, as for other autoimmune diseases, including infectious diseases[39], other autoimmune diseases[40], and lifestyle factors, such as smoking[41] and a high-fat diet[42]. The research group from Newcastle-upon-Tyne in the UK, conducted a population-based case–control study[43] in northeast England to investigate risk factors for PBC susceptibility through the use of postal self-completion questionnaires covering medical history and lifestyle. Results from 100 patients with PBC and 223 controls showed no significant association with events in pregnancy, or past infections; on the other hand, an association with past smoking habit was observed.

To identify environmental factors leading to PBC, our group performed an epidemiological study on 1032 patients with PBC and 1041 controls[16]. Patients were referred from 20 tertiary referral centres in the US and were representative of all but two States (Delaware and Hawaii). Controls were matched to cases for sex, age, and geographical origin (using the random-digit dialling system). The survey instrument was based on over 180 standardized questions, including subquestions, largely drawn from NHANES and investigating demographics, personal and family medical history, reproductive history, and lifestyle factors. The instrument was based on a preliminary study[44] that had

Table 3 Results of multiple logistic regression models from our epidemiological study on 1032 patients with PBC and 1041 controls[16]

	β	*OR*	*95% CI*	*p*
Medical/family history				
Family history of PBC	1.1868	10.736	4.227–27.268	<0.0001
Family history of SLE	0.4019	2.234	1.261–3.957	0.005
Family history of Sjögren's	0.8801	5.814	1.279–26.435	0.0227
History of urinary tract infections	0.2065	1.511	1.192–1.915	0.0006
Lifestyle factors				
Ever smoked > 100 cigarettes	0.2252	1.569	1.292–1.905	<0.0001
No passive smoke at work/	–0.2368	0.820	0.582–1.155/	0.0005
Don't have a job	0.2574	1.369	1.095–1.712	
Uses of nail polish/year	0.00177	1.002*	1.000–1.003	0.0136
Number of cigarettes smoked	–0.00109	0.999	0.998–1.000	0.0031
Each smoker in household	–0.6776	0.5078	0.3167– 0.8143	0.0041
*Reproductive history***				
Ever used hormonal replacement	0.2185	1.548	1.273–1.882	<0.0001
Never pregnant	–0.4906	0.6118	0.4489–0.8338	0.0012
Age of first pregnancy	–0.0470	0.9541	0.9331–0.9755	<0.0001

*Calculated for each additional use of nail polish/year.

** For female cases and controls only.

SLE, systemic lupus erythematosus.

highlighted limitations in the first version of the questionnaire that were corrected for the present study. The analysis of data included the creation of three models (medical history, lifestyle, and reproductive factors, respectively) with social and demographic factors included in all models. All patients studied were within 7 years of their diagnosis. The prevalence of all investigated factors was evaluated in an unadjusted fashion as well as within multiple logistic regression models (Table 3). Data indicate that having a first-degree relative with PBC (adjusted odds ratio [AOR] 10.736; 95% confidence interval 4.227–27.268), a history of urinary tract infections (AOR 1.511, 95% CI 1.192–1.915), past smoking (AOR 1.569, 95% CI 1.292–1.905), or the use of hormone replacement therapies (AOR 1.548, 95% CI 1.273–1.882) were significantly associated with increased risk of PBC[16]. The frequent use of nail polish slightly increased the risk of having PBC. Other autoimmune diseases coexisted with PBC in 32% of patients while only 13% of controls had one autoimmune disease (excluding PBC). Importantly, no significant differences were observed for race and ethnicities, although few African American patients were referred for our study.

CONCLUSIONS AND FUTURE DIRECTIONS

Factors leading to the breakdown of immune tolerance in PBC are still largely unknown; however, a better understanding of such factors is the critical step towards improving our diagnostic and therapeutic tools. Epidemiological studies of PBC should not be disregarded by researchers involved in basic aspects of the disease. Data from rigorous epidemiological studies might indicate potential sources for environmental factors triggering PBC in susceptible individuals, as well as providing an accurate picture of the social and economic weight of the disease. We are convinced that the interpretation of epidemiological data on PBC is not of secondary importance, and should not be overlooked either by clinicians or by laboratory scientists. Several clues to disease pathogenesis and to novel treatment options have been obtained in the past from epidemiological observations and experimental findings, respectively. For this reason we encourage all researchers interested in PBC to better their knowledge of genetic and geoepidemiology data in the search for new ideas that might shed light on the mysteries of PBC. The common *adagio* that states that findings and observations must be transferred 'from bench to bedside' should become a two-way street in PBC. Future epidemiological studies should use uniform methods and case-finding criteria to allow a better comparison of data from different populations. This approach will provide clues to the unanswered questions on PBC and the disease will finally evolve into 'a matter of (well-known) bad genes and bad luck'[45].

References

1. Talwalkar JA, Lindor KD. Primary biliary cirrhosis. Lancet. 2003;362:53–61.
2. Prince M, Chetwynd A, Newman W, Metcalf JV, James OF. Survival and symptom progression in a geographically based cohort of patients with primary biliary cirrhosis: follow-up for up to 28 years. Gastroenterology. 2002;123:1044–51.
3. Gluud C, Christensen E. Ursodeoxycholic acid for primary biliary cirrhosis. Cochrane Database Syst Rev 2002:CD000551.
4. MacQuillan GC, Neuberger J. Liver transplantation for primary biliary cirrhosis. Clin Liver Dis. 2003;7:941–56, ix.
5. Brind AM, Bray GP, Portmann BC, Williams R. Prevalence and pattern of familial disease in primary biliary cirrhosis. Gut. 1995;36:615–17.
6. Hamlyn AN, Macklon AF, James O. Primary biliary cirrhosis: geographical clustering and symptomatic onset seasonality. Gut. 1983;24:940–5.
7. Myszor M, James OF. The epidemiology of primary biliary cirrhosis in north-east England: an increasingly common disease? Q J Med. 1990;75:377–85.
8. James OF, Bhopal R, Howel D, Gray J, Burt AD, Metcalf JV. Primary biliary cirrhosis once rare, now common in the United Kingdom? Hepatology. 1999;30:390–4.
9. Bach N, Schaffner F. Familial primary biliary cirrhosis. J Hepatol. 1994;20:698–701.
10. Danielsson A, Boqvist L, Uddenfeldt P. Epidemiology of primary biliary cirrhosis in a defined rural population in the northern part of Sweden. Hepatology. 1990;11:458–64.
11. Floreani A, Naccarato R, Chiaramonte M. Prevalence of familial disease in primary biliary cirrhosis in Italy. J Hepatol. 1997;26:737–8.
12. Tsuji K, Watanabe Y, Van De Water J et al. Familial primary biliary cirrhosis in Hiroshima. J Autoimmun. 1999;13:171–8.
13. Kim WR, Lindor KD, Locke GR 3rd et al. Epidemiology and natural history of primary biliary cirrhosis in a US community. Gastroenterology. 2000;119:1631–6.
14. Jones DE, Watt FE, Metcalf JV, Bassendine MF, James OF. Familial primary biliary cirrhosis reassessed: a geographically-based population study. J Hepatol. 1999;30:402–7.
15. Olafsson S, Gudjonsson H, Selmi C et al. Antimitochondrial antibodies and reactivity to N. aromaticivorans proteins in Icelandic patients with primary biliary cirrhosis and their relatives. Am J Gastroenterol. 2004;99:2143–6.
16. Gershwin ME, Selmi C, Worman HJ et al. Risk factors and comorbidities in primary biliary cirrhosis: a controlled interview-based study of 1032 patients. Hepatology. 2005;42: 1194–202.

17. MacGregor AJ, Snieder H, Rigby AS et al. Characterizing the quantitative genetic contribution to rheumatoid arthritis using data from twins. Arthritis Rheum. 2000;43:30–7.
18. Greco L, Romino R, Coto I et al. The first large population based twin study of coeliac disease. Gut. 2002;50:624–8.
19. Chohan MR. Primary biliary cirrhosis in twin sisters. Gut. 1973;14:213–14.
20. Kaplan MM, Rabson AR, Lee YM, Williams DL, Montaperto PA. Discordant occurrence of primary biliary cirrhosis in monozygotic twins. N Engl J Med. 1994;331:951.
21. Selmi C, Mayo MJ, Bach N et al. Primary biliary cirrhosis in monozygotic and dizygotic twins: genetics, epigenetics, and environment. Gastroenterology. 2004;127:485–92.
22. Abu-Mouch S, Selmi C, Benson GD et al. Geographic clusters of primary biliary cirrhosis. Clin Dev Immunol. 2003;10:127–31.
23. Arbour L, Field L, Ross P, Erikson A, Yoshida E. The mystery of primary biliary cirrhosis in British Columbia's First Nations people. Int J Circumpolar Health. 2004;63(Suppl. 2):185–8.
24. Ala A, Stanca CM, Bu-Ghanim M et al. Increased prevalence of primary biliary cirrhosis near Superfund toxic waste sites. Hepatology. 2006;43:525–31.
25. Metcalf JV, Bhopal RS, Gray J, Howel D, James OF. Incidence and prevalence of primary biliary cirrhosis in the city of Newcastle upon Tyne, England. Int J Epidemiol. 1997;26:830–6.
26. Triger DR. Primary biliary cirrhosis: an epidemiological study. Br Med J. 1980;281:772–5.
27. Ala A, Ahmado I, Stanca C et al. The prevalence of primary biliary cirrhosis is increased near toxic waste sites in New York City. Hepatology. 2004;40:205A.
28. Kaplan MM. Primary biliary cirrhosis. N Engl J Med.1996;335:1570–80.
29. Hislop WS, Main G, Robertson AJ. Primary biliary cirrhosis: experience in a non-referral centre. Lancet. 1981;2:144.
30. Goudie B, MacFarlane G, Boyle P. Epidemiology of antimitichondrial antibody seropositivity and primary biliary cirrhosis in west of Scotland. Gut. 1987;28:A1346.
31. Eriksson S, Lindgren S. The prevalence and clinical spectrum of primary biliary cirrhosis in a defined population. Scand J Gastroenterol. 1984;19:971–6.
32. Lofgren J, Jarnerot G, Danielsson D, Hemdal I. Incidence and prevalence of primary biliary cirrhosis in a defined population in Sweden. Scand J Gastroenterol. 1985;20:647–50.
33. Remmel T, Remmel H, Uibo R, Salupere V. Primary biliary cirrhosis in Estonia, with special reference to incidence, prevalence, clinical features, and outcome. Scand J Gastroenterol. 1995;30:367–71.
34. Triger DR, Berg PA, Rodes J. Epidemiology of primary biliary cirrhosis. Liver. 1984;4:195–200.
35. Shibata M, Onozuka Y, Morizane T et al. Prevalence of antimitochondrial antibody in Japanese corporate workers in Kanagawa prefecture. J Gastroenterol. 2004;39:255–9.
36. Watson RG, Angus PW, Dewar M, Goss B, Sewell RB, Smallwood RA. Low prevalence of primary biliary cirrhosis in Victoria, Australia. Melbourne Liver Group. Gut. 1995;36:927–30.
37. Sood S, Gow PJ, Christie JM, Angus PW. Epidemiology of primary biliary cirrhosis in Victoria, Australia: high prevalence in migrant populations. Gastroenterology. 2004;127:470–5.
38. Witt-Sullivan H, Heathcote J, Cauch K et al. The demography of primary biliary cirrhosis in Ontario, Canada. Hepatology. 1990;12:98–105.
39. Rose NR. The role of infection in the pathogenesis of autoimmune disease. Semin Immunol. 1998;10:5–13.
40. Culp KS, Fleming CR, Duffy J, Baldus WP, Dickson ER. Autoimmune associations in primary biliary cirrhosis. Mayo Clin Proc. 1982;57:365–70.
41. George J, Levy Y, Shoenfeld Y. Smoking and immunity: an additional player in the mosaic of autoimmunity. Scand J Immunol. 1997;45:1–6.
42. Fernandes G. Dietary lipids and risk of autoimmune disease. Clin Immunol Immunopathol. 1994;72:193–7.
43. Howel D, Fischbacher CM, Bhopal RS, Gray J, Metcalf JV, James OF. An exploratory population-based case–control study of primary biliary cirrhosis. Hepatology. 2000;31: 1055–60.

44. Parikh-Patel A, Gold EB, Worman H, Krivy KE, Gershwin ME. Risk factors for primary biliary cirrhosis in a cohort of patients from the United States. Hepatology. 2001;33:16–21.
45. He XS, Ansari AA, Gershwin ME. Xenobiotic considerations for the development of autoimmune liver disease: bad genes and bad luck. Rev Environ Health. 2001;16:191–202.

6
Interface hepatitis in primary biliary cirrhosis: prognostic marker and therapeutic target

R. POUPON, L. ABENAVOLI, N. RABAHI, Y. CHRÉTIEN, O. CHAZOUILLERES, D. WENDUM and C. CORPECHOT

INTRODUCTION

Interlobular bile duct is the primary target in primary biliary cirrhosis (PBC). PBC may progress over 10–20 years to cirrhosis by two different pathways having in common an interface hepatitis. One is the so-called biliary or ductular piecemeal necrosis (PMN). In contrast, in other patients the progression seems to result by the same mechanisms as in chronic viral hepatitis or as in autoimmune hepatitis (AIH). They show the typical features of lymphocytic PMN.

PMN as proposed by Popper et al.[1] is defined by the destruction of the liver cells at the limiting plate of hepatocytes. Because the lesion occurs at the interface between parenchymal cells and connective tissue it is currently termed interface hepatitis.

INTERFACE HEPATITIS IN PBC: NATURE, SEVERITY AND LINK TO FIBROGENESIS

The typical ductular reaction associated with biliary PMN is shown in Figure 1. Lymphocytes, macrophages and reactive ductules extend between liver plates. Hepatocytes degenerate but swelling and feathery degeneration, at least in the early stages, is less marked than in AIH. Figure 2 is another example of a liver biopsy specimen from a patient with a massive biliary PMN. The periportal parenchyma is disturbed and a mixture of inflammatory cells and loose fibrous tissue separates the liver cell plates, some of which tend to show tubular transformation.

The lymphocytic PMN observed in PBC is an inflammatory destruction of a group of hepatocytes associated with lymphohistiocytic cells. In the case depicted in Figure 3 there is obvious close contact between clusters of

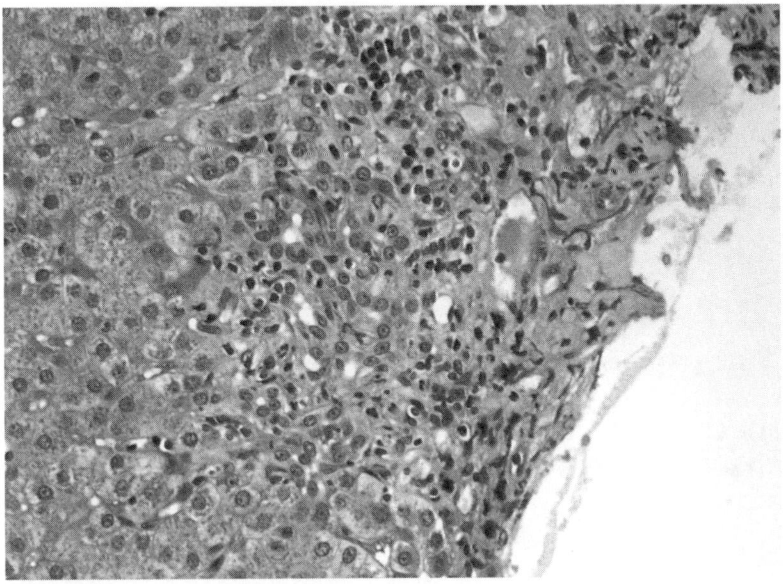

Figure 1 Typical ductular reaction associated with loose fibrotic tissue, macrophages, polymorphonuclears and lymphocytes

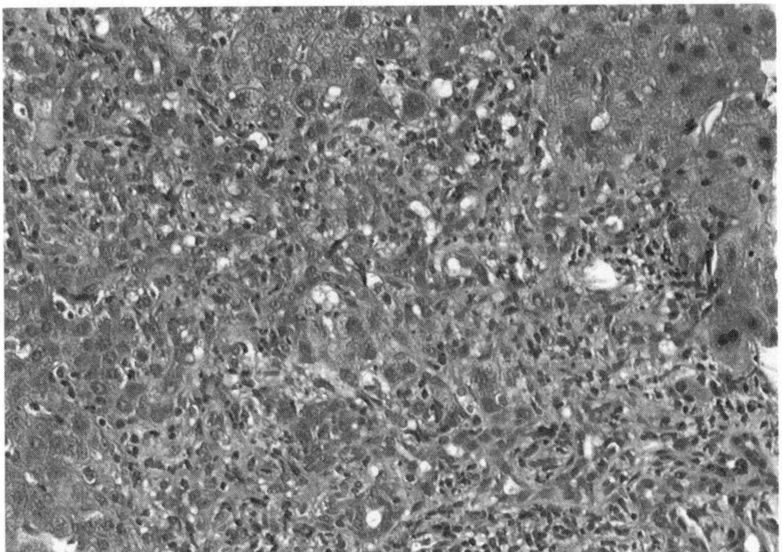

Figure 2 Another example of massive biliary piecemeal necrosis

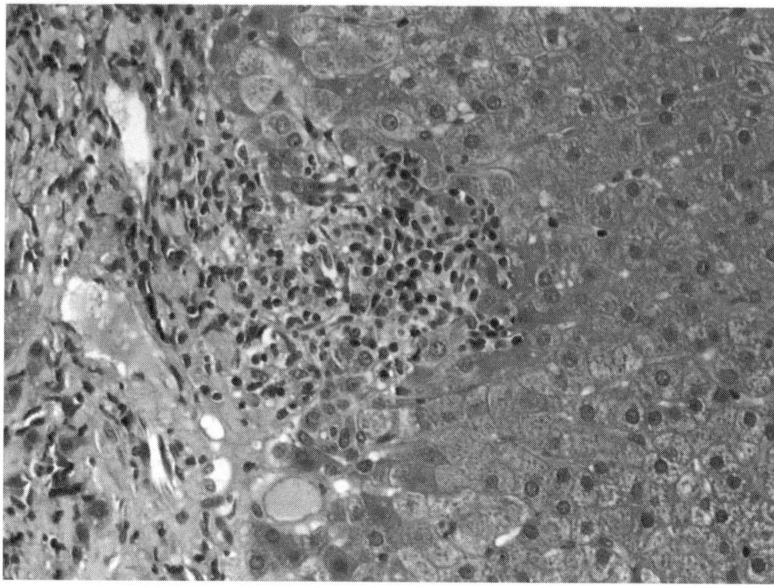

Figure 3 The lymphocytic piecemeal necrosis observed in PBC: lymphohistiocytic cells spreading out the portal tract and killing the periportal hepatocytes

lymphocytes and hepatocytes. In other cases lymphocytes invade the hepatocyte. Liver cells show degenerative features, swelling, shrinkage, or apoptosis, while other cells seem to survive, become hyperplastic and grow in a tubular fashion (so-called resetting). As a result the limiting plates are replaced by newly formed connective tissue.

Key features of lymphocytic PMN are the following: (a) peripolesis, which describes close contact between $CD4^{-}$ CD8 lymphocytes with the hepatocyte membrane, thus forming an immunological synapse; (b) emperipolesis, which is the internalization of the synapse into the lymphocyte followed by transport to a lysozome of the lymphocyte for digestion. By this mechanism the target cell is slowly diminished in size and gradually disappears[2].

Grading the severity of PMN was initially performed in a way roughly similar to that proposed by Knodell[3] for chronic hepatitis. According to this grading the prevalence of the lymphocytic PMN and biliary PMN has been reported to be respectively 40% and 60% of the cases studied by Portmann et al.[4] and Nakanuma et al.[5] in 1984 and 1990, respectively. Surprisingly, only few patients had both types of interface hepatitis and this is also our experience. During the past 10 years we have graded PMN in a different manner. We prefer the grading according to the scoring proposed by Bedossa and Poynard for hepatitis C[6] because it is very simple and highly reproducible, i.e. (a) grade minimal, defined by a focal alteration of the periportal plate in some portal tracts; (b) moderate, defined by a diffuse alteration of the periportal plate in some portal tracts or focal lesions around all the portal tracts; (c) severe, defined by a diffuse alteration of the periportal plate in all the portal tracts. As expected, we have found that the intensity of either the ductular reaction and/or the lymphocytic PMN correlated tightly with the severity of fibrosis in untreated PBC patients, confirming that these two lesions herald not only chronicity but also progression towards cirrhosis[7].

Do the fundamental liver lesions of PBC, bile duct paucity, lymphocytic PMN, biliary PMN and fibrosis have a biochemical counterpart? We studied this point in untreated PBC patients. We did not find major differences according their severity. For example, patients with a mild PMN had lower IgM, lower gammaglobulins, and lower AST levels than those with a high-grade PMN. However, there was a large overlap between patients with or without significant lymphocytic PMN or paucity. The same observation was made in patients receiving ursodeoxycholic acid (UDCA). So, to date, the only way to correctly identify these lesions is to perform a liver biopsy. This is, however, not true for the diagnosis of fibrosis; now we have a device, the fibroscan (Echosens, Paris). With this device we are able to diagnose patients with fibrous septa and early cirrhosis, and this might be helpful in the future to assess the efficacy of new drugs in PBC.

THE IMPACT OF UDCA TREATMENT ON INTERFACE HEPATITIS

Is the severity of the interface hepatitis still prognostic in UDCA-treated patients? The effect of UDCA treatment on histology in PBC is depicted in Figure 4. These data come from our clinical trial published in 1991[8]. UDCA was shown to stabilize or improve significantly the grades of both ductular proliferation and lymphocytic PMN. This result was confirmed by a combined analysis of four large trials[9]. Because 2 years of treatment is too short a period

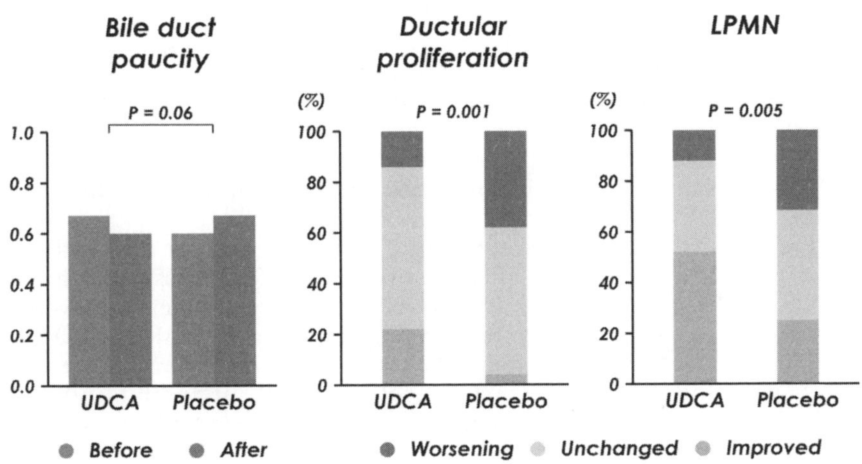

Figure 4 The effect of UDCA treatment on histology in PBC

we developed a Markov model to determine the possible impact of UDCA on fibrosis[10]. UDCA produced a five-fold reduction of liver fibrosis development when compared to placebo or patients receiving D-penicillamine. However, and this a weakness of UDCA, UDCA is not able to produce regression of fibrosis, suggesting that other more powerful drugs are desirable. In this context we aimed to determine the actual incidence and predictive factors for the development of cirrhosis[11]. We validated a Markov model for this purpose. We studied 437 liver biopsies from 187 non-cirrhotic patients during 655 patient-years of follow-up. By univariate analysis the following parameters were found to be significantly associated with progression towards cirrhosis: the histological stage at diagnosis, the grade of interlobular bile duct paucity, lymphocytic PMN, and the serum bilirubin level measured after 1 year of UDCA therapy. For example, given a patient with a stage 2 disease at onset, the average risk of development rose from 3% to 19% at 5 years of follow-up depending on the grade of lymphocytic PMN. By multivariate analysis, in addition to the histological stage at onset of treatment, serum bilirubin and lymphocytic PMN were the two major parameters associated with the development of cirrhosis.

WHAT IS THE FUTURE IN TERMS OF THERAPEUTIC MANAGEMENT OF PBC?

Before dealing with this question it is necessary to discover what PBC really is. How does UDCA work? PBC is a disease characterized by unresolved inflammation and cholestasis. Inflammation and cholestasis are responsible for cell necrosis, further immune activation and fibrosis.

The inflammatory process involves mainly T cells, CD40 activation and interferon-γ, but also many other cytokines and cells, macrophages, polymorphonuclears and fibroblasts. This process is characterized by an intense crosstalk between all these cells[12]. Because of the multiple players, it may be anticipated that anti-inflammatory drugs with broad specificity should be more relevant than highly specific targeted drugs.

In PBC, cholestasis is both a ductal and canalicular cholestasis. Cholestasis, at least in the early stages, is due to the action of cytokines present around the interlobular bile duct and the periportal canaliculi. In advanced stages, cholestasis results from both inflammation and mechanical obstruction of the biliary canals.

UDCA has multiple biological effects[13]. At least in part for this reason, we ignore the subtle mechanisms of action of UDCA. Nonetheless the following scenario is the most probable according to the present state of knowledge. UDCA relieves cholestasis by three means: it improves both ductular and canalicular cholestasis by promoting the targeting of biliary transporters to the canaliculus and ATP canalicular secretion; finally it inhibits bile acid resorption. The resulting combination of stimulation of biliary secretion together with biliary diversion has several consequences that tend to restore normal physiological patterns, by reducing immune activation, apoptosis and fibrogenesis. In addition, UDCA by itself directly abrogates interferon-induced HLA class II hyperexpression and inhibits cytokine-induced apoptosis.

TARGETING STRATEGIES TO IMPROVE UDCA ACTION

Theoretically many pathomechanisms could be targeted to improve UDCA action. We will comment on three of them: (1) blockers of immune activation, (2) inducers of ductal choleresis, (3) nucleoreceptor modulators.

As mentioned above, the inflammatory process involves many players. Therefore our belief is that we have to use properly anti-inflammatory drugs with broad specificity rather than specific targeted weapons. Two studies[14,15] have shown that budesonide combined with UDCA is superior to UDCA alone for treating PBC with non-cirrhotic PBC: the combination is better in terms of biochemistries, immune parameters and very importantly in terms of grade of inflammation and fibrosis. This is an important step forward. The next step is to demonstrate that this combination is equally efficient in the subset of patients not responding adequately to UDCA alone.

A second approach is represented by inducers of ductal choleresis. Glybenclamide[16], sulindac[17] and norUDCA[18] belong to this class of drugs. NorUDCA is the C23 homologue of UDCA. NorUDCA does not undergo significant conjugation and generates a bicarconate-rich choleresis. In mdr2 knockout mice norUDCA markedly improved liver tests and histology, significantly reduced fibrosis and proliferating cholangiocytes and was far more effective than UDCA[19].

A third approach is represented by the nucleoreceptor modulators. The main aim of the biliary function is to irreversibly eliminate from the body a series of harmful lipophilic substances. Export of the compounds is determined by canalicular transporters which are transcriptionally regulated by nucleoreceptors.

When bile acid concentrations increase in hepatocytes, bile acids bind to FXR and PXR. Subsequently transcription of a series of gene products aimed to extrude bile acids from the cell are activated (Figure 4). Accordingly, hepatocytes adapt by reducing import, increasing efflux, decreasing synthesis and increasing biotransformation of bile acids.

Table 1

	FXR	*PXR*	*LXR*	*PPAR*
Targets	CYP3A4 BSEP MDR3 MRP2	CYP3A4 MDR3 MRP2	Reverse chol. transp. Inflammation ABCG5/8	Fatty acid oxidation Inflammation MDR3
Inducers	CDCA GW-4064 6ECDCA Fexaramine	RIF Statin Dexa Troglitazone LCA	Oxysterols	Fatty acids Eicosanoids Fibrates Statin Thiazolidinediones

chol. transp., cholesterol transport

Target genes and inducers of nucleoreceptors are shown in Table 1. UDCA is not an agonist of these receptors *in vitro*; moreover, UDCA behaves as an antagonist *in vivo*. Accordingly it is very tempting to imagine that some agonists of the nucleoreceptors could augment the therapeutic effect of UDCA. Peroxisome proliferator-activated receptor (PPAR) agonists might be of use in cholestasis for two reasons: first, PPARγ is down-regulated in cholangiocytes in PBC[20], second, PPARγ agonists inhibit stellate cell activation and may counteract liver fibrosis[21].

Rifampicin, a PXR agonist, is widely used for the treatment of cholestasis, in particular in childhood, mainly because of its beneficial effect on pruritus. Very rare cases of side-effects have been reported. There are no data assessing the effects of UDCA–rifampicin combination in adulthood or childhood. However, in our experience the vast majority of PBC patients presenting with pruritus respond completely with this combination.

This may be explained by the study of Marshall et al.[22] carried out in patients with gallstones. They found that rifampicin, unlike UDCA, has the unique property to markedly stimulate CYP3A activity and the formation of hydroxylated bile salt, an effect not shared by UDCA. Another argument suggesting that rifampicin could have a long-term benefit is that activation of PXR inhibits the expression of several fibrosis-related genes as well as proliferation-related genes[23].

The potential utility of FXR agonist therapy in cholestasis is based on the fact that the expression of most of the major enzymes and membrane transporters that control the synthesis and enterohepatic circulation of bile acids are regulated by FXR. GW-4014 and 6 ECDCA are 50–100-fold more potent than UDCA in transactivating FXR. Both have been shown to protect the hepatocyte during bile acid accumulation induced experimentally[24,25]. GW-4064 affords marked protection against a-naphthyl isothiocyanate (ANIT). ANIT-treated rats show large areas of parenchymal necrosis, which was reduced but still notable in tauroursodeoxycholic acid (TUDCA)/ANIT-treated rats and absent in rats receiving the agonist. In these rats serum bile acids and bilirubin were markedly reduced when compared to sham animals or TUDCA-treated rats. Analysis of gene expression in livers from GW-4064-treated cholestatic rats revealed decreased expression of bile acid biosynthetic genes and increased genes involved in bile acid transport, including mdr2.

It should, however, be noted that FXR-mediated stimulation of canalicular bile flow via induction of the bile salt export pump could aggravate liver injury as observed in UDCA-fed mice with obstructive cholestasis[26]. Adult cholangiopathies, PBC in advanced stage, i.e. with marked ductopenia and primary sclerosing cholangitis, have a significant obstructive component, implying cautious use of FXR agonist in advanced disease stages.

FXR ligands have in addition recently been demonstrated to have antifibrotic properties through the FXR/SHP cascade that regulates negatively hepatic stellate cell activation and also through activation of PPARγ[26].

In summary, many novel therapies based on a strong rationale could be available in the next few years; however, it remains to be seen if they are free of side-effects. As in other clinical situations, and because of the number of genes and pathomechanisms involved in cholangiopathies, we can also imagine targeted combination therapies based on liver genetic background characterization and molecular profiling using microarray technologies.

References

1. Popper H, Paronetto F, Schaffner F. Immune processes in the pathogenesis of liver disease. Ann NY Acad Sci. 1965;124:781–99.
2. Wang MX, Morgan T, Lungo W, Wang L, Sze GZ, French SW. 'Piecemeal' necrosis: renamed troxis necrosis. Exp Mol Pathol. 2001;71:137–46.
3. Knodell RG, Ishak KG, Black WC et al. Formulation and application of a numerical scoring system for assessing histological activity in asymptomatic chronic active hepatitis. Hepatology. 1981;1:431–5.
4. Portmann B, Popper H, Neuberger J, Williams R. Sequential and diagnostic features in primary biliary cirrhosis based on serial histologic study in 209 patients. Gastroenterology. 1985;88:1777–90.
5. Nakanuma Y, Saito K, Unoura M. Semiquantitative assessment of cholestasis and lymphocytic piecemeal necrosis in primary biliary cirrhosis: a histologic and immunohistochemical study. J Clin Gastroenterol. 1990;12:357–62.
6. Bedossa P, Poynard T. An algorithm for the grading of activity in chronic hepatitis C. The METAVIR Cooperative Study Group. Hepatology. 1996;24:289–93.
7. Degott C, Zafrani ES, Callard P, Balkau B, Poupon RE, Poupon R. Histopathological study of primary biliary cirrhosis and the effect of ursodeoxycholic acid treatment on histology progression. Hepatology. 1999;29:1007–12.
8. Poupon RE, Balkau B, Eschwege E, Poupon R. A multicenter, controlled trial of ursodiol for the treatment of primary biliary cirrhosis. UDCA-PBC Study Group. N Engl J Med. 1991;324:1548–54.
9. Poupon RE, Lindor KD, Pares A, Chazouilleres O, Poupon R, Heathcote EJ. Combined analysis of the effect of treatment with ursodeoxycholic acid on histologic progression in primary biliary cirrhosis. J Hepatol. 2003;39:12–16.
10. Corpechot C, Carrat F, Bonnand AM, Poupon RE, Poupon R. The effect of ursodeoxycholic acid therapy on liver fibrosis progression in primary biliary cirrhosis. Hepatology. 2000;32:1196–9.
11. Corpechot C, Carrat F, Poupon R, Poupon RE. Primary biliary cirrhosis: incidence and predictive factors of cirrhosis development in ursodiol-treated patients. Gastroenterology. 2002;122:652–8.
12. Fava G, Glaser S, Francis H, Alpini G. The immunophysiology of biliary epithelium. Semin Liver Dis. 2005;25:251–64.
13. Paumgartner G, Beuers U. Ursodeoxycholic acid in cholestatic liver disease: mechanisms of action and therapeutic use revisited. Hepatology. 2002;36:525–31.
14. Leuschner M, Maier KP, Schlichting J et al. Oral budesonide and ursodeoxycholic acid for treatment of primary biliary cirrhosis: results of a prospective double-blind trial. Gastroenterology. 1999;117:918–25.
15. Rautiainen H, Karkkainen P, Karvonen AL et al. Budesonide combined with UDCA to improve liver histology in primary biliary cirrhosis: a three-year randomized trial. Hepatology. 2005;41:747–52.
16. Nathanson MH, Burgstahler AD, Mennone A, Dranoff JA, Rios-Velez L. Stimulation of bile duct epithelial secretion by glybenclamide in normal and cholestatic rat liver. J Clin Invest. 1998;101:2665–76.
17. Bolder U, Trang NV, Hagey LR et al. Sulindac is excreted into bile by a canalicular bile salt pump and undergoes a cholehepatic circulation in rats. Gastroenterology. 1999;117:962–71.
18. Yoon YB, Hagey LR, Hofmann AF, Gurantz D, Michelotti EL, Steinbach JH. Effect of side-chain shortening on the physiologic properties of bile acids: hepatic transport and effect on biliary secretion of 23-nor-ursodeoxycholate in rodents. Gastroenterology. 1986; 90:837–52.

19. Fickert P, Wagner M, Marschall HU et al. 24-norUrsodeoxycholic acid is superior to ursodeoxycholic acid in the treatment of sclerosing cholangitis in Mdr2 (Abcb4) knockout mice. Gastroenterology. 2006;130:465–81.
20. Harada K, Isse K, Kamihira T, Shimoda S, Nakanuma Y. Th1 cytokine-induced downregulation of PPARgamma in human biliary cells relates to cholangitis in primary biliary cirrhosis. Hepatology. 2005;41:1329–38.
21. Fiorucci S, Rizzo G, Antonelli E et al. Cross-talk between farnesoid-X-receptor (FXR) and peroxisome proliferator-activated receptor gamma contributes to the antifibrotic activity of FXR ligands in rodent models of liver cirrhosis. J Pharmacol Exp Ther. 2005;315:58–68.
22. Marschall HU, Wagner M, Zollner G et al. Complementary stimulation of hepatobiliary transport and detoxification systems by rifampicin and ursodeoxycholic acid in humans. Gastroenterology. 2005;129:476–85.
23. Haughton EL, Tucker SJ, Marek CJ et al. Pregnane X receptor activators inhibit human hepatic stellate cell transdifferentiation *in vitro*. Gastroenterology. 2006;131:194–209.
24. Liu Y, Binz J, Numerick MJ et al. Hepatoprotection by the farnesoid X receptor agonist GW4064 in rat models of intra- and extrahepatic cholestasis. J Clin Invest. 2003;112:1678–87.
25. Fiorucci S, Rizzo G, Antonelli E et al. A farnesoid X receptor-small heterodimer partner regulatory cascade modulates tissue metalloproteinase inhibitor-1 and matrix metalloprotease expression in hepatic stellate cells and promotes resolution of liver fibrosis. J Pharmacol Exp Ther. 2005;314:584–95.
26. Fickert P, Zollner G, Fuchsbichler A et al. Ursodeoxycholic acid aggravates bile infarcts in bile duct-ligated and Mdr2 knockout mice via disruption of cholangioles. Gastroenterology. 2002;123:1238–51.

Section III
Viral hepatitis (Part 1)

Chair: G PAUMGARTNER and HC THOMAS

7
Pathobiology of hepatitis B virus infection

K. KOIKE

INTRODUCTION

Hepatitis B virus (HBV) is a small enveloped DNA virus, one of the hepatotropic and highly species-specific virus family which is able to infect in culture only well-differentiated primary hepatocytes of their specific hosts[1–4]. The specificity of virus attachment, uncoating and entry has not yet been well characterized, as many efforts to develop a small animal model to study the HBV infection[5] are hampered by self-limiting infection and could not yet establish a chronic carrier state[6]. After a non-specific uptake of hepadnavirus genomes, later steps in the viral life cycle are not strictly host-restricted, as is shown by transgenic HBV-producing mice[7].

CHRONIC INFECTION

Usually, outcomes of HBV infections are acute infection, fulminant hepatitis and chronic infection.

1. Acute infection: more than 90% of HBV infection is grouped in this category with clearance of virus particles from the liver and blood and patients acquired lasting immunity to reinfection.
2. Fulminant infection is a rare case (nearly 1%) but with severe liver failure its mortality is 90%.
3. Chronic hepatitis (around 10%) is a persistent HBV carrier state, most of which is neonatal infection from carrier mother to infant. Chronic HBV infection is more common in Asian and sub-Saharan African countries. In addition to vertical transmission the virus can be transmitted horizontally by subcutaneous and transmucosal exchange of infected blood and body fluid. Chronic infection is a frequent cause of liver cirrhosis and hepatocellular carcinoma. With more than 350 million chronically infected carriers (around 5% of the total world population), persistent infection with

HBV is a major health problem worldwide. The number of HBV carriers is increasing due to the lack of molecular pathobiology of HBV infection and fully effective antiviral treatments[8].

VIRAL GENES AND EXPRESSION

HBV is a spherical particle 42 nm in diameter, and has any of seven virus genotypes; A, B, C, D, E, F, G. The genome of HBV is a circular DNA duplex of about 3.2 kilobase pairs (kbp) that replicates by reverse transcription of a pre-genome RNA and contains a single-stranded gap region of variable length in one strand. Molecular biological studies on HBV particles and/or genomes revealed mechanisms of DNA replication, transcription of viral genes, or virion assembly to some extent. However, mechanisms of the virus–cell interaction, such as cellular receptor, uncoating, entry into target cells, or HBV genome integration into the host cell genome, are still not well characterized, although viral sequences implicated in the virion interaction with the cellular membrane are located in a variable region of the aminoterminal part of the pre-S1 domain.

The nucleotide (nt) sequences of HBV DNA and deduced amino acid sequences revealed a compact, partially overlapping coding organization of the entire genome (Figure 1). The genome of HBV contains four functionally distinct ORF on one strand in the same transcriptional porality. Two of these, S and C, are assigned to the viral structural proteins of envelope antigen (HBsAg) with 226 amino acids, and nuclear core protein (HBcAg) with 183 amino acids, respectively.

The smallest gene, X, codes for a regulatory protein of 154 amino acids, and is known to exhibit a transcriptional transactivation function through X-responsive elements for many viral and cellular genes[9–14] and implicated in the carcinogenecity of this virus. In spite of the basic property of the X protein, it has no DNA-binding activity; therefore its transactivation and/or oncogenic function is probably manifested through protein–protein interaction with cellular factor(s).

The largest fourth gene P, consists of 843 amino acids, and overlaps with other coding regions on different frames. It is indispensable for virus replication with its activities of DNA polymerase, reverse transcriptase, and ribonuclease H. This overlapping of the P region is a characteristic feature and, with its unusual overlapping of coding and regulatory sequences (such as promoters, two specific enhancers, and a poly-A termination signal), makes this virus unique. In addition to the central homologous region, two homology blocks in the X protein might correspond to a conservative selective pressure for functional activity of the X as a *trans*-regulator, and the viral ribonuclease H, encoded by C-terminal sequences of P, which overlaps with the 5′ end of the X protein.

Another important characteristic of HBV is a high mutation rate due to the lack of a proofreading function of the reverse transcriptase that replicates its genome. This may increase the production of mutated variants during persistent infection, which enables the virus to escape from the host immune

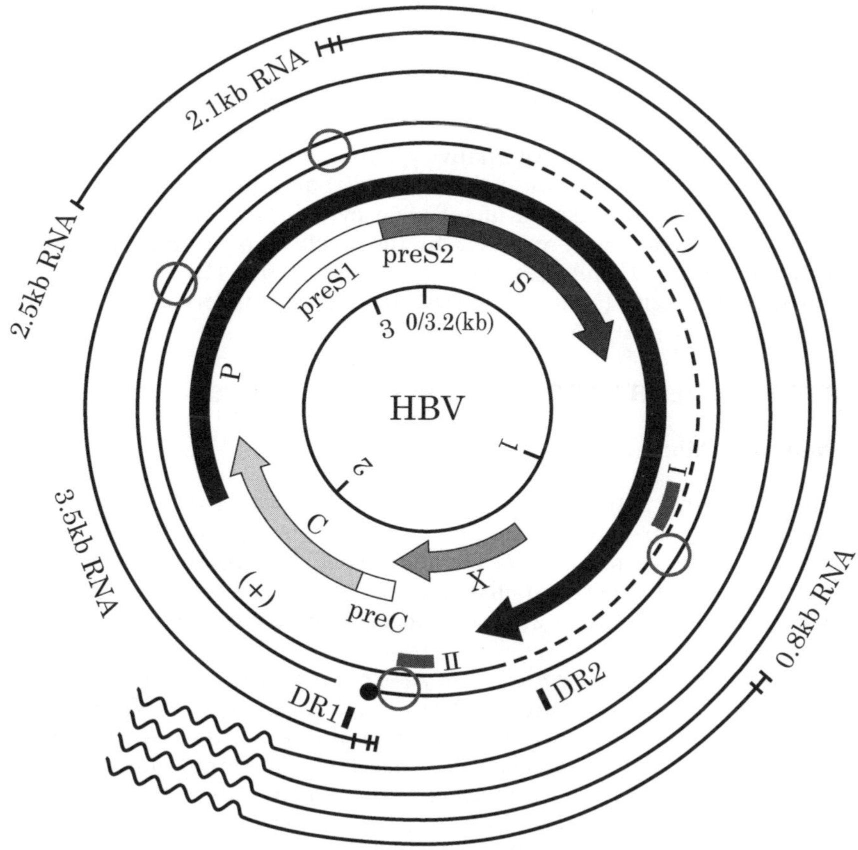

Figure 1 Genetic organization and transcription map of the hepatitis B virus genome. The HBV DNA was derived from virus of the genotype C, with a nucleotide sequence of 3215 bp. The positions of promoters and two enhancers are shown by open circle and (I) and (II), respectively. The four different transcripts are indicated with a poly(A) tail (wavy line). The terminal protein domain of polymerase (●) is attached to the 5′ end of the minus strand (–). DR1 and DR2 represent 11-bp direct repeats

response or enables a growth advantage of mutant viral genomes. Substantial nucleotide and amino acid sequence heterogeneity in different isolates of HBV, plus specific mutations in the HBV genome in individual patients, have been observed. However, there is no experimental evidence that these defective HBV genomes might harbour oncogenic potential, or that any particular HBV genotype might be more oncogenic than others. Comparison of the HBV genome with those of animal hepadnaviruses reveals extensive homologies among mammalian hepadnaviruses (WHV, GSHV), which share a basically identical genomic organization.

The antibody response to HBV HBsAg is well characterized to be T-cell-dependent, and plays an important role in clearance of the circulating virus and preventing their attachment and uptake by susceptible hosts. Most HBV infections are cleared by an effective immune response. Persistent infection of HBV is mainly caused by a weak antiviral immune response that cannot kill or cure all of the infected cells. Alternatively, chronic hepatitis may be caused by an inappropriate T-cell response that can kill but cannot cure the infected hepatocyles without production of sufficient cytokines to clear the infection. The role of the antibody response to the nucleocapsid (HBcAg and its derivative in serum, HBeAg, the best marker for active HBV infection) and non-structural proteins (X and P) in HBV pathogenesis is less clear.

HEPATOCELLULAR CARCINOMA (HCC)

Chronic hepatitis

In chronic hepatitis patients HBV appear to cause repeated cycles of cell death and regeneration in hepatocytes. In these cases malignant transformation usually occurs in the presence of cirrhosis. HCC is a common fatal cancer with $\sim 5 \times 10^5$ deaths per year in Southeast Asia and China, but is rare in Western countries (less than 2% fatal cancers) where most cases are alcohol-related.

More than 80% of the world's liver cancers per year are caused by viral infection, with ~60% associated with HBV and ~40% with HCV. Co-factors such as aflatoxins and nitrosamines can induce PHC-like disease in experimental animals; so the chance of developing a chronic infection, and therefore the risk of liver cancer, depends to some extent on the way people become infected with this virus. Vertical transmission from mother to child is known to play a major role in the spread of HBV in Asian countries; 72% of patients chronically infected with HBV become negative for HBeAg (the best marker for active HBV infection) within 10 years[15].

HBV X protein

Independent studies show that HBV, and the X gene, have been implicated in the carcinogenicity of this virus because of its ability to transform rodent cells[16–18]. Moreover, the development of HCC in at least one strain of transgenic mouse has been shown[19]. Although there was an absence of tumours in other X transgenic mice so far analysed, some of these mice exhibited an increased susceptibility to chemical carcinogens[20,21]. These results suggest that HBV X protein (HBx) acts in cooperation with cellular oncogenic factors. However, pathogenesis and carcinogenesis by HBV infection occur over years of chronic infection, and are associated with hepatocyte death and regeneration.

The X gene product is known to transactivate a number of viral and cellular genes[12,22]. In recent years the transactivation function of X protein has been extensively studied by many investigators. Some reports suggested that the X protein transactivates transcription via protein–protein interaction with cellular factors that bind to their recognition sequences[23,24]. Some other reports suggested a possible role of X protein in the regulation of signal transduction pathways[25–28]. In addition, a Kunitz inhibitor (KI) domain-like region was shown to be in dispensable for the transactivation function of X protein[29,30].

It has been reported that the X protein is essential for the viral replication[31,32]. The X protein itself is unable to bind to DNA directly, but is a potent transcriptional co-activator through multiple *cis*-acting elements, including AP-1, AP-2, ATF/CREB, NF-κB, NF-AT, c-myc, and C/EBP[33]. The X protein also mediates several signal transduction cascades, including the RAS/MAPK-, JNK-, NF-κB-, and Src-dependent pathways[25,27,34–36], as schematically shown in Figure 2.

The mechanism of transactivation by the X protein is still under debate[33]. It has been proposed that the X protein may stimulate transcription by interacting with several cellular sequence-specific transcription factors or with the basal transcriptional machinery[23,37,38]. These functions, interacting with transcription factors and signalling cascades, are associated with two regions of the X protein, amino acids 61–69 and amino acids 105–140, which act together to cause cell proliferation in cooperation[39].

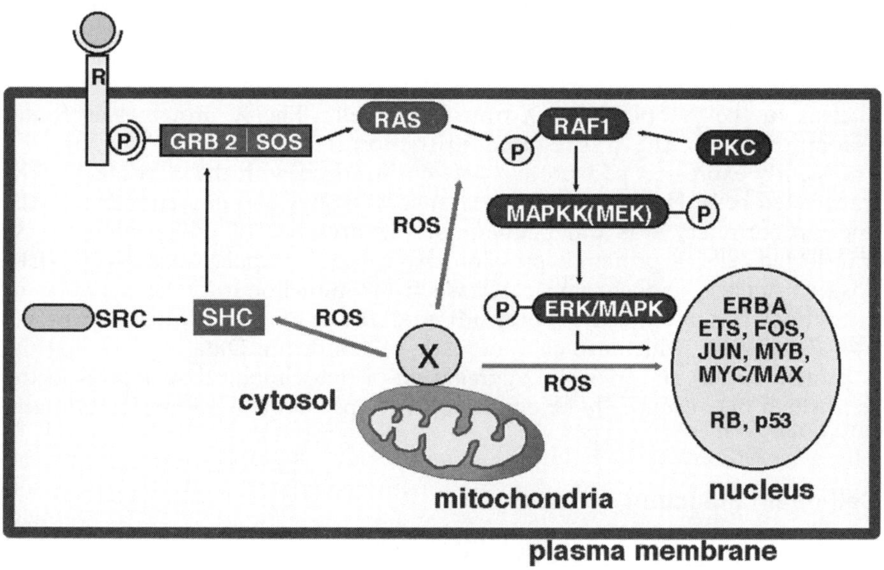

Figure 2 Cellular location of some oncogene and suppressor oncogene products and HBV X protein in relation to the signal transduction pathway. Open circled X: HBV X protein; ROS: reactive oxygen species

Furthermore, association of the X protein with mitochondria causes loss of the mitochondrial membrane potential and subsequently causes cell death[40], suggesting that the amino acids 68–104 region of the X protein may be sufficient for association with mitochondria and loss of mitochondrial membrane potential. Thus, the X protein has independent proliferative and cell death-promoting activities.

From these observations it has been suggested that the X protein may facilitate HCC through transactivating cellular oncogenes and signalling cascades that stimulate cell proliferation.

Interaction with mitochondria and p53

It is known that normal and most p53 mutant proteins are located in the nucleus[41], whereas the X protein is found in the cytoplasm around the nuclear periphery[25]. In the previous study[42] it was observed that the X protein translocates endogenous or transiently expressed p53 protein from the nucleus to the cytoplasm, and that the p53 protein was found associated with the aggregated structures in the cytoplasm. This change was dependent on a direct or indirect interaction between the carboxyterminal region of p53 protein and the transcriptional activation domain (KI domain-like region) of the X protein. p53-independent cell death caused by X protein expression has also been reported[43].

To determine the exact location of the X protein in X gene-transfected cells a fluorescent staining technique was employed, as well as biochemical analyses. Aggregation of mitochondrial structures became evident at the periphery of nucleus in the cytoplasm of X-transfected cells. The X protein was found associated with the aggregated mitochondrial structures (Figure 2). Transiently expressed p53 protein was co-localized with the X protein in X-transfected cells. However, the appearance of aggregated mitochondria at the nuclear periphery was independent of the presence of p53 protein in X-transfected cells. X protein expression also caused an appearance of TUNEL-positive nucleus, cytochrome c release from mitochondria, the decrease of mitochondrial membrane potential and the membrane blebbing of X-transfected cells, which are characteristic of cell death. Data suggest that the X protein causes an abnormal aggregation of mitochondrial structures in the cell, which may eventually be connected with p53 protein co-localization and cell death.

Cell death induction

In contrast to proliferative effects, several studies[39,44,45] showed that the X protein induces cell death or makes cells susceptible to apoptotic stimuli such as tumour necrosis factor (TNF). The X protein is of importance in controlling apoptosis in HBV-infected cells. Expression of the X protein either during virus infection or alone has been found to sensitize cells to apoptosis and to increase sensitivity to TNF-induced killing. This response involves activation of both c-Myc- and MEKK-dependent pathways and requires p53. But in the absence of p53, the X protein is growth-promoting and appears to play a role in the oncogenicity of HBV. Thus, like adenovirus E1A and polyomavirus Large T antigen, stimulation of the proliferative response activates p53-dependent apoptotic pathways. This implies a multi-step development for HCC, where loss of p53 is an important stage[46].

Previous reports have revealed that expression of the X gene brought some apoptotic features to cells, such as a TUNEL-positive nucleus, decreased mitochondrial membrane potential, and membrane blebbing. It was reported that the ectopically expressed X protein is located in mitochondria[39]. Recently, many reports have focused on mitochondria as important players in apoptosis and programmed cell death, where many cell death signals reach mitochondria. Mitochondria undergo major changes in membrane integrity before cell death signals are obvious. These changes concern both inner and outer mitochondrial membranes. Outer mitochondrial membrane permeabilization causes the redistribution of numerous apoptogenic factors such as cytochrome *c*, AIF, and Smac/DIABLO in the intermembrane space[47,48]. The redistribution of these apoptogenic factors is an important step in cell death. For instance, cytochrome *c* release is essential for the generation of the apoptosome, the complex containing cytochrome *c*, Apaf-1, and caspase-9, and this apoptosome complex initiates effector caspase activation[49–51]. On the other hand, inner mitochondrial membrane permeabilization causes disruption of mitochondrial membrane potential ($\Delta\Psi m$). The $\Delta\Psi m$ is indispensable for driving the ATP synthase, which phosphorylates ADP to ATP. Therefore the loss of mitochondrial membrane potential means loss of ATP generation.

The mechanism of X protein-related cell death was therefore examined, using enhanced green fluorescent protein (EGFP) fusion constructs of the X protein. The region required for its mitochondrial localization was mapped to amino acids 68–117, which is essential for cell death but not for transactivation function[40,49] (Figure 3). *In vitro* binding analysis supported the notion that the recombinant X (rX) protein associates with isolated mitochondria through the region of amino acids 68–117 without causing redistribution of cytochrome *c* and apoptosis-inducing factor (AIF). Cytochemical analysis revealed that mitochondrial membrane potential was decreased by X protein association with mitochondria, suggesting that the X protein induces dysfunction of the permeability transition pore (PTP) complex. Furthermore, PTP inhibitors, reactive oxygen species (ROS) scavengers and Bcl-xL, which are known to stabilize mitochondrial membrane potential, prevented X protein-induced cell death. Collectively, the present results suggest that appearance of the X protein in mitochondria of hepatitis B virus-infected cells causes loss of mitochondrial membrane potential and subsequently induces mitochondria-dependent cell death.

DNA cleavage and mutation

To understand the mechanism of X protein-induced cell death and/or transformation, effective introduction of DNA and expression of X-related protein in the cell was essential. The DNA transfection method involves problems such as over-expression or low efficiency of introduction. A protein transduction method was employed which enables rX protein and its mutants

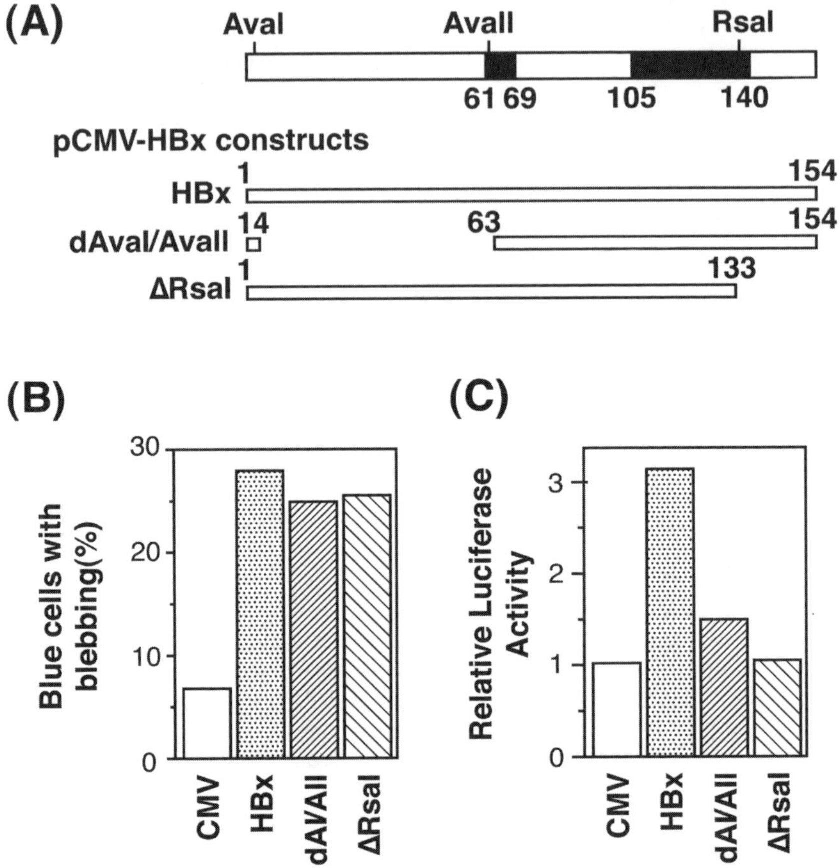

Figure 3 Transactivation activity of HBV X gene is not connected with cell death induction. **A**: Schematic presentation of wild-type X protein (HBx) gene and mutants. The regions necessary for transactivation activity are indicated as black boxes. Mutants lacking one of the regions lose transactivation ability. **B**: Cell death inducibility was examined by DNA transfection experiment using transactivation-deficient mutants, pCMVX-dAvaI/AvaII (dAI/AII) and pCMVX-ΔRsaI (ΔRsaI). Both mutants showed similar cell death induction ability with the HBx gene. **C**: Transactivation ability of the deficient mutants measured by luciferase assay

to be entered across the cell membrane, and allows us to directly analyse their cytopathic effects on cells. Using rX protein and its deletion or truncation mutant, XΔ(5-67) or X(1-67), respectively, the function of the X protein was analysed. Only rX protein and XΔ(5–67), which associated with mitochondria, inhibited cell proliferation and decreased the cellular ATP level, while unassociated mutant X(1-67) did not show such a function (Figure 4). The mutagenic effect of rX protein or XΔ(5–67) was then examined by the Big Blue

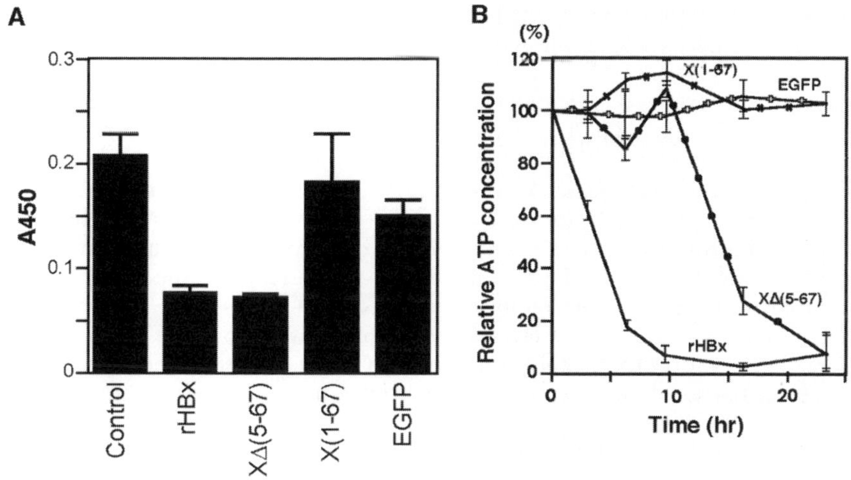

Figure 4 HBV X protein induces cell death and decreases the level of cellular ATP. **A**: WST-1 assay was carried out to demonstrate the cytopathic effect of wild-type recombinant X proteins (rHBx) transduced into HuH7 cells. Concentrations of proteins: control without protein, 1 μM rHBx, 3 μM XΔ(5–67), 3 μM X(1–67) and 3 μM EGFP. The rHBx and XΔ(5–67) proteins, which associated with mitochondria, inhibited cell proliferation. **B**: Cellular ATP levels were decreased by addition of the rHBx and XΔ(5–67) proteins. HuH7 cells were incubated with protein, then cells were harvested and the cellular ATP level analysed by measurement of ATP-driven luciferin luciferase activity. Cellular ATP level was normalized with untreated cells as 100%

rat cell line method. The rX protein transduction increased the mutation rate more than 4-fold, but was counteracted by the addition of ROS scavenger, Tiron or *N*-acetylcystein (NAC). These data suggest that the X protein causes DNA damage by increasing mutation frequency via ROS generation in mitochondria as a result of loss of membrane potential.

CONCLUDING REMARKS

Tumorigenesis is triggered by chemical carcinogens, radiation, ultraviolet light and viral infection. Irrespective of triggers, it is widely accepted that tumour formation progresses through multiple processes, most of which consist of a mutation followed by clonal selection of the mutated cells. Tumour cells consequently accumulate a number of mutations in oncogenes, tumour-suppressor genes and their related genes. Indeed, cancer has been shown to result from a series of mutations in specific oncogenes and tumour-suppressor genes[52]. Our study indicated that the X protein directly causes DNA mutation through ROS generation, and that the X protein directly induces transformation through the increase of mutant frequency. However, our observations do not exclude the possibility that tumorigenicity of the X protein is in part attributable to an activation of a transcription factor or signalling cascade.

Cancer progression is frequently a pathological consequence of extensive and sustained damage related to free-radical stress. Mutations in cancer-related genes or post-translational modifications of proteins by nitration, nitrosation, phosphorylation, acetylation or poly-ADP-ribosylation by free radicals or lipid peroxidation byproducts are some of the key events that can increase the cancer risk. In addition, free radicals can modulate cell growth and tumour promotion by activating signal-transduction pathways, which result in the transcriptional activation of proto-oncogenes, such as c-FOS, c-JUN and c-Myc[53]. Inflammation as a result of HBV infection is thought to cause DNA damage. In our observation rX protein stimulated ROS generation in the mitochondria due to collapse of the membrane potential, and consequently caused DNA damage. This strongly suggests that the X protein has the ability to induce DNA cleavage and mutation by itself.

Until recently the possibility that some viral proteins could be directly engaged in DNA damage was not seriously considered. However, in the past 10 years the number of viral proteins has increased to such an extent that they can no longer be considered as meaningless curiosities. It has occasionally been discussed that the X protein is essential for viral replication, although the X protein itself is unable to interact with DNA directly. The physiological significance of the X protein expression is partly known, but there is some clue that the X protein acts as an 'arrangement protein' for HBV replication. From an evolutionary viewpoint it is noteworthy that several arrangement proteins have been found in viruses.

Acknowledgements

We thank Drs S. Takada, Dept. of Biochemistry and Molecular Biology, M.D. Anderson Cancer Center, University of Texas, USA and Y. Shirakata, Patent Examination Dept., Japan Patent Office, Tokyo for their collaboration; and N. Kaneniwa, Dept. of Biochemistry, the Cancer Institute (JFCR), Tokyo and M. Kobayashi, Kitasato University for their technical help. The work was supported in part by a grant-in-aid from the Ministry of Education, Culture, Sports, Science and Technology, Japan.

References

1. Aldrich CE, Coates L, Wu TT et al. *In vitro* infection of woodchuck hepatocytes with woodchuck hepatitis virus and ground squirrel hepatitis virus. Virology. 1989;172:247–52.
2. Galle PR, Hagelstein J, Kommerell B, Volkmann M, Schranz P, Zentgraf H. *In vitro* experimental infection of primary human hepatocytes with hepatitis B virus. Gastroenterology. 1994;106:664–73.
3. Gripon P, Diot C, Theze N et al. Hepatitis B virus infection of adult human hepatocytes cultured in the presence of dimethyl sulfoxide. J Virol. 1988;62:4136–43.
4. Tuttleman JS, Pugh JC, Summers JW. *In vitro* experimental infection of primary duck hepatocyte cultures with duck hepatitis B virus. J Virol. 1986;58:17–25.
5. Schaefer S, Tolle TK, Lottmann S, Gerlich WH. Animal models and experimental systems in hepatitis B virus research. In: Koshy J, Caselmann WH, editors. Hepatitis B Virus:

Molecular Mechanisms in Disease and Novel Strategies for Therapy. London: Imperial College Press, 1998:51–74.

6. Walter E, Keist R, Niederost B, Pult I, Blum HE. Hepatitis B virus infection of tupaia hepatocytes *in vitro* and *in vivo*. Hepatology. 1996;24:1–5.
7. Guidotti LG, Matzke B, Schaller H, Chisari FV. High-level hepatitis B virus replication in transgenic mice. J Virol. 1995;69:6158–69.
8. Blum HE, Wieland S, Walter E, von Wiezsacker F, Offensperger WB, Moradpour D. Natural course of HBV infection. In: Koshy R, Caselmann WH, editors. Hepatitis B Virus: Molecular Mechanisms in Disease and Novel Strategies for Therapy. London: Imperial College Press, 1998:75–92.
9. Koike K, Shirakata Y, Yaginuma K et al. Oncogenic potential of hepatitis B virus. Mol Biol Med. 1989;6:151–60.
10. Aufiero B, Schneider RJ. The hepatitis B virus X-gene product *trans*-activates both RNA polymerase II and III promoters. EMBO J. 1990;9:497–504.
11. Hu JC, O'Shea K, Kim P, Sauer RT. Sequence requirements for coiled-coils: analysis with lambda repressor-GCN4 leucine zipper fusions. Science. 1990;250:1400–3.
12. Spandau DF, Lee C-H. *trans*-Activation of viral enhancers by the hepatitis B virus X protein. J Virol. 1988;62:427–34.
13. Twu JS, Schloemer RH. Transcriptional *trans*-activating function of hepatitis B virus. J Virol. 1987;61:3448–53.
14. Zahm P, Hofschneider PH, Koshy R. The HBV X-ORF encodes a *trans*-activator: a potential factor in viral hepatocarcinogenesis. Oncogene. 1988;3:169–77.
15. McMahon BJ. Holck P, Balkow L, Snowball M. Serologic and clinical outcomes of 1536 Alaska natives chronically infected with hepatitis B virus. Ann Intern Med. 2001;135:759–68.
16. Shirakata Y, Kawada M, Fujiki Y et al. The X gene of hepatitis B virus induced growth stimulation and tumorigenic transformation of mouse NIH3T3 cells. Jpn J Cancer Res. 1989;80:617–21.
17. Hohne M, Schaefer S, Scifer M, Feitelson MA, Paul D, Gerlich WH. Malignant transformation of immortalized transgenic hepatocyte after transfection with hepatitis B virus DNA. EMBO J. 1990;9:1137–45.
18. Rakotomahanina CK, Hilger C, Fink T, Zentgraf H, Schroder CH. Biological activities of a putative truncated hepatitis B virus X gene product fused to a polylysin stretch. Oncogene. 1994;9:2613–21.
19. Kim CM, Koike K, Saito I, Miyamura T, Jay G. HBx gene of hepatitis B virus induces liver cancer in transgenic mice. Nature. 1991;351:317–20
20. Slagle BL, Lee TH, Medina D, Finegold MJ, Butel JS. Increased sensitivity to the hepatocarcinogen diethylnitrosamine in transgenic mice carrying the hepatitis B virus X gene. Mol Carcinogen. 1996;15:261–9.
21. Madden CR, Finegold MJ, Slagle BL. Hepatitis B virus X protein acts as a tumor promoter in development of diethylnitrosamine-induced preneoplastic lesions. J Virol. 2001;75:3851–8.
22. Rossner MT. Review: hepatitis B virus X-gene product: a promiscuous transcriptional activator. J Med Virol. 1992;36:101–17.
23. Maguire HF, Hoeffler JP, Siddiqui A. HBV X protein alters the DNA binding specificity of CREB and ATF-2 by protein-protein interactions. Science. 1991;252:842–4.
24. Natoli G, Avantaggiati ML, Chirillo P et al. Induction of the DNA-binding activity of c-jun/c-fos heterodimers by the hepatitis B virus transactivator pX. Mol Cell Biol. 1994;14: 989–98.
25. Benn J, Schneider RJ. Hepatitis B virus HBx protein activates Ras-GTP complex formation and establishes a Ras, Raf, MAP kinase signaling cascade. Proc Natl Acad Sci USA. 1994;91:10350–4.
26. Cross JC, Wen P, Rutter WJ. Transactivation by hepatitis B virus X protein is promiscuous and dependent on mitogen-activated cellular serine/threonine kinases. Proc Natl Acad Sci USA. 1993;90:8078–82.
27. Kekule AS, Lauer U, Weiss L, Luber B, Hofschneider PH. Hepatitis B virus transactivator HBx uses a tumour promoter signalling pathway. Nature. 1993;361:742–5.

28. Natoli G, Avantaggiati ML, Chirillo P et al. Ras- and Raf-dependent activation of c-jun transcriptional activity by the hepatitis B virus transactivator pX. Oncogene. 1994b;9: 2837–43.
29. Takada S, Kido H, Fukutomi A, Mori T, Koike K. Interaction of hepatitis B virus X protein with a serine protease, tryptase TL2 as an inhibitor. Oncogene. 1994;9:341–8.
30. Takada S, Koike K. X protein of hepatitis B virus resembles a serine protease inhibitor. Jpn J Cancer Res. 1990;81:1191–4.
31. Zoulim F, Saputelli J, Seeger C. Woodchuck hepatitis virus X protein is required for viral infection *in vivo*. J Virol. 1994;68:2026–30.
32. Bouchard MJ, Wang LH, Schneider RJ. Calcium signaling by HBx protein in hepatitis B virus DNA replication. Science. 2001;294:2376–8.
33. Yen TS. Hepadnaviral X protein: review of recent progress. J Biomed Sci. 1996;3:20–30.
34. Luber B, Lauer U, Weiss L, Hohne M, Hofschneider PH, Kekule AS. The hepatitis B virus transactivator HBx causes elevation of diacylglycerol and activation of protein kinase C. Res Virol. 1993;144:311–21.
35. Lee YH, Yun Y. HBx protein of hepatitis B virus activates Jak1-STAT signaling. J Biol Chem. 1998;273:25510–15.
36. Waris G, Huh KW, Siddiqui A. Mitochondrially associated hepatitis B virus X protein constitutively activates transcription factors STAT-3 and NF-kappa B via oxidative stress. Mol Cell Biol. 2001;21:7721–30.
37. Lin Y, Nomura T, Cheong J, Dorjsuren D, Iida K, Murakami S. Hepatitis B virus X protein is a transcriptional modulator that communicates with transcription factor IIB and the RNA polymerase II subunit 5. J Biol Chem. 1997;272:7132–9.
38. Shamay M, Borak O, Doitsh G, Ben-Dor I, Shaul Y. Hepatitis B virus pX interacts with HBXAP, a PHD finger protein to coactivate transcription. J Biol Chem. 2002;277:9982–8.
39. Takada S, Shirakata Y, Kaneniwa N, Koike K. Association of hepatitis B virus X protein with mitochondria causes mitochondrial aggregation at the nuclear periphery, leading to cell death. Oncogene.1999;18:6965–73.
40. Shirakara Y, Koike, K. Hepatitis B virus X protein induces cell death by causing loss of mitochondrial membrane potential. J Biol Chem. 2003;278:22071–8.
41. Shaulsky G, Goldfinger N, Tosky MS, Levine AJ, Rotter V. Nuclear localization is essential for the activity of p53 protein. Oncogene. 1991;6:2055–65.
42. Takada S, Kaneniwa N, Tsuchida N, Koike K. Cytoplasmic retention of the p53 tumor suppressor gene product is observed in the hepatitis B virus X gene-transfected cells. Oncogene. 1997;15:1895–901.
43. Terradillos O, Pollicino T, Lecoeur H et al. p53-Independent apoptotic effects of the hepatitis B virus HBx protein *in vivo* and *in vitro*. Oncogene. 1998;17:2115–23.
44. Su F, Schneider RJ. Hepatitis B virus HBx protein sensitizes cells to apoptotic killing by tumor necrosis factor alpha. Proc Natl Acad Sci USA. 1997;94:8744–9.
45. Bergametti F, Prigent S, Luber B et al. The proapoptotic effect of hepatitis B virus HBx protein correlates with its transactivation activity in stably transfected cell lines. Oncogene. 1999;18:2860–71.
46. Roulston, A. Marucellus RC, Branton PE. Viruses and apoptosis. Annu Rev Microbiol. 1999;53:577–628.
47. Green DR, Reed JC. Mitochondria and apoptosis. Science. 1998;281:1309–12.
48. Kroemer G, Reed JC. Mitochondrial control of cell death. Nat Med. 2000;6:513–19.
49. Li P, Nijhawan D, Budihardjo I et al. Cytochrome c and dATP-dependent formation of Apaf-1/caspase-9 complex initiates an apoptotic protease cascade. Cell. 1997;91:479–89.
50. Zou H, Li Y, Liu X, Wang X. An APAF-1.cytochrome c multimeric complex is a functional apoptosome that activates procaspase-9. J Biol Chem. 1999;274:11549–56.
51. Wang X. The expanding role of mitochondria in apoptosis. Genes Dev. 2001;15:2922–33.
52. Vogelstein B, Fearon ER, Hamilton SR et al. Genetic alterations during colorectal-tumor development. N Engl J Med. 1988;319:525–32.
53. Hussain SP, Hofseth LJ, Harris CC. Radical causes of cancer. Nat Rev Cancer. 2003;3:276–85.

8
Individualized treatment of hepatitis C

S. ZEUZEM

INTRODUCTION

Today at least six different hepatitis C virus (HCV) genotypes and more than 100 subtypes have been reported with different geographical distribution[1]. In Western countries, genotypes 1, 2, and 3 are primarily found, with a predominance of genotype 1. Genotype 4 is widely distributed in North Africa and the Middle East, whereas genotype 5 is found in South Africa, and genotype 6 in Hong Kong and more recently in Australia.

Therapy for HCV infection was first reported in the late 1980s when patients with so-called non-A non-B hepatitis were treated with interferon[2]. After isolation and characterization of HCV it became obvious that sustained virological response rates defined as non-detectable HCV RNA 24 weeks after cessation of therapy were markedly different with regard to the HCV genotype. In large multicentre studies sustained virological response rates were 2% and 7% for patients infected with genotype 1 treated for 24 and 48 weeks, respectively. In contrast, for patients infected with genotypes 2/3 a sustained virological response was achieved in 16% (24 weeks) and 29–33% (48 weeks)[3,4].

The introduction of combination therapy with interferon alpha (IFN-α) and the nucleoside analogue ribavirin for 48 weeks in the late 1990s substantially improved treatment outcome with mean sustained virological response rates of 41%. However, differences between genotype 1- and genotype 2- and 3-infected patients remained evident. Sustained virological response rates for patients with genotypes 1, 4–6 were 28–36% compared to 61–79% for patients with genotype 2 and 3 infection[3–6]. The development of pegylated (PEG) interferons with a sustained absorption, a slower rate of clearance, and a longer half-life than unmodified interferons, led to further improvement of sustained virological response rates, especially for patients infected with genotype 1[5,6]. Again, sustained virological response rates were significantly higher in genotype 2 and 3 (76–82%) as compared to genotype 1-infected patients (42–52%) treated with pegylated IFN-α in combination with ribavirin.

ANTIVIRAL THERAPY OF PATIENTS WITH HCV GENOTYPE 1

In the different guidelines of European countries and the US recommendation for standard treatment of patients with chronic hepatitis C genotype 1 infection is the application of PEG-IFN-α in combination with ribavirin for 48 weeks. At present two types of PEG-IFN-α are approved for the treatment of chronic hepatitis C: PEG-IFN-α2a and PEG-IFN-α2b, which differ in size and form of the linked polyethylene glycol molecule (40 vs 12 kDa). Due to pharmacokinetics the 40 kDa PEG-IFN-α2a is given independently from body weight with 180 μg, whereas for the 12-kDa PEG-IFN-α2b, a dose of 1.0–1.5 μg/kg body weight is approved. Both PEG-interferons are injected subcutaneously once a week. For HCV genotype 1-infected patients ribavirin is given adjusted to the body weight between 800 and 1400 mg orally per day.

Prediction of virological response at baseline is possible by different host- and virus-related factors. Host-related factors associated with a reduced virological response to (PEG)-interferon/ribavirin combination therapy are: older age, male gender, African-American ethnicity, elevated body mass index, advanced fibrosis or cirrhosis, liver steatosis, and elevated gammaglutamyltranspeptidase (γ-GT) levels[7]. As already highlighted in the Introduction, the most import viral factor for prediction of virological response is the HCV genotype. HCV genotype 2- or 3-infected patients generally show higher sustained response rates than genotype 1-infected patients. In addition, a high baseline viral load is also associated with a reduced probability of virological response[7]. Furthermore, different HCV proteins have been associated with inhibition of IFN-α signal transduction pathways[8]. While the underlying mechanisms of HCV-related IFN-α treatment resistance are still not fully understood, clinically, a clear association of multiple mutations within the so-named interferon sensitivity determining region (ISDR) within the HCV NS5A protein and sustained virological response has been shown in numerous studies and a recent meta-analysis[9–11].

During antiviral therapy early discontinuation in virological non-responders is possible at certain time-points based on HCV RNA kinetics. In different studies a high predictive value for virological non-response (98–100%) was observed for HCV genotype 1-, 4–6-infected patients with a decline of HCV RNA serum concentration of less than 2 log steps between baseline and week 12 of (pegylated)-IFN-α/ribavirin combination therapy[5,12,13]. In addition, at week 24 of treatment it is recommended to discontinue therapy on the basis of detectable HCV RNA in serum by qualitative PCR-based assays, again with a predictive value for virological non-response of 98–100%[6,12–14].

Results of HCV RNA concentrations determined by different quantitative HCV RNA assays can differ up to 0.5 log steps despite a standardization of all commercially available test systems to IU[15]. Therefore, it is important to use the same assay at baseline and week 12 to take correct decisions on treatment discontinuation based on the 2 log decline rule. For assessment of detectable HCV RNA at week 24 a sensitive assay with a detection limit $\leqslant$50 IU/ml should be used[5,6,14]. Whether the presence of low viral titres at week 24 (>5–10 but <50 IU/ml) can be used for selection of patients which may benefit from prolonged therapy is unknown today.

INDIVIDUALIZED TREATMENT DURATION OF PATIENTS WITH HCV GENOTYPE 1

Most investigations on early viral decay and treatment response focused on developing treatment algorithms for discontinuation of therapy in patients with HCV genotype 1 infection and little or no chance for achieving a sustained virological response. However, the slopes of initial decline of HCV RNA concentrations differ widely, and in individual patients HCV RNA may become undetectable as early as week 4 in some patients and only at week 24 in others. Therefore, in different clinical trials early HCV RNA decline data at week 4 were used to identify patients with rapid virological response in whom standard therapy duration of 48 weeks might be over-treatment and shortening to 24 weeks may be possible. One recent prospective clinical trial investigated shortening of treatment duration from 48 weeks to 24 weeks in patients with HCV genotype 1 infection and low baseline viral load ⩽600 000 IU/ml. However, the overall sustained virological response rate was 50% in this study in comparison with 71% in a matched historical control of patients which also had a baseline viral load ⩽600 000 IU/ml, but were treated for 48 weeks. In a retrospective analysis it was then shown that shortening of treatment duration in HCV genotype 1-infected patients with low baseline viral load is possible only on the basis of a rapid virological response (RVR) defined as HCV RNA negativity (⩽29 IU/ml) at week 4. In this subgroup of patients a sustained virological response rate after 24 weeks of therapy with PEG-IFN-α2b and ribavirin was 89% in comparison to 85% in the historical control of patients with low baseline viral load and RVR treated for 48 weeks[16].

Recently one retrospective analysis and one ongoing prospective clinical trial showed similar results for the treatment with PEG-IFN-α2a and ribavirin[17,18]. Jensen et al. analysed the different treatment durations within the trial conducted by Hadziyannis et al.[22] for HCV genotype 1-infected patients with RVR defined as HCV RNA below 50 IU/ml at week 4. In this subgroup of patients sustained virological response rates were similar after 24 and 48 weeks with 88% and 91%, respectively[17]. Interestingly, flat dosing of ribavirin with 800 mg/day in patients with RVR also led to high sustained virological response rates (89%) after 24 weeks of therapy. Low baseline HCV RNA concentration (<200 000 vs >600 000 IU/ml) and HCV subtype (1b vs 1a) were independent predictors of a rapid virological response[17]. In a prospective study by Ferenci et al, shortening of treatment duration from 48 to 24 weeks was performed in HCV genotype 1-infected patients with RVR (<50 IU/ml HCV RNA at week 4) and an interim analysis of patients who already completed therapy was presented recently. In this study sustained virological response rates were significantly higher in patients with a low baseline viral load (⩽600 000 IU/ml) in comparison to those with high baseline viral load (>600 000 IU/ml) (98% vs 74%). HCV RNA at week 4 was assessed both by a standard HCV RNA assay with a lower detection limit of 50 IU/ml and a real-time PCR-based assay with a sensitivity of 10 IU/ml. Interestingly, up to 30% of patients with a negative result by the standard assay (<50 IU/ml) turned out to be still HCV RNA-positive with the more sensitive assay[18]. It is highly conceivable that these differences will be of importance for the probability of subsequent sustained virological response, and recommendations for shortening of treatment durations must carefully take into account the sensitivity of the applied HCV RNA assay.

Similar to shortening of treatment duration in patients with a rapid virological response, prolongation of therapy from 48 to 72 weeks may be suitable in patients with a very slow decline of HCV RNA levels. Two large multicentre studies have addressed this question[19,20]. While treatment prolongation in all patients was not associated with an overall increase of sustained virological response rates, both studies show that a subgroup of patients with slow virological response may benefit from prolongation of therapy from 48 to 72 weeks. In the study by Berg et al. a significant improvement of sustained virological response was observed in patients with a 2 log decline but detectable HCV RNA at week 12, which then became HCV RNA negative at week 24. In this subgroup of patients sustained virological response rates increased from 17% to 29% by prolongation of combination therapy from 48 to 72 weeks[19]. Similar results were observed in the study by Sanchez-Tapias et al, showing a significant improvement of sustained virological response rates in slow virological responders (defined as patients with detectable HCV RNA at week 4) when treated for 72 weeks instead of 48 weeks[20].

Taken together, for HCV genotype 1-infected patients individualization of treatment duration on the basis of HCV RNA concentration at baseline and decline during therapy seems to be possible, and prospective trials are under way to prove this concept with different treatment durations of 24, 30, 36, 42, 48, 60 and 72 weeks. At present, shortening of treatment with PEG-IFN-α2b and α2a plus ribavirin from 48 to 24 weeks is approved in the EU for HCV genotype 1-infected patients with a low baseline viral load (<600 000 IU/ml and <800 000IU/ml, respectively) and rapid virological response (HCV RNA <29 IU/ml or <50 IU/ml, respectively at week 4).

ANTIVIRAL THERAPY OF PATIENTS INFECTED WITH HCV GENOTYPES 2 AND 3

In the first pivotal trials using PEG-IFN-α in combination with ribavirin, antiviral therapy was administered for 48 weeks independently of HCV genotype. Patients with genotype 2 or 3 infection treated with PEG-IFN-α2b/ribavirin or PEG-IFN-α2a/ribavirin achieved sustained virological response rates of 82% and 76%, respectively, confirming the favourable results achieved by standard IFN/ribavirin combination treatment[5,6]. Subsequent studies showed that similar sustained virological response rates (78–81%) can be achieved in patients treated for 24 weeks compared with those treated for 48 weeks[21,22]. Consequently, the 24-week treatment duration with PEG-IFN-α and ribavirin has been established as the standard of care for first-line therapy in patients with genotype 2 or 3 infection.

A fixed dose of 800 mg ribavirin per day independently of body weight has been proven as effective as a body weight-adapted ribavirin schedule[22]. Current studies even investigate lower ribavirin doses (e.g. 400 mg per day) in the treatment of HCV genotype 2-, or 3-infected patients.

In previous studies genotype 2- and 3-infected patients were always reported together. In recent studies differences in the sustained virological response rates were well described between patients infected with genotypes 2 and 3. In general, patients infected with genotype 2 have even more favourable sustained virological response rates compared to those infected with HCV genotype 3 (93% vs 79%)[21]. By analyses of patient characteristics before initiation of antiviral therapy, apparent differences between genotypes 2 and 3 were mainly attributed to a subgroup of genotype 3-infected patients with a high baseline HCV RNA concentration of more than 6×10^5 IU/ml[21]. In this subgroup a high relapse rate after cessation of combination treatment was observed as compared with relapse rates in genotype 2- and genotype 3-infected patients with low baseline viral load (23% vs 5–8%)[21]. Furthermore, histologically proven liver steatosis, which is frequently observed in genotype 3-infected patients, has been shown to be a significant negative prognostic factor for achieving a sustained virological response. Although the fibrosis stage has not been identified as a statistically independent negative predictive factor for sustained response in this study, it is worthwhile noting that patients who have no fibrosis achieve higher sustained virological response rates compared with those who have bridging fibrosis or cirrhosis (97% vs 75%)[21].

INDIVIDUALIZED TREATMENT OF PATIENTS INFECTED WITH HCV GENOTYPE 2 AND 3

Recently four independent studies have investigated whether treatment duration of PEG-IFN-α2a/2b and ribavirin in patients with genotype 2 and 3 infection can be further reduced from 24 weeks to 16, 14, and 12 weeks, respectively, without compromising the sustained virological response rates[23–26].

In a first-published non-randomized trial, 122 treatment-naive Norwegian patients were treated with PEG-IFN-α2b 1.5 μg/kg body weight and ribavirin (800–1400 mg adjusted to body weight)[24]. Patients with an early virological response defined as undetectable HCV RNA (<50 IU/ml) at week 4 and 8 after initiation of treatment were treated for 14 weeks (n = 95). The remaining patients received 24 weeks of treatment (n = 27). A control group receiving standard treatment of 24 weeks independent of the early virological response at week 4 and 8 was not included in this study. The overall sustained virological response was comparable to previous studies (82%). Patients with an early virological response achieved a sustained virological response rate of 90%, whereas patients without early virological response at week 4 showed sustained response rates of only 56%. As there was no control arm in this study no data on sustained virological response rates of patients with early virological response who received combination therapy for a standard 24-week duration are available. In this study independent factors associated with a sustained virological response included younger age, treatment according to protocol, undetectable HCV RNA at treatment week 4, and lower viral load at baseline. Indeed, patients with genotype 3 and a baseline HCV RNA concentration of less than 6×10^5 IU/ml had higher sustained virological response rates as compared to those with a higher HCV RNA concentration (98% vs 79%). There was no difference in sustained virological response rates between patients infected with HCV genotype 2 or 3. However, only 23 patients infected with genotype 2 had been enrolled in this trial.

In a second study 283 patients were randomized to receive antiviral treatment with PEG-IFN-α2b 1.0 μg/kg body weight and ribavirin (1000–1200 mg) for 24 weeks (standard duration group, $n = 70$) or, according to early virological response at week 4 (variable-duration group, $n = 213$) for either 24 weeks (HCV RNA $\geqslant 50$ IU/ml at week 4, $n = 80$) or 12 weeks (HCV RNA < 50 IU/ml at week 4, $n = 133$)[23]. In the standard duration group the sustained virological response rate was 76% compared with 77% in the variable duration group. Patients in the variable duration group who were HCV RNA-negative at week 4 achieved a sustained response rate of 85% compared with 64% in those who did not show an early virological response. In the standard duration group the sustained virological response rates were 91% and 48% with regard to the presence and absence of an early virological response, respectively. In this study no independent baseline factor was significantly associated with early virological response or virological relapse. The overall rate of sustained virological response was 80% among patients infected with genotype 2 and 66% among those infected with genotype 3 ($p < 0.001$).

A third study reported virological response rates in 153 patients who were treated with PEG-IFN-α2a 180 μg/week and 800–1200 mg/day ribavirin based on body weight[25]. Patients with undetectable HCV RNA by quantitative RT-PCR after 4 weeks of treatment (rapid virological responder, <600 IU/ml) were randomized for a total duration of 16 weeks ($n = 71$) or 24 weeks ($n = 71$). Patients without rapid virological response at week 4 were treated for 24 weeks ($n = 11$). The sustained virological response rates in early responders who were treated for 16 or 24 weeks were similar, at 82% and 80%, respectively. However, only 36% of patients with detectable HCV RNA at week 4 cleared the virus. Generally, patients with HCV genotype 2 infection had higher sustained virological response rates compared to patients with genotype 3 infection (92% vs 73%). Infection with HCV genotype 2 was confirmed as an independent factor for sustained virological response in a multivariate analysis. Other factors associated with sustained virological response were low gammaglutamyltransferase levels at baseline and low pretreatment HCV RNA concentration. Interestingly, when only those patients who showed an early virological response ($n = 142$) were stratified by pretreatment HCV RNA concentration ($\leqslant 800\,000$ IU/ml vs $> 800\,000$ IU/ml), it was shown that patients with HCV genotype 2 infection achieved sustained virological response rates of 100% and 93% independent of treatment duration. However, in patients with HCV genotype 3 infection, treatment outcome was different with regard to the pretreatment HCV RNA concentration. Sustained virological response was observed in 85% of patients with low pretreatment HCV RNA concentration whereas only 59% of patients with high HCV RNA concentration became sustained virological responders ($p = 0.003$).

Most recently, results of a large international multicentre trial on shortening of treatment duration in HCV genotype 2- or 3-infected patients (ACCELERATE) were presented at the EASL meeting in Vienna in April

2006[26]. In this study patients were randomized to receive PEG-IFN-α2a (180 μg q.w.) in combination with a fixed dose of 800 mg ribavirin for 16 or 24 weeks. Overall, sustained virological response rates were significantly lower in the 16-week treatment group in comparison with treatment for 24 weeks (62% vs 70%, respectively). Also, in the subgroup of patients with a rapid virological response (< 50 IU/ml at week 4) shortening of treatment duration from 24 to 16 weeks was still associated with a significantly reduced sustained virological response rate (82% vs 90%)[26].

Several open questions need to be answered in the future. While in the three pilot studies ribavirin was dosed according to body weight (800–1200 mg), in the ACCELERATE trial a flat dose of 800 mg/day was given to all patients. Thus, for shortening of treatment duration below 24 weeks, higher ribavirin doses may be required, and this question should be prospectively addressed in future studies. In addition, HCV genotype 2- and 3-infected patients should be analysed separately and further grouped according to baseline HCV RNA concentrations. Furthermore, the cutoff levels between low and high baseline viraemia must be revisited. Although not statistically significant in all studies, the relapse rates of patients who were treated for 12, 14 and 16 weeks, respectively, were always higher compared to the relapse rates of those treated for 24 weeks, including those patients who achieved an early virological response at week 4. The optimal duration of therapy has to be determined according to HCV genotype, baseline viral load, initial virological response, and most probably also according to liver fibrosis stage and other relevant host parameters for response prediction. Moreover, for patients without an early virological response at week 4, an extension of the current standard therapy duration beyond 24 weeks needs to be investigated in future trials (Table 1).

INDIVIDUALIZED TREATMENT OF PATIENTS WITH HCV GENOTYPE 4, 5, AND 6

Data on individualized treatment durations in patients infected with HCV genotype 4, 5 or 6 are sparse[32–35] and summarized in Table 1. In general, patients should be treated with PEG-IFN-α plus ribavirin for 48 weeks.

Future antiviral strategies

New antiviral enzyme inhibitors directly targeting the HCV NS3/4A protease or the virus-encoded NS5B RNA-dependent RNA-polymerase are currently under clinical investigation in phase I/II studies. They are expected to fundamentally expand the treatment options for patients infected with chronic hepatitis C. However, due to variable amino acid sequences of the NS3 and NS5B proteins, the antiviral efficacy of enzyme inhibitors may differ between HCV genotypes. In a first proof-of-principle study of the NS3 protease inhibitor BILN-2061 a rapid and sharp decline of HCV RNA levels within 2 days of treatment was observed in HCV genotype 1-infected patients. However, it was shown that antiviral efficacy was less pronounced and more variable in patients with HCV genotype 2 or 3 infection compared with results obtained in

Table 1 Individualized treatment duration in patients with chronic hepatitis C

HCV genotype	*Pretreatment viral load*	*Definition of initial virological response*	*Data support the following treatment duration (weeks)*	*References*
HCV-1	<600 000 IU/ml	<50 IU/ml (week 4)	24	Zeuzem et al.[16]
	Independent	<2 $\log_{10}$ (week 12)	48	Fried et al.[5] Davis et al.[12]
	Independent	50–6000 IU/ml (week 12)	72	Berg et al.[19]
HCV-2	Independent	<50 IU/ml (week 4)	14	Dalgard et al.[24]
	Independent	<50 IU/ml (week 4)	12	Mangia et al.[23]
	Independent	<600 IU/ml (week 4)	16	v. Wagner et al.[25]
	<400 000 IU/ml	Independent	16	Shiffman et al.[26]
HCV-3	<800 000 IU/ml	<600 IU/ml (week 4)	16	v. Wagner et al.[25]
	<400 000 IU/ml	Independent	16	Shiffman et al.[26]
	>800 000 IU/ml		24	Hadziyannis et al.[22] Zeuzem et al.[16]
	Independent	>600 IU/ml	36–48 (?)	No data available
HCV-4	<800 000 IU/ml	<50 IU/ml or 2 $\log_{10}$ Decline (week 12)	36	Kamal et al.[32]
	>800 000 IU/ml		48	Diago et al.[33] Hasan et al.[34] Kamal et al.[32]
HCV-5	Independent	Independent	48	Bonny et al.[35]
HCV-6	Independent	Independent	48 (?)	No data available

patients infected with HCV genotype 1[27,28]. A lower affinity of BILN-2061 for the NS3 protease of HCV genotypes 2 and 3 has been suggested as a major contributor to these findings. Further development of BILN-2061 was halted due to cardiotoxic side-effects in monkeys.

Two other HCV protease inhibitors are currently in phase II studies (VX-950 and SCH503034) which showed a 2–5 $\log_{10}$ decline in HCV RNA during a 2-week dosing period in HCV genotype 1-infected patients[29,30]. Similar to other viruses (e.g. HBV, HIV) a high antiviral efficacy of an antiviral drug may be associated with the rapid selection of resistant isolates and subsequent virological breakthrough. The emergence of resistant isolates harbouring specific mutations within the HCV NS3 protease gene has already been shown for the HCV protease inhibitor VX-950[31]. However, optimal dosing, a sharp initial decline of HCV RNA levels, and combination with PEG-IFN-α (and ribavirin) were associated with a continuous decline of HCV RNA without the apparent development of resistance. Future strategies of HCV therapy will most probably continue to be optimized according to HCV genotype, baseline viraemia and early virological response criteria.

References

1. Simmonds P, Bukh J, Combet C et al. Consensus proposals for a unified system of nomenclature of hepatitis C virus genotypes. Hepatology. 2005;42:962–73.
2. Hoofnagle JH, Mullen KD, Jones DB et al. Treatment of chronic non-A, non-B hepatitis with recombinant human alpha interferon. A preliminary report. N Engl J Med. 1986;315: 1575–8.
3. McHutchison JG, Gordon SC, Schiff ER et al. Interferon alfa-2b alone or in combination with ribavirin as initial treatment for chronic hepatitis C. Hepatitis Interventional Therapy Group. N Engl J Med. 1998;339:1485–92.
4. Poynard T, Marcellin P, Lee SS et al. Randomised trial of interferon alpha2b plus ribavirin for 48 weeks or for 24 weeks versus interferon alpha2b plus placebo for 48 weeks for treatment of chronic infection with hepatitis C virus. International Hepatitis Interventional Therapy Group (IHIT). Lancet. 1998;352:1426–32.
5. Fried MW, Shiffman ML, Reddy KR et al. Peginterferon alfa-2a plus ribavirin for chronic hepatitis C virus infection. N Engl J Med. 2002;347:975–82.
6. Manns MP, McHutchison JG, Gordon SC et al. Peginterferon alfa-2b plus ribavirin compared with interferon alfa-2b plus ribavirin for initial treatment of chronic hepatitis C: a randomised trial. Lancet. 2001;358:958–65.
7. Mihm U, Herrmann E, Sarrazin C, Zeuzem S. Review article: Predicting response in hepatitis C virus therapy. Aliment Pharmacol Ther. 2006;23:1043–54.
8. Hofmann WP, Zeuzem S, Sarrazin C. Hepatitis C virus-related resistance mechanisms to interferon alpha-based antiviral therapy. J Clin Virol. 2005;32:86–91.
9. Sarrazin C, Berg T, Lee JH et al. Improved correlation between multiple mutations within the NS5A region and virological response in European patients chronically infected with hepatitis C virus type 1b undergoing combination therapy. J Hepatol. 1999;30:1004–13.
10. Enomoto N, Sakuma I, Asahina Y et al. Mutations in the nonstructural protein 5A gene and response to interferon in patients with chronic hepatitis C virus 1b infection. N Engl J Med. 1996;334:77–81.
11. Pascu M, Martus P, Hohne M et al. Sustained virological response in hepatitis C virus type 1b infected patients is predicted by the number of mutations within the NS5A-ISDR: a meta-analysis focused on geographical differences. Gut. 2004;53:1345–51.
12. Davis GL, Wong JB, McHutchison JG, Manns MP, Harvey J, Albrecht J. Early virologic response to treatment with peginterferon alfa-2b plus ribavirin in patients with chronic hepatitis C. Hepatology. 2003;38:645–52.
13. Berg T, Sarrazin C, Herrmann E et al. Prediction of treatment outcome in patients with chronic hepatitis C: significance of baseline parameters and viral dynamics during therapy. Hepatology. 2003;37:600–9.
14. Mihm U, Kronenberger B, Hofmann WP, Zeuzem S, Sarrazin C. Highly sensitive HCV RNA detection assays for decision of treatment (dis)continuation in patients with chronic hepatitis C. J Hepatol. 2005;42:605–6.
15. Sarrazin C, Gärtner B, Sizmann D et al. Comparison of conventional PCR with real-time PCR and branched DNA-based assays for hepatitis C virus RNA quantification and clinical significance for genotypes 1 to 5. J Clin Microbiol. 2006;44:729–37.
16. Zeuzem S, Buti M, Ferenci P et al. Efficacy of 24 weeks treatment with peginterferon alfa-2b plus ribavirin in patients with chronic hepatitis C infected with genotype 1 and low pretreatment viremia. J Hepatol. 2006;44:97–103.
17. Jensen DM, Morgan TR, Marcellin P et al. Early identification of HCV genotype 1 patients responding to 24 weeks peginterferon alpha-2a (40 kd)/ribavirin therapy. Hepatology. 2006;43:954–60.
18. Ferenci P, Bergholz U, Laferl H et al. 24 Week treatment regimen with peginterferon alpha-2a (40KD) (Pegasys) plus ribavirin (Copegus) in HCV genotype 1 or 4 super-responders. J Hepatol. 2006;44(Suppl. 2):6A.
19. Berg T, von Wagner M, Nasser S et al. Extended treatment duration for hepatitis C virus type 1: comparing 48 versus 72 weeks of peginterferon-alfa-2a plus ribavirin. Gastroenterology. 2006;130:1086–97.
20. Sanchez-Tapias JM, Diago M, Escartin P et al. for the TeraViC-4 Study Group: Peginterferon-alfa2a plus ribavirin for 48 versus 72 weeks in patients with detectable

hepatitis C virus RNA at week 4 of treatment. Gastroenterology. 2006;131:451–60. Erratum in: Gastroenterology. 2006;131:1363.
21. Zeuzem S, Hultcrantz R, Bourliere M et al. Peginterferon alfa-2b plus ribavirin for treatment of chronic hepatitis C in previously untreated patients infected with HCV genotypes 2 or 3. J Hepatol. 2004;40:993–9.
22. Hadziyannis SJ, Sette H Jr, Morgan TR et al. Peginterferon-alfa2a and ribavirin combination therapy in chronic hepatitis C. A randomized study of treatment duration and ribavirin dose. Ann Intern Med. 2004;140:346–55.
23. Mangia A, Santoro R, Minerva N et al. Peginterferon alfa-2b and ribavirin for 12 vs 24 weeks in HCV genotype 2 or 3. N Engl J Med. 2005;352:2609–17.
24. Dalgard O, Bjoro K, Hellum KB et al. Treatment with pegylated interferon and ribavarin in HCV infection with genotype 2 or 3 for 14 weeks: a pilot study. Hepatology. 2004;40:1260–5.
25. von Wagner M, Huber M, Berg T et al. Peginterferon-alpha-2a (40KD) and ribavirin for 16 or 24 weeks in patients with genotype 2 or 3 chronic hepatitis C. Gastroenterology. 2005; 129:522–7.
26. Shiffman ML, Pappas S, Nyberg L et al. Peginterferon alpha-2a (Pegasys) plus ribavirin (Copegus) for 16 or 24 weeks in patients with HCV genotype 2 or 3. Final results of the ACCELERATE trial. J Hepatol. 2006;44(Suppl. 2):271A.
27. Hinrichsen H, Benhamou Y, Wedemeyer H et al. Short-term antiviral efficacy of BILN 2061, a hepatitis C virus serine protease inhibitor, in hepatitis C genotype 1 patients. Gastroenterology. 2004;127:1347–55.
28. Reiser M, Hinrichsen H, Benhamou JP et al. Antiviral efficacy of NS3-serine protease inhibitor BILN-2061 in patients with chronic genotype 2 and 3 hepatitis C. Hepatology. 2005;41:832–5.
29. Sarrazin C, Rouzier R, Wagner F et al. SCH 503034, a novel hepatitis C virus protease inhibitor, plus pegylated interferon alpha-2b for genotype 1 nonresponders. Gastroenterology. 2007;132:1270–8.
30. Reesink HW, Zeuzem S, Weegink CJ et al. Rapid decline of viral RNA in hepatitis C patients treated with VX-950: a phase Ib, placebo-controlled, randomized study. Gastroenterology. 2006;131:997–1002.
31. Sarrazin C, Kieffer T, Bartels D et al. Dynamic hepatitis C virus genotypic and phenotypic changes in patients treated with protease inhibitor telaprevir. Gastroenterology. 2007;132:1767–77.
32. Kamal SM, El Tawil AA, Nakano T et al. Peginterferon α-2b and ribavirin therapy in chronic hepatitis C genotype 4: impact of treatment duration and viral kinetics on sustained virological response. Gut. 2005;54:858–66.
33. Diago M, Hassanein T, Rodes J, Ackrill AM, Sedarati F. Optimized virologic response in hepatitis C virus genotype 4 with peginterferon-alpha2a and ribavirin. Ann Intern Med. 2004;140:72–3.
34. Hasan F, Asker H, Al-Khaldi J et al. Peginterferon alfa-2b plus ribavirin for the treatment of chronic hepatitis C genotype 4. Am J Gastroenterol. 2004;99: 1733–7.
35. Bonny C, Roche C, Fontaine H et al. Efficacy of interferon (standard or pegylated) plus ribavirin in naive patients with hepatitis C virus genotype 5. A French national study. J Hepatol. 2005;42(Suppl. 2):200.

Section IV
Viral hepatitis (Part 2)

Chair: P BERK and P FERENCI

9
The future in the treatment of hepatitis C

J.-M. PAWLOTSKY

INTRODUCTION

The current standard of care for patients with chronic hepatitis C is the combination of pegylated interferon (IFN) alpha and ribavirin. This treatment is effective in approximately 80% of patients with hepatitis C virus (HCV) genotype 2 or 3 infection, but less than 50% of those with HCV genotype 1[1]. Many new HCV drugs are at the preclinical developmental stage and several are in clinical development. Novel HCV therapies currently in development schematically belong to four categories: novel IFN, alternatives to ribavirin, immune therapies, and specific and non-specific inhibitors of the HCV life cycle.

NOVEL INTERFERONS

New IFN-α molecules are currently being developed. They are expected to yield more potent antiviral effects, and eventually more potent immunomodulatory effects, with improved pharmacokinetic and pharmacodynamic properties and, if possible, better tolerance. The potential interest in non-alpha IFN is also being studied.

Alpha interferons

Albumin-linked IFN-α

Albumin-linked IFN-α (Albuferon®, Human Genome Sciences and Novartis) is an IFN-α-2b molecule attached to a human albumin moiety which has a prolonged half-life that allows dosing at intervals of 2–4 weeks. Albumin-linked IFN-α is able to induce an antiviral response in previous non-responders to pegylated IFN and ribavirin combination. Results presented at the 42nd Annual Meeting of the European Association for the Study of the Liver (EASL) have shown no significant difference in the rates of sustained virological response between patients receiving pegylated IFN-α-2a and ribavirin or various doses of albumin-linked IFN-α administered every 2 or 4 weeks with ribavirin.

Gene-shuffled IFN-α

Novel IFN-α variants have been derived experimentally through DNA shuffling. In this technique natural IFN-α genes are fragmented and then reassembled through recombination, a laboratory process that mimics evolution. Gene-shuffled IFN variants may have increased antiviral and antiproliferative activity[2]. A library of clones has been generated by Maxygen, and two clones have been selected because of their improved specific *in-vitro* activity over human IFN-α. One of them has been recently pegylated by Roche and will soon enter clinical evaluation.

Consensus IFN-α

IFN alphacon-1 (Infergen, Amgen and InterMune-Yamanouchi) is a synthetic recombinant 'consensus' IFN (cIFN) created by scanning the sequences of several natural alpha IFN and assigning the most frequently observed amino acid in each corresponding position to the recombinant molecule[3]. Intrinsic antiviral activity of this IFN appears to be greater than that of other standard IFN. However, there is no clear evidence that cIFN is superior to other alpha IFN when both are given in equivalent doses in terms of sustained virological response. The future of cIFN will depend on the development of a pegylated or otherwise pharmacologically modified form of the drug.

Other alpha IFN in development

Multiferon (Viragen) is a highly purified, multi-subtype natural human IFN-α derived from human leucocytes which has already been approved for use in HCV therapy in several countries. Medusa® (Flamel Technologies) is a self-assembled poly-amino acid nanoparticles system that can be used as a protein carrier for the development of novel long-acting native protein drugs. Medusa IFN-α-2a and Medusa IFN-β are currently in early clinical development. Various types of orally administered IFN-α are also currently being developed. Whether similar efficacy as with subcutaneous administration can be achieved is under study.

Non-alpha interferons

Interferon gamma

In spite of *in-vitro* results that would support a potentiating effect of IFN-γ on IFN-α[4], the proof of concept that IFN-γ could have a utility in the treatment of chronic hepatitis C still remains to be made.

Interferon omega

IFN-ω, like IFN-α or IFN-β, is a type 1 IFN. It shares 70% homology with IFN-α and binds to the same receptor. A stable, glycosylated form of IFN-ω has been developed (Intarcia) and has been administered to patients with chronic hepatitis C in phase I and II trials. The results of a phase II trial presented at the 42nd EASL Annual Meeting have shown sustained virological response rates (HCV RNA below 50 international units (IU)/ml at week 12 post-therapy) of 36% in the IFN-ω–ribavirin combination arm versus 6% in patients treated with IFN-ω alone. The manufacturer plans to develop an implantable infusion pump that will release a steady amount of IFN-ω for approximately four weeks.

Interferon lambda

IFN-λ-1 (interleukin 29 or IL-29) is a member of a novel family of cytokines that are distantly related to the IL-10 family and type 1 IFNs. IFN-λ-1 exhibits dose- and time-dependent inhibition of HCV replication in various models, independent of types 1 and 2 IFN receptors and induced pathways[5]. A pegylated form of IFN-λ will soon enter clinical evaluation.

ALTERNATIVES TO RIBAVIRIN

Because of its side-effects, principally haemolytic anaemia, ribavirin administration is exposed to dose reductions and drug discontinuations that are associated with breakthroughs during and relapses after therapy. Thus, alternatives to ribavirin are needed that would have the same effects as ribavirin on infection in combination with pegylated IFN or other IFN molecules without its haemolytic properties. However, the challenge for discovering such molecules is hampered by the fact the antiviral mechanisms of ribavirin remain poorly understood. Various approaches have already been abandoned because they were inferior to ribavirin in combination with pegylated IFN-α. They included: inhibitors of the enzyme inosine monophosphate dehydrogenase (IMPDH), such as mycophenolate mofetil (Cellcept) or merimepodib (VX-497, Vertex Pharmaceuticals), an oral IMPDH inhibitor; levovirin, an L-enantiomer of ribavirin that has similar immunomodulatory activity as ribavirin but does not induce haemolytic anaemia.

Taribavirin (Valeant Pharmaceuticals) is an amidine prodrug of ribavirin converted into ribavirin by adenosine deaminases, which are primarily present in hepatocytes. Taribavirin is thus preferentially taken up in the liver where it serves for ribavirin delivery to the major site of HCV replication, whereas it is not transported efficiently into red blood cells[6]. In two recent phase III trials in combination with pegylated IFN-α-2a and IFN-α-2b, respectively, taribavirin at a flat dose of 600 mg twice a day failed to achieve similar efficacy as weight-based dosed ribavirin in patients with chronic hepatitis C of various genotypes. The incidence of haemolytic anaemia was, however, significantly lower with taribavirin. New trials with higher doses of taribavirin have been planned.

IMMUNE THERAPIES

Non-specific immunomodulatory agents

Various non-specific immunomodulatory agents, including thymosin-α_1 (Thymalphasin, SciClone), IL-10 or histamine (Maxym Pharmaceuticals) have been administered to patients with chronic hepatitis C, with little success.

Therapeutic vaccines

Vaccines can induce $CD4^+$ and $CD8^+$ T-cell responses to HCV. Preclinical and early human studies indicate that therapeutic vaccines using different forms of recombinant HCV proteins together with various adjuvants could up-regulate both cellular and humoral immune responses in patients with chronic hepatitis C[7]. However, there is currently no evidence that therapeutic vaccines alone can result in changes in HCV-RNA levels. It remains to be determined whether therapeutic vaccines could be useful in combination with potent antiviral molecules.

SPECIFIC AND NON-SPECIFIC INHIBITORS OF THE HCV LIFE CYCLE

Every step of the HCV life cycle constitutes a potential target for specific or non-specific antiviral molecules. Many drugs are at the preclinical developmental stage and several are in clinical development, but initial trials using some of these inhibitors alone have raised concerns about their tolerability and the development of viral resistance. A number of specifically targeted therapies are now also being tested in combination with pegylated IFN-α with or without ribavirin.

Inhibitors of viral attachment, entry and fusion

Specific antibodies

Inhibition of HCV entry can be based on the use of specific antibodies that neutralize infectious particles and prevent their attachment to the receptor molecules. Polyclonal hepatitis C immune globulins (Civacir®, NABI Biopharmaceuticals) have been administered to prevent HCV infection in HCV-infected liver transplant recipients with little success[8]. New trials are ongoing. Anti-HCV monoclonal antibodies have been shown to have high-affinity HCV neutralizing properties *in vitro*[9] and to reduce viral replication by 0.3–1.0 log[10,11]. Other HCV-specific monoclonal antibodies are at the preclinical developmental stage, and attempts to develop mimotope-based therapeutic vaccine strategies are aimed at achieving HCV viral particle neutralization[12,13].

Small molecule entry inhibitors

Small molecule entry inhibitors can theoretically belong to two groups of drugs: molecules that specifically fix onto the HCV surface structures and neutralize the virus, or molecules that compete with infectious viral particles at the receptor level. Unfortunately, our understanding of HCV entry mechanisms remains rudimentary, hampering the development of such molecule inhibitors; it should, however, rapidly improve with the recent development of a productive cell culture system. New vaccine strategies are aimed at achieving HCV viral particle neutralization[14–16]. Fusion could also become an interesting target for novel therapies when its mechanisms are better understood.

HCV-RNA translation inhibitors

Nucleic acid-based strategies

Several nucleic acid-based strategies have been tested; all of them have been shown to potently inhibit HCV translation *in vitro*. The phosphorothioate antisense oligodeoxynucleotide ISIS 14803 (Isis Pharmaceuticals) progressed to phase II clinical trials, but development was halted due to lack of antiviral efficacy and transient serum alanine aminotransferase elevations[17,18]. A synthetic phosphorodiamidate morpholino oligomer (PMO), AVI-4065 (AVI Biopharma) has induced no significant reductions in HCV-RNA levels at an interim analysis. Higher doses and longer duration of treatment will be explored. Ribozymes early phase clinical trials were halted because of animal toxicity. RNA interference initiated by small interfering RNAs (siRNA) or short hairpin RNAs (shRNA) is very specific and offers a potential to be used as antiviral against HCV. However, because of their size and chemical composition, siRNAs and shRNAs currently are not orally bioavailable and require parenteral administration. Methods for effective *in-vivo* delivery are being explored.

Small molecule inhibitors of HCV translation

One limitation of sequence-based nucleic acid drug design is that it is strictly dependent on the predicted secondary structure of the target RNA sequence. The target RNA sequence may, however, not be fully accessible in the infected cell *in vivo*. Alternative strategies currently target the three-dimensional functional internal ribosome entry site (IRES) complexed with ribosomal subunits and viral and cellular proteins. VGX-410C (VGX Pharmaceuticals), a molecule that specifically blocks binding of HCV IRES and the 40S ribosomal subunit to eukaryotic factor 3, subsequently inhibiting ribosome formation and polyprotein translation, has entered phase II clinical evaluation.

Inhibitors of HCV post-translational processing

Peptidomimetic inhibitors of NS3/4A serine proteinase

Highly selective, potent peptidomimetic inhibitors of HCV NS3/4A proteinase have been derived from the viral substrate of the enzyme using structure-based drug design techniques. VX-950 or telaprevir (Vertex Pharmaceuticals) and SCH 503034 (Schering-Plough Corporation) have now advanced to phase II clinical trials. Patients who received 750 mg of telaprevir alone every 8 h experienced a median 4.4-log reduction in HCV RNA levels[19]. The drug is well tolerated over short-term administration. However, viral breakthroughs occur during the second week of telaprevir administration in patients with low exposure to the drug. These are due to selection of telaprevir-resistant variants. Combination with pegylated IFN-α with or without ribavirin could theoretically at least partly prevent telaprevir resistance. This hypothesis is currently under study in two phase II trials, PROVE 1 and PROVE 2. SCH 503034 appears to have less potent antiviral properties than telaprevir *in vivo* at the doses used to date[20]. In combination, the antiviral effect of SCH 503034 appears to be additive to that of pegylated IFN-α-2b[20]. A phase II clinical trial is ongoing in combination with pegylated IFN-α and ribavirin, in which high doses of SCH503034 are administered to treatment-naive patients. ITMN 191 (InterMune), a macrocyclic inhibitor of NS3/4A, will soon enter phase I clinical evaluation.

Inhibitors of NS4A-NS3 proteinase interaction

Other approaches have been developed to inhibit the NS3/4A serine proteinase function. ACH-806/GS-9132 (Achillion Pharmaceuticals and Gilead Sciences) inhibits binding of NS4A to the NS3 proteinase, therefore inhibiting polyprotein processing by preventing the formation of the active proteinase complex. ACH-806/GS-9132 has potent *in-vitro* activity against HCV genotype 1 and administration of 300 mg twice daily for 5 days resulted in an average change in HCV-RNA level from baseline of –0.9 log[21]. The development of this drug has been halted because of nephrotoxicity[21], but the study has demonstrated that inhibition of NS4A fixation to NS3 is a target for new NS3/4A proteinase inhibitors and other molecules are being developed.

Inhibitors of HCV replication

RNA-dependent RNA polymerase inhibitors

Inhibitors of the RNA-dependent RNA polymerase (RdRp) belong to two categories: nucleoside/nucleotide inhibitors, that target the catalytic site of the enzyme, and non-nucleoside inhibitors that target allosteric sites of the RdRp. Three RdRp inhibitors have been administered to patients in clinical trials, including two nucleoside and one non-nucleoside inhibitor. NM283 or valopicitabine (Idenix Pharmaceuticals and Novartis) is a nucleoside analogue that induces a dose-dependent HCV-RNA decline in HCV genotype-1-infected patients, up to –1.2 log on average, and has a dose-dependent additive effect to that of pegylated IFN-α[22]. A few cases of valopicitabine resistance have been observed after several months of monotherapy. Frequent gastrointestinal side-effects have led to a dose reduction in ongoing trials. R1626 (Roche Products) is another nucleoside analogue that induces a dose-dependent HCV RNA level reduction[23]. At very high doses, HCV-RNA level decrease reaches more than 3 logs, but side-effects are frequent. R1626 in combination with pegylated IFN-α and ribavirin has recently progressed into phase II of clinical development.

HCV-796 (ViroPharma and Wyeth Pharmaceuticals) is a non-nucleoside inhibitor of HCV RdRp. The patients with high exposure to HCV-796 achieve a peak mean HCV RNA level reduction of –1.4 log IU/ml[24]. However, most patients relapse after a few days of administration due to the selection of resistance mutations. In combination with pegylated IFN-α, the HCV-RNA level reduction is greater than with pegylated IFN alone, and no relapse due to resistance selection is observed over 14 days of administration[25]. Other drugs have entered phase I clinical development, such as XTL 2125 (XTL Pharmaceuticals), a non-nucleoside HCV RdRp inhibitor, or R7128 (Pharmasset), an oral cytidine analogue.

Cyclophilin B inhibitors

The HCV RdRp has been reported to bind cyclophilin B, a cellular peptidyl-prolyl *cis trans* isomerase that apparently regulates HCV replication through modulation of the RNA-binding capacity of RdRp. Synthetic, non-immunosuppressive cyclophilin B inhibitors have been developed and are being tested in patients with chronic HCV infection. DEBIO-025 (DebioPharm) has been administered to treatment-naive patients coinfected with human immunodeficiency virus and induced an HCV-RNA level decrease of more than 2 log upon administration, with a maximal HCV-RNA level reduction of –3.6 log, independent of the HCV genotype[26]. Side-effects were however frequent, and led to treatment withdrawal in several cases[26]. Another cyclophilin B inhibitor, NIM 811 (Novartis), is currently being tested in a phase I trial[27,28].

NS5A inhibitors

So-called 'NS5A inhibitors' have been tested in various models *in vitro*. Since the function of NS5A in the HCV life cycle remains unknown, the three-dimensional structure of the full protein has not been determined and no functional assay is available, the hypothesis of an effect of these drugs on NS5A is based on the observation of an inhibitory effect in replicon assays and selection of amino acid substitutions in the NS5A protein sequence. It is difficult to know, however, if such selection occurred as a result of a direct effect of the drugs on NS5A or of an adaptation of the replicons to cellular changes induced by the administration of the drug.

Inhibitors of virus assembly and release

Iminosugars have been suggested to be able to cross cellular membranes and concentrate in the endoplasmic reticulum where they could competitively inhibit envelope proteins glycosylation and interfere with viral assembly[29]. UT-231B (United Therapeutics), an alpha-glucosidase inhibitor, has no proven significant antiviral efficacy. MX-3253 or celgosivir (Migenix) has a modest antiviral effect on HCV in monotherapy. It is currently administered in combination with pegylated IFN-α and ribavirin in a phase II clinical trial.

References

1. NIH Consensus Statement on Management of Hepatitis C: 2002. NIH Consens State Sci Statements. 2002;19:1–46.
2. Chang CC, Chen TT, Cox BW et al. Evolution of a cytokine using DNA family shuffling. Nat Biotechnol. 1999;17:793–7.
3. Keeffe EB, Hollinger FB. Therapy of hepatitis C: consensus interferon trials. Hepatology. 1997;26:101S–7S.
4. Larkin J, Jin L, Farmen M, Venable D, Huang Y, Tan SL, Glass JI. Synergistic antiviral activity of human interferon combinations in the hepatitis C virus replicon system. J Interferon Cytokine Res. 2003;23:247–57.
5. Marcello T, Grakoui A, Barba-Spaeth G et al. Interferons alpha and lambda inhibit hepatitis C virus replication with distinct signal transduction and gene regulation kinetics. Gastroenterology. 2006;131:1887–98.
6. Wu JZ, Lin CC, Hong Z. Ribavirin, viramidine and adenosine-deaminase-catalysed drug activation: implication for nucleoside prodrug design. J Antimicrob Chemother. 2003;52: 543–6.
7. Leroux-Roels G, Batens AH, Desombere I et al. Immunogenicity and tolerability of intradermal administration of an HCV E1-based vaccine candidate in healthy volunteers and patients with resolved or ongoing chronic HCV infection. Hum Vaccin. 2005;1:61–5.
8. Davis GL, Nelson DR, Terrault N et al. A randomized, open-label study to evaluate the safety and pharmacokinetics of human hepatitis C immune globulin (Civacir) in liver transplant recipients. Liver Transplant. 2005;11:941–9.
9. Eren R, Landstein D, Terkieltaub D et al. Preclinical evaluation of two neutralizing human monoclonal antibodies against hepatitis C virus (HCV): a potential treatment to prevent HCV reinfection in liver transplant patients. J Virol. 2006;80:2654–64.
10. Schiano TD, Charlton M, Younossi Z et al. Monoclonal antibody HCV-AbXTL68 in patients undergoing liver transplantation for HCV: results of a phase 2 randomized study. Liver Transplant. 2006;12:1381–9.
11. Galun E, Terrault NA, Eren R et al. Clinical evaluation (phase I) of a human monoclonal antibody against hepatitis C virus: safety and antiviral activity. J Hepatol. 2007;46:37–44.
12. Roccasecca R, Folgori A, Ercole BB et al. Mimotopes of the hypervariable region 1 of the hepatitis C virus induce cross-reactive antibodies directed against discontinuous epitopes. Mol Immunol. 2001;38:485–92.
13. Zucchelli S, Roccasecca R, Meola A et al. Mimotopes of the hepatitis C virus hypervariable region 1, but not the natural sequences, induce cross-reactive antibody response by genetic immunization. Hepatology. 2001;33:692–703.
14. Lindenbach BD, Evans MJ, Syder AJ et al. Complete replication of hepatitis C virus in cell culture. Science. 2005;309:623–6.
15. Wakita T, Pietschmann T, Kato T et al. Production of infectious hepatitis C virus in tissue culture from a cloned viral genome. Nat Med. 2005;11:791–6.
16. Zhong J, Gastaminza P, Cheng G et al. Robust hepatitis C virus infection *in vitro*. Proc Natl Acad Sci USA. 2005;102:9294–9.
17. McHutchison JG, Patel K, Pockros P et al. A phase I trial of an antisense inhibitor of hepatitis C virus (ISIS 14803), administered to chronic hepatitis C patients. J Hepatol. 2006;44:88–96.

18. Soler M, McHutchison JG, Kwoh TJ, Dorr FA, Pawlotsky JM. Virological effects of ISIS 14803, an antisense oligonucleotide inhibitor of hepatitis C virus (HCV) internal ribosome entry site (IRES), on HCV IRES in chronic hepatitis C patients and examination of the potential role of primary and secondary HCV resistance in the outcome of treatment. Antivir Ther. 2004;9:953–68.
19. Reesink HW, Zeuzem S, Weegink CJ et al. Rapid decline of viral RNA in hepatitis C patients treated with VX-950: a phase Ib, placebo-controlled, randomized study. Gastroenterology. 2006;131:997–1002.
20. Sarrazin C, Rouzier R, Wagner F et al. SCH 503034, a novel hepatitis C virus protease inhibitor, plus pegylated interferon α-2b for genotype 1 non-responders. Gastroenterology. 2007 (In press).
21. Pottage JC, Lawitz E, Mazur D et al. Short-term antiviral activity and safety of ACH-806 (GS-9132), an NS4A antagonist, in HCV genotype 1 infected individuals. J Hepatol. 2007 (In press).
22. Afdhal NH, O'Brien C, Godofsky E et al. Valopicitabine (NM283), alone or with peg-interferon, compared to peg-interferon/ribavirin (PEGIFN/RBV) retreatment in hepatitis C patients with prior non-response to PEGIFN/RBV: week 24 results. J Hepatol. 2006;44 (Suppl. 2):S19.
23. Roberts S, Cooksley G, Dore G et al. Results of a phase 1b, multiple dose study of R1626, a novel nucleoside analogue targeting HCV polymerase in chronic HCV genotype 1 patients. Hepatology. 2006;44(Suppl. 1):692A.
24. Chandra P, Raible D, Harper D, Speth J, Villano S, Bichier G. Antiviral activity of the non-nucleoside polymerase inhibitor, HCV-796, in patients with chronic hepatitis C virus: preliminary results from a randomized, double-blind, placebo-controlled, ascending multiple dose study. Gastroenterology. 2006;130(Suppl. 2):A748.
25. Villano SA, Raible D, Harper D et al. Antiviral activity of the non-nucleoside polymerase inhibitor, HCV-796, in combination with pagylated interferon alfa-2b in treatment-naive patients with chronic HCV. J Hepatol. 2007 (In press).
26. Flisiak R, Orban A, Kierkus J et al. The cyclophilin inhibitor DEBIO-025 has a potent dual anti-HIV and anti-HCV activity in treatment-naœve HIV/HCV co-infected subjects. Hepatology. 2006;44(Suppl. 1):609A.
27. Ma S, Boerner JE, Tiong-Yip C et al. NIM811, a cyclophilin inhibitor, exhibits potent *in vitro* activity against hepatitis C virus alone or in combination with alpha interferon. Antimicrob Agents Chemother. 2006;50:2976–82.
28. Goto K, Watashi K, Murata T, Hishiki T, Hijikata M, Shimotohno K. Evaluation of the anti-hepatitis C virus effects of cyclophilin inhibitors, cyclosporin A, and NIM811. Biochem Biophys Res Commun. 2006;343:879–84.
29. Durantel D, Carrouee-Durantel S, Branza-Nichita N, Dwek RA, Zitzmann N. Effects of interferon, ribavirin, and iminosugar derivatives on cells persistently infected with noncytopathic bovine viral diarrhea virus. Antimicrob Agents Chemother. 2004;48:497–504.

10
Treatment of interferon non-responder patients with chronic hepatitis C in Asia

M. NISHIOKA

INTRODUCTION

Interferon (IFN) has been widely used in patients with chronic hepatitis C (CHC) as an effective antiviral drug. IFN-based antiviral therapy seems to be effective in about half of the patients with CHC who undergo IFN therapy, as judged by sustained virological response. However, not all patients with CHC are suitable candidates for IFN therapy. IFN often gives rise to severe adverse effects. Therefore, alternative therapeutic agents are necessary to prevent progressive liver diseases.

In Asia, medicinal herbs have been commonly used in patients with chronic liver diseases since early times. At the present time glycyrrhizin, ursodeoxycholic acid (UDCA) and some medicinal herbs such as Sho-saiko-to are used for treatment of patients with chronic liver disease, including HBV and HCV infections, in Japan and other Asian countries. They seem to work by suppressing inflammatory processes, reflected by reduction of serum alanine aminotransferase (ALT) levels, thereby preventing the development of hepatocellular carcinoma (HCC)[1–3]. These drugs are now used as an alternative therapy for patients with CHC who fail to respond to IFN.

CURRENT THERAPY FOR CHC IN ASIA

IFN-based antiviral therapy including pegylated IFN does not seem to be adopted in a high percentage of prescriptions: only 10% of patients in Japan, 20–30% in China, 1% in Korea, and 3–5% in Indonesia. Glycyrrhizin preparations based on an extract of licorice are, on the other hand, used in 30% of patients in Japan, 20–30% in China, and 3–5% in Indonesia. Several natural medicines such as Sho-saiko-to, silymarin, DDB (dimethyl diphenyl bicarboxylate – a natural substance-related synthetic molecule) and other hepatoprotectors such as UDCA, are also common in Asian countries. The

Table 1 Current status of therapy for chronic hepatitis C patients in Asia (estimation)

	IFN PEG-IFN	*Glycyrrhizin*	*Medicinal herbs, liver protectives*
Japan[a]	9.4%	30%	60% UDCA, Sho-saiko-to, glycyron, prohepal, Juzentaiho-to, etc.
Korea[b]	1%	–	99% DDB (dimethyl diphenyl bicarboxylate), silymarin, FAD, UDCA, ornithine aspartate, etc.
China[c] (in central areas)	20–30%	20–30%	40–60% UDCA, silibinin, domperidone, vitamin A, E, Sui-bon-so, etc.
Indonesia[d]	3–5%	3–5%	90–95% DDB, curcumin, Sho-saiko-to, UDCA, silymarin, etc.

[a]Chugai Pharmaceutical Co., Ltd; [b]HanAll Pharmaceutical Co., Ltd; [c]Shanghai Public Hygiene Center; [d]Indonesian Top Doctors.

use of such agents represents the majority of prescriptions, with 60% in Japan, 99% in Korea, 40–60% in China and 90–95% in Indonesia (Table 1).

IFN has made an impact by introducing an antiviral therapy, and has changed the picture greatly. However, non-antiviral drugs have never been replaced by IFN. The therapeutic strategy for hepatitis management now focuses on improvement of prognosis of CHC patients who cannot be treated with IFN. Most non-antiviral drugs have remained drugs of choice in parallel with IFN.

Silymarin, which is derived from the milk thistle plant, has been used as a natural remedy for diseases of the liver and the biliary tract throughout the world[4], and has also been known in Japan as a folk medicine. The name 'milk thistle' was derived from the milk of the Virgin Mary. Silymarin and its active constituent silybin have been reported to work as antioxidants scavenging free radicals.

DDB is commonly used in China, Korea, Indonesia and Egypt. This compound seems to have anti-oxidative effects and an effect of lowering serum ALT levels in patients with various liver diseases[5].

In order to assess the beneficial and harmful effects of medicinal herbs for hepatitis C virus (HCV) infection, 13 randomized trials (n = 818) evaluated 14 medicinal herbs[6]. Some of them seem to be effective. Oxymatrine showed positive effects on HCV-RNA clearance, compared with vitamins[7]. Yi Er Gan Tang was more effective on normalizing serum ALT, compared with silymarin plus glucurolactone[8]. The herbal mixture Bing Gan Tang plus IFN-α was more effective on clearance of HCV-RNA and normalization of serum ALT than IFN-α alone[9]. Silybin showed significant effects on reducing AST, compared with placebo[10]; however, herbs were also associated with adverse events. Consequently, there was no firm evidence supporting the efficacy of medicinal

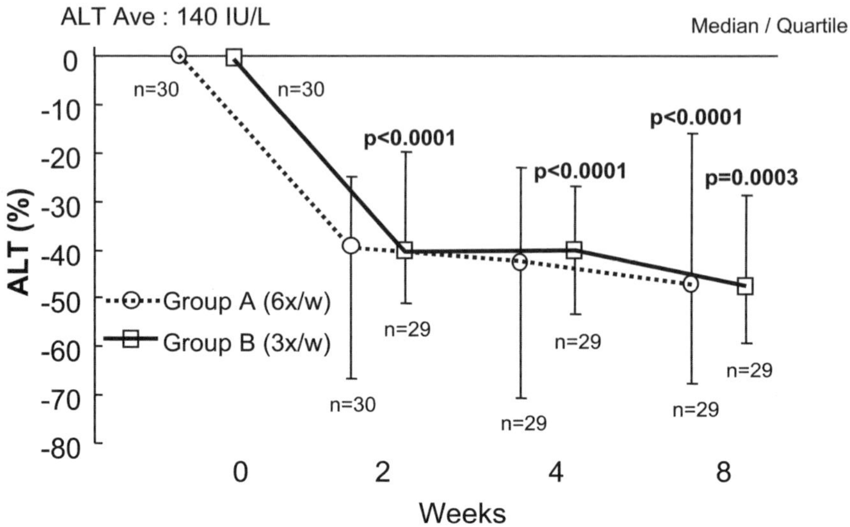

Figure 1 Glycyrrhizin therapy for chronic hepatitis C patients in Indonesia

herbs for HCV infection[6]. In spite of the fact that we may not yet have a plausible pharmacological reason for use of herbs, such use has been widespread. Certain medicinal herbs contain substances that may interact with virus and/or prevent liver injury. Further randomized trials are needed.

SHO-SAIKO-TO FOR HCV INFECTION

Sho-saiko-to, a crude extract from seven herbs, is a Chinese medicine used widely in Japan for treatment of liver disorders. It is well known that Sho-saiko-to is effective in protecting hepatic necrosis caused by CCl_4 in rats, and in preventing hepatic fibrosis caused by dimethylnitrosamine. Recently, Sakaida et al. reported that Sho-saiko-to prevents liver fibrosis in rat liver cirrhosis induced by a choline-deficient L-amino acid-defined diet[11].

Sho-saiko-to was shown to lower serum ALT levels in patients with chronic hepatitis B[12]. In addition to ALT levels, serum concentrations of the 7S domain of type IV collagen and the aminoterminal propeptide of type III procollagen were significantly decreased by Sho-saiko-to in CHC patients with fibrosis[13].

Importantly, effects of Sho-saiko-to in CHC patients non-responsive to IFN were reported[14]. Serum transaminase levels decreased significantly in patients given Sho-saiko-to at a daily dose of 7.5 g orally for 24 weeks (Figure 3). Some other reports indicate that Sho-saiko-to helps to prevent the development of HCC in patients with cirrhosis, particularly in cirrhosis patients without HBs antigen[3].

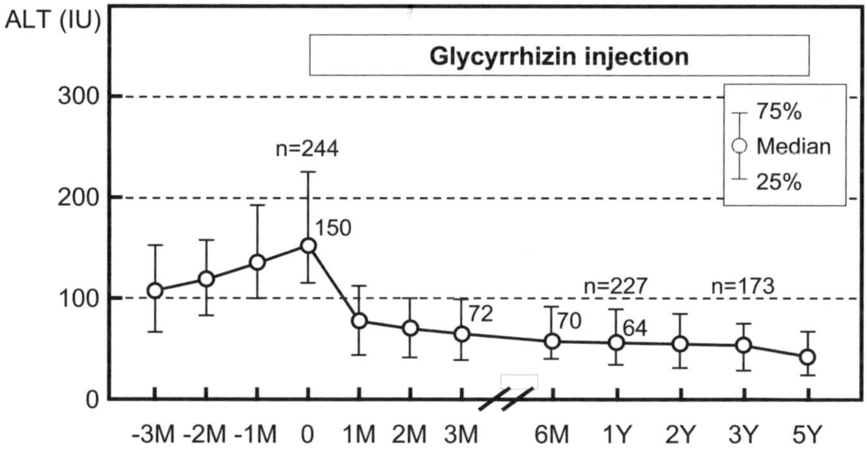

Figure 2 Change of ALT levels in patients with IFN-resistant chronic hepatitis C during glycyrrhizin therapy

URSODEOXYCHOLIC ACID IN HCV INFECTION

Ursodeoxycholic acid (UDCA), a choleretic bile acid, has been known as a potential therapeutic agent in patients with liver diseases. Proposed pharmacological mechanisms are direct cytoprotection, detergent effects on toxic hydrophobic bile acid, induction of hypercholeresis and immunomodulatory effects.

UDCA is commonly used in patients with primary biliary cirrhosis and primary sclerosing cholangitis. Puoti C et al. first reported the efficacy of UDCA in lowering liver enzymes such as ALT and gammaglutamyl transpeptidase (γ-GTP) in patients with HCV-associated liver diseases[15].

A multicentre randomized controlled dose study of UDCA for CHC showed a significant decrease of serum ALT and γ-GTP in patients administered 600 and 900 mg of UDCA[16]. Interestingly, UDCA and IFN combined therapy significantly prolongs the period for which serum ALT levels remain within the normal range after discontinuation of IFN[17].

Furthermore, to assess the efficacy and safety of combination therapy using UDCA with glycyrrhizin in patients with CHC, Tsubota et al.[18] conducted a prospective randomized controlled trial of glycyrrhizin, compared with glycyrrhizin plus UDCA. Serum ALT levels were significantly decreased during treatment in both groups. Levels of γ-GTP fell significantly more in the combination group and decreased to normal levels in more cases at the end of the 24-week trial (Figure 4). Concentrations of HCV-RNA did not change during the trial in either group. No adverse effects have been noticed. It is suggested that combined therapy with UDCA is an alternative to IFN in patients with CHC.

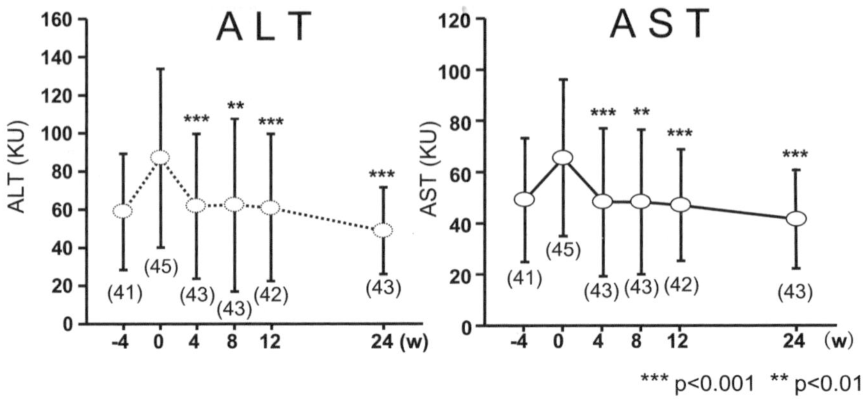

Figure 3 Effect of Sho-saiko-to on chronic hepatitis C patients with non-response to interferon

EFFECTS OF GLYCYRRHIZIN ON HCV INFECTION

An aqueous extract from the licorice root (*Glycyrrhiza radix*) has been used in China since early times as an anti-inflammatory agent[19]. Glycyrrhizin is effective in improving abnormal liver function tests in patients with chronic hepatitis[1]. A subsequent study showed a significant improvement in histological findings in patients with CHC[20]. Therapeutic effects of glycyrrhizin (SNMC: glycyrrhizin product, Stronger Neo-Minophagen C) at different doses on chronic hepatitis and cirrhosis were studied[21]; 52.3% of patients given 100 ml of SNMC and 26.1% of patients given 40 ml of SNMC improved, showing a dose–effect relationship and suggesting that a higher dose is more effective.

Since then, glycyrrhizin has been used widely in CHC on the basis of its anti-inflammatory effect on liver cells. In addition, long-term administration of glycyrrhizin for patients with CHC is effective in preventing liver carcinogenesis[2].

A recent study from China showed an effect of glycyrrhizin in combination with IFN in CHC patients[22]. The patients were given IFN for 24 weeks. At the end of the trial the patient group which additionally received glycyrrhizin to IFN showed significantly higher normalization of serum ALT levels and clearance of HCV-RNA compared with the group which received IFN alone (Table 2). Normalization rates of serum ALT in the IFN + glycyrrhizin group were also higher 24 weeks after the end of treatment, compared with the IFN group. These data suggest that addition of glycyrrhizin to IFN therapy is beneficial.

Sulaiman et al.[23] investigated the benefit and safety of glycyrrhizin for IFN non-responders with CHC in Indonesia. In patients with ALT concentration 2 times or more above the upper limit of normal, glycyrrhizin was administered

Table 2 Interferon and glycyrrhizin combined therapy in treatment of chronic hepatitis C

	No of cases	*Normalization of ALT level (%)*	*HCV-RNA negativity (%)*
Group A: IFN-α-2b + SNMC	30	66.7* (46.7*)	63.3* (30.0)
Group B: IFN-α-2b	26	45.2 (30.8)	38.5 (26.9)

Figures in parentheses indicate after 24-week treatment.

*$p < 0.05$ vs group B.

six times per week (group A) or three times per week (group B) for 8 weeks. Glycyrrhizin lowered ALT levels in both groups without side-effects during treatment (see Figure 1).

EFFECTS OF LONG-TERM GLYCYRRHIZIN THERAPY

Long-term glycyrrhizin therapy is extremely useful in patients with CHC[2]. Recent studies have revealed that long-term glycyrrhizin injection therapy reduces liver carcinogenesis rates in patients with IFN-resistant active CHC[24]. In this cohort study, among 346 patients with high ALT levels, 244 patients received intravenous glycyrrhizin injection and 102 patients did not. Proportional hazard analysis revealed that the glycyrrhizin injection therapy significantly decreased the incidence of HCC in patients with IFN-resistant active CHC. Long-term therapy also lowered serum ALT levels over a long period of time (Figure 2).

COMBINATION THERAPY

A sustained low ALT level is able to reduce the incidence of HCC, even if HCV cannot be eradicated from the liver. High serum ALT levels are associated with rapid development of HCC in patients with HCV-associated cirrhosis, and suppression of serum ALT levels by treatment with anti-inflammatory drugs prevents hepatocarcinogenesis[25].

Recently, Tarao et al.[26] reported that combination therapy of anti-inflammatory drugs such as glycyrrhizin, Sho-saiko-to, UDCA, etc., prolongs the interval from HCV-associated cirrhosis to the development of HCC. It is suggested that such combination therapy is a choice for CHC patients with IFN non-response.

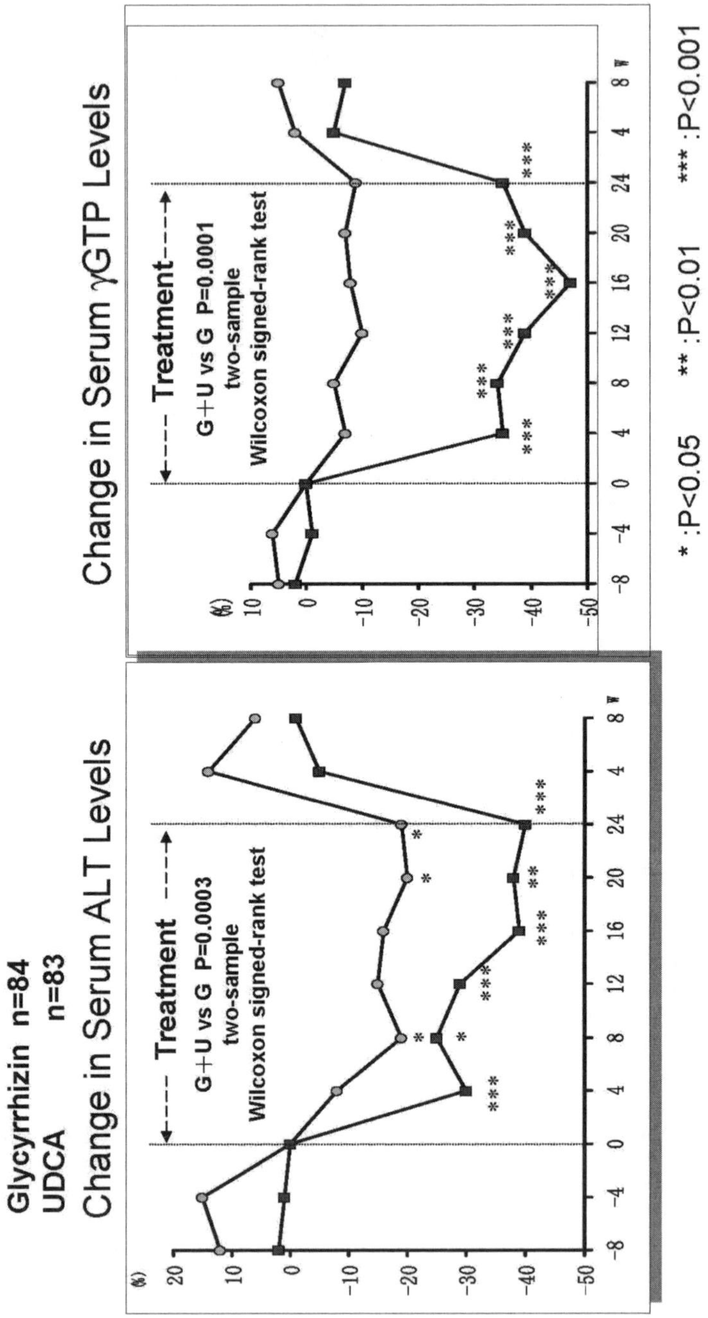

Figure 4 Combined ursodeoxycholic acid and glycyrrhizin therapy for chronic hepatitis C virus infection

CONCLUSIONS

Suppression of liver inflammation is very important. Non-antiviral agents, particularly of herbal origin, have been in common use for the treatment of liver diseases in Asia. Glycyrrhizin, UDCA and herbal medicines such as Sho-saiko-to are effective and safe anti-inflammatory drugs among known compounds, and they can prevent liver fibrosis and reduce the development of HCC through suppression of liver inflammation.

Furthermore, long-term glycyrrhizin therapy, combination therapy using anti-inflammatory drugs or using both IFN and anti-inflammatory drugs are also useful in IFN non-responders or patients who are not eligible for IFN therapy.

References

1. Suzuki H, Ohta Y, Takino T et al. Effects of glycyrrhizin on biochemical tests in patients with chronic hepatitis–double blind trial. Asian Med J. 1983;26:324–38.
2. Arase Y, Ikeda K, Murashima N, The long term efficacy of glycyrrhizin in chronic hepatitis C patients. Cancer. 1997;79:1494–500.
3. Oka H, Yamamoto S, Kuroki T et al. Prospective study of chemoprevention of hepatocellular carcinoma with Sho-saiko-to (TJ-9). Cancer. 1995;76:743–9.
4. Flora K, Hahn M, Rosen H et al. Milk thistle (*Silybum marianum*) for the therapy of liver disease. Am J Gastroenterol. 1998;93:139–43.
5. Fu H, Jing X, Yu S et al. Quantitative risk assessment for lung cancer from exposure to metal ore dust. Biomed Environ Sci. 1992;5:221–8.
6. Liu J, Manheimer E, Tsutani K et al. Medicinal herbs for hepatitis C virus infection: a Cochrane hepatobiliary systematic review of randomized trials. Am J Gastroenterol. 2003; 98:538–44.
7. Li JQ, Li CQ, Zeng MD et al. Preliminary study on therapeutic effect of oxymatrine in treating patients with chronic hepatitis C. Chin J Integrated Trad West Med. 1998;18:227–9.
8. Yu WJ. Yi Er Gan decoction for treatment of 40 cases of hepatitis C. NJ Trad Chin Med. 1995;27:46–7.
9. Pei ZG, Liu YL, Zhao SM. Therapeutic observation on Binggan decoction combined with interferon in 46 cases of hepatitis C. Pract J Integr Chin Mod Med. 1996;9:444.
10. Buzzelli G, Moscarella S, Giusti A et al. A pilot study on the liver protective effect of silybin–phosphatidylcholine complex (IdB1016) in chronic active hepatitis. Int J Clin Pharmacol Ther Toxicol. 1993;31:456–60.
11. Sakaida I, Matsumura Y, Akiyama S et al. Herbal medicine Sho-saiko-to (TJ-9) prevents liver fibrosis and enzyme-altered lesions in rat liver cirrhosis induced by a choline-deficient L-amino acid-defined diet. J Hepatol. 1998;28:298–306.
12. Tajiri H, Kozaiwa K, Ozaki Y et al. Effects of Sho-saiko-to (Xiao-Chai-Hu-Tang) on HBeAG clearance in children with chronic hepatitis B virus infection and with sustained liver disease. Am J Chin Med. 1991;19:121–9.
13. Shimizu I. Sho-saiko-to. Herbal Medicine and Molecular Basis in Health and Disease Management. New York: Dekker, 2003:573–93.
14. Kumada H. Effects of Sho-saiko-to in chronic hepatitis C with non-response to interferon. Rinsho Seijinbyou. 1994;24:1103–9.
15. Puoti C, Pannullo A, Annovazzi G et al. Ursodeoxycholic acid and chronic hepatitis C infection. Lancet. 1993;341:1413–14.
16. Takano S, Ito Y, Yokosuka O et al. A multicenter randomized controlled dose study of ursodeoxycholic acid for chronic hepatitis C. Hepatology. 1994;20:558–64.
17. Boucher E, Jouanolle H, Andre P et al. Interferon and ursodeoxycholic acid combined therapy in the treatment of chronic viral C hepatitis: results from a controlled randomized trial in 80 patients. Hepatology. 1995;21:322–7.

18. Tsubota A, Kumada H, Arase Y et al. Combined ursodeoxycholic acid and glycyrrhizin therapy for chronic hepatitis C virus infection: a randomized controlled trial in 170 patients. Eur J Gastroenterol Hepatol. 1999;11:1077-83.
19. Finney RS, Somers GF. The anti-inflammatory activity of glycyrrhetinic acid and derivatives. J Pharm Pharmacol. 1958;10:613–20.
20. Hino K, Miyakawa H, Takahashi A et al. The effects on hepatohistological patterns on massive administration of stronger Neo-Minophagen C for the treatment on chronic active hepatitis. Kan-Tan-Sui. 1986;13:797–807.
21. Iino S, Tango T, Matsushima T et al. Therapeutic effects of stronger neo-minophagen C at different doses on chronic hepatitis and liver cirrhosis. Hepatol Res. 2001;19:31–40.
22. Wu WF. Biochemical and sustained viral response at the end of treatment in patients with chronic hepatitis C. China Pharmacy. 2004;15(9).
23. Sulaiman A, Lesmana L, Budihusodo U et al. Experience of using stronger Neo-Minophagen C (SNMC) for chronic hepatitis C in Indonesia. Indonesian group (APASL 2002). J Gastroenterol Hepatol. 2002;17:A1061–82.
24. Ikeda K, Arase Y, Kobayashi M et al. A long-term glycyrrhizin injection therapy reduces hepatocellular carcinogenesis rate in patients with interferon-resistant active chronic hepatitis C: a cohort study of 1249 patients. Dig Dis Sci. 2006;5:603–9.
25. Tarao K, Rino Y, Ohkawa S et al. Association between high serum alanine aminotransferase levels and more rapid development and higher rate of incidence of hepatocellular carcinoma in patients with hepatitis C virus-associated cirrhosis. Cancer. 1999;86:589–95.
26. Tarao K, Tamai S, Okawa S et al. Studies on improvement of serum transaminase by administration by anti-inflammatory drugs in patients with hepatitis C virus associated cirrhosis. Rinsho to kenkyu. 1998;75:1858–67.

Section V
Autoimmune hepatitis

Chair: S MATERN and LW POWELL

11
Update on autoimmune hepatitis

C. P. STRASSBURG and M. P. MANNS

DESCRIPTION AND DEFINITION OF AUTOIMMUNE HEPATITIS

Autoimmune hepatitis (AIH) represents a chronic immune-mediated disease of the liver targeted at the hepatocyte leading to inflammation, fibrosis, cirrhosis and eventually liver failure[1]. The most distinguishing diagnostic features are hypergammaglobulinaemia, female gender and the presence of serum autoantibodies. Historically, hypergammaglobulinaemia and the detection of antinuclear antibodies have led to the definition of this disease entity by Waldenström and Mackay. Subsequently it became increasingly evident that AIH is characterized by serological heterogeneity best reflected in the diversity of autoantigen targets defined by different circulating autoantibodies[2]. Apart from nuclear autoantigens, actin, microsomal drug-metabolizing enzymes and soluble cytoplasmic proteins are targets of serologically detectable B cell autoimmunity. Autoantibodies directed against the endoplasmic reticulum, in particular against members of the cytochrome P450 (CYP) superfamily of proteins, also occur as markers of serological autoimmunity in hepatitis C and hepatitis D virus infections, as well as frequent transient markers of drug-mediated allergic hepatic disease, and even in the context of genetically determined autoimmune disease such as the autoimmune polyglandular syndrome (APS) type 1. The exact immunological basis of AIH still remains unresolved despite awareness of its serological features and considerable research efforts invested into the identified autoantigen targets. When AIH is diagnosed the disease is usually not in its early stages, and the initiating events are therefore not available for detailed analysis. The parallel serological features of virus-associated autoimmunity and genuine autoimmune hepatitis, however, have led to the hypothesis of an external trigger (infectious or chemical) for this disease. This suggests that the susceptibility inherent to the host is a required co-factor. The epidemiology of AIH unfortunately offers very little information suggesting genetic susceptibility, mainly because familial risk is understudied. Apart from individual reports, and in the absence of twin studies, AIH itself does not appear to segregate in families, but the prevalence of other autoimmune diseases, including autoimmune thyroid disease, coeliac disease, ulcerative colitis, etc., is increased in relatives. This feature is the basis of inclusion of first-degree relatives with autoimmune diseases into the revised international scoring system for AIH, a score that describes – for scientific purposes – the likelihood of the presence of AIH. The hypothesis of trigger and genetic susceptibility is strengthened by a significant body of evidence linking major histocompatibility complex (MHC) genes with AIH[3]. MHC class I and II antigens are critical players in T cell immunity in their ability to present short antigenic peptides for recognition by antigen-specific T cells. Variants of MHC-encoded proteins therefore influence the precise interplay of T cell receptor and HLA molecule, including the possibility of determining immunological susceptibility and resistance. The study of autoantibodies and autoantigens on the one hand offers a window to identify the relevant antigenic determinants involved in the loss of tolerance; the study of genetic associations, on the other hand, leads to the definition of the permissive genetic profile, and the combination represents the stage upon which the pathophysiology of AIH unfolds.

UPDATE ON CLASSICAL AUTOANTIBODIES IN AIH AND THEIR DETECTION

The serological hallmarks of AIH are autoantibodies detected by serological *in vitro* assays[4]. For screening purposes, antinuclear antibodies (ANA), smooth muscle antibodies (SMA), liver–kidney microsomal antibodies (LKM) and antimitochondrial antibodies (AMA) continue to be determined by indirect immunofluorescence on rodent liver, stomach and kidney tissue. When these autoantibodies are undetectable, and the suspicion of autoimmune hepatitis remains, autoantibodies against cytosolic antigens, in particular the SLA/LP antigen, are helpful. These SLA/LP autoantibodies are detected by radioimmunoassay and enzyme-linked immunosorbent assays, because they cannot be observed by immunofluorescence. In approximately 10% of patients with AIH, SLA autoantibodies are present as the only marker in the absence of the above-mentioned autoantibodies. In addition, many cases of AIH display autoantibodies that show reactivity towards the asialoglycoprotein receptor (ASGPR) and some reactivity with the liver cytosolic antigen 1 (LC-1), although these autoantibodies are not routinely determined. LC1 autoantibodies are found in AIH type 2, and have been characterized as recognizing formiminotransferase cyclodeaminase. For scientific purposes, autoantibody profiles have been utilized as a means of subclassification of AIH[2].

Antinuclear antibodies (ANA)

These show specificity for both functional and structural proteins in the cell nucleus. They can be directed against nuclear membranes or even DNA[5,6]. Unfortunately, for diagnostic reasons, the target antigens are heterogeneous and incompletely defined, and heterogeneous even within a single disease such as AIH. For the diagnostic work-up of AIH the molecular characterization of target antigen specificity does not supply important additional information to increase diagnostic precision. Screening determinations of ANA are routinely performed by indirect immunofluorescence on cryostat sections of rat liver and on hepatitis 2 cell slides. Most commonly, a homogeneous or speckled immunofluorescence pattern is demonstrable. ANA have been found to be reactive with an array of antigens which include centromeres, ribonucleoproteins, and cyclin A. They represent the most common autoantibody in AIH and occur in high titres usually exceeding 1:160. However, the titre does not correlate with disease course, prognosis, disease progression, requirement of transplantation or disease activity.

Anti-smooth muscle antibodies (SMA)

SMA are directed against cytoskeletal proteins such as actin, troponin and tropomyosin[7,8]. They frequently occur in high titres in association with ANA. They are not highly specific for AIH and have been shown to occur in advanced liver diseases of other aetiologies, in infectious diseases and rheumatic disorders. In these cases titres are often lower than 1:80. SMA autoantibodies are also determined by indirect immunofluorescence on cryostat sections of rat stomach. In paediatric patients SMA autoantibodies may be the only marker of AIH type 1. When present in very young patients with AIH type 1 the titres may be as low as 1:40.

Autoantibodies against antigens of the endoplasmic reticulum

The endoplasmic reticulum (ER) of the cell harbours two important enzyme families, which are the main players in phase I and phase II metabolism: the cytochrome P450 (CYP) and the UDP-glucuronosyltransferases (UGT)[9]. Phase I metabolism leads to the oxidative modification of compounds, usually by the addition of functional groups such as hydroxylation. Phase II metabolism leads to conjugation with polar prosthetic groups such as glucuronic acid (glucuronidation)[10]. In the case of the UGT, glucuronidation leads to a water-soluble glucuronide which is targeted for renal or biliary elimination. Both enzyme families are preferred targets of a B cell response in autoimmune liver diseases. The characterization of autoantibodies directed against CYP and UGT has led to the definition of specificities which span a number of immune-mediated diseases. These include AIH, drug-induced hepatitis[11,12], autoimmune polyendocrine syndrome type 1 (APS-1)[9], and chronic hepatitis C (HCV) and D (HDV)[13]. LKM autoantibodies are visualized by indirect immunofluorescence on rodent cryostat sections. Subclassification is achieved by enzyme-linked immunosorbent assay (ELISA) and Western blot, preferably using recombinant antigens. AIH type 2 is characterized by the presence of LKM-1 autoantibodies against CYP 2D6. In 10%, LKM-3 autoantibodies against UDP-glucuronosyltransferases are also present[13]. In contrast to AIH type 1, additional organ-specific autoantibodies are present such as antithyroid, antiparietal cell, and anti-Langerhans cell autoantibodies. The number of extrahepatic immune syndromes such as diabetes, vitiligo and autoimmune thyroid disease is also more prevalent in AIH type 2 relative to AIH type 1. Serum immunoglobulin levels are moderately elevated with a reduction of IgA. AIH type 2 is a rare disorder which affects 20% of AIH patients in Europe but only 4% in the USA[14,15].

Autoantibodies against soluble liver antigen

Autoantibodies which target a soluble liver and pancreas antigen (anti-SLA/LP) exhibit a high specificity for AIH. They are detectable in about 10–30% of all patients with AIH[16–19]. In 1992 specific autoantibodies were identified in patients with a severe form of autoimmune chronic active hepatitis. These antibodies precipitated a UGA suppressor serine tRNA–protein complex, which is likely to be involved in co-translational selenocysteine incorporation in human cells[16]. The screening of cDNA expression libraries confirmed these findings and identified a previously unknown amino acid sequence, which presumably encodes the UGA-suppressor tRNA-associated protein, as target antigen of SLA antibodies[18,19]. Additionally, it was demonstrated that anti-SLA and the independently described liver–pancreas autoantibody (anti-LP) are identical, and the designation anti-SLA/LP has replaced anti-SLA and anti-LP. Recent data confirmed the previous finding that patients with anti-SLA/LP display a more severe course of autoimmune hepatitis[20,21]. Regarding disease specificity, anti-SLA/LP may be linked to the pathogenesis of the autoimmune process; however, the exact function and its role in autoimmunity are so far unclear.

UPDATE ON THE AUTOREACTIVITY AGAINST MICROSOMAL PROTEINS AND LESSONS FROM DRUG-INDUCED IMMUNE-MEDIATED LIVER DISEASE

LKM autoantibodies against CYP1A2 and CYP2A6 are found in patients with autoimmune polyglandular syndrome type 1 (APS-1) and hepatic involvement[22]. Anti-CYP 2A6 autoantibodies also occur in HCV infection[23]. LM autoantibodies, which are characterized by an immunofluorescence pattern selectively staining the hepatocellular cytoplasm but not kidney, have been found to be directed against CYP1A2[24–26]. These autoantibodies are also found in APS-1 syndrome with hepatic involvement and occur in dihydralazine-induced hepatitis[27]. A second type of LKM autoantibodies, LKM-2, are directed against CYP2C9 and are induced in ticrynafen-associated hepatitis[11,28]. A third group of LKM autoantibodies, LKM-3, were identified in 6–10% of patients with chronic hepatitis D virus infection (HDV) by Crivelli et al.[29]. These autoantibodies are directed against family 1 UDP-glucuronosyltransferases (UGT1A)[13]. LKM-3 autoantibodies have been identified in HDV infection but also in AIH type 2 patients[13]; they can also occur in LKM-1-negative and ANA-negative AIH.

A small percentage of patients treated with therapeutic drugs can develop severe hepatitis, which is characterized by lymphocytic liver infiltrations and autoantibodies directed against hepatic proteins. It is believed that drug-metabolizing enzymes, mainly CYP, create reactive metabolites, which in turn modify either the metabolizing CYP enzyme itself and/or other hepatic proteins[30]. In susceptible patients these modified proteins induce an immune response resulting in severe ‘drug-induced hepatitis’[31]. Modified proteins preferentially include CYP, which in themselves are often targets for autoantibodies. As typical examples, tienilic acid-induced hepatititis[32], dihydralazine hepatitis[27], halothane hepatitis[33] and anticonvulsant-induced hepatitis[34] have been characterized. It is debated whether alcoholic liver disease is caused in part by an autoimmune reaction against hepatic proteins, directed against both acetaldehyde- and hydroxyethyl-modified hepatic proteins. It is suggested that metabolism of ethanol by CYP2E1 generates hydroxyethyl radicals, which can represent targets of autoimmunity[35].

UPDATE ON THE AUTOEPITOPE SPECIFICITY OF CYTOCHROME P450 2D6 IN AIH AND HEPATITIS C

Hepatitis C virus (HCV) is associated with an array of extrahepatic manifestations, including mixed cryoglobulinaemia, membranoproliferative glomerulonephritis, polyarthritis, porphyria cutanea tarda, Sjögren's syndrome and autoimmune thyroid disease[36]. Not surprisingly, numerous autoantibodies are found to be associated with chronic hepatitis C. Similar to AIH antinuclear, SMA, LKM and antithyroid antibodies are found with a high prevalence.

The examination of LKM autoantibodies in HCV patients revealed that, although anti-CYP2D6 titres are similar to titres in AIH type 2, differences exist regarding the epitopes recognized by LKM autoantibodies[37–40]. In patients with AIH type 2 the epitope of amino acids 257–269 is recognized with a significantly higher prevalence than in chronic hepatitis C[39]. In addition, the immune reaction seems to be more heterogeneous than in autoimmune hepatitis, as indicated by recognized protein targets of 59 kDa and 70 kDa[41].

When the CYP2D6 molecule is analysed in detail for epitope reactivity of circulating autoantibodies in patients with AIH and HCV four regions of interest emerge[42]. All sera from AIH as well as HCV patients which are positive for LKM-1 autoantibodies upon standard screening tests recognize the full-length protein. The major region of reactivity resides between amino acids 257 and 269. With respect to this epitope a differential pattern is observed characterized by the binding of 64% of sera from AIH type 2 patients and only 24% of HCV patients. A second major epitope was found between amino acids 321 and 351 with a less convincing differential recognition pattern (AIH 20% and HCV 12%). Two minor epitopes are located between amino acids 373–389 and 419–429, recognized by under 10% of sera. Taken together none of these epitopes accounted for the entire reactivity measurable with the full-length protein, most probably owing to the fact that these were *Escherichia coli* or similarly expressed linear fragments assayed with denaturing techniques such as Western blot. In a recent analysis[42] we found that elimination of sequence from the amino terminus of the molecule up to amino acid 351 virtually eliminated all binding activity which was not reconstituted by any of the individual fragments located upstream of amino acid 351. This indicates that conformation-dependent epitopes are most probably responsible for the remainder of immunoreactivity with the CYP2D6 molecule. This was also substantiated by the finding that a larger fragment spanning amino acids 321–376 reacted with 76% of all AIH type 2 and HCV sera representing the highest number of all epitope definitions found to date. The three-dimensional rendering of the CYP2D6 molecule was able to show that this novel epitope was located on the surface of the molecule distant from the previous major linear epitope between amino acids 257 and 269[42]. A sequence search and comparison analysis further showed that this area of the CYP molecule overlaps with the previously defined epitopes recognized by autoantibodies directed against CYP1A2, CYP2C9, CYP3A1 and CYP21B in different drug-mediated autoimmune reactions, as well as in adrenal autoimmunity (CYP21B). This interesting finding points to the usefulness of epitope analyses to elucidate common immunological processes involved in the generation of autoimmunity. The model of drug-induced hepatitis is therefore linked to the patterns found in genuine autoimmune hepatitis at the B cell level.

From a practical point of view the differential recognition of autoepitopes in HCV and AIH type 1 has clinical implications. LKM autoantibodies in chronic hepatitis C seem to indicate an increased risk of exacerbation of the disease[43,44]. Dalekos et al. studied antibody titres and performed epitope mapping of LKM-1-positive sera from patients with chronic hepatitis C. Interestingly, a patient with a high LKM-1 titre and autoantibodies directed against an epitope of amino acids 257–269, which is preferentially recognized by patients with AIH type 2, showed exacerbation of the disease under interferon treatment. In contrast to other patients with HCV infection, this patient further recognized a rarely detected epitope on the C-terminal third of the protein.

Recently another autoantibody was detected in patients infected with HCV. About 2% of HCV-positive sera in general, and 7.5% of LKM-1 positive HCV sera, recognize CYP2A6. This autoantibody appears to occur more frequently in HCV-infected patients with LKM-1 autoantibodies. Interestingly anti-CYP2A6 autoantibodies are not detected in patients with AIH type 2, who exhibit high titres of LKM-1 autoantibodies. The clinical relevance of this finding remains to be determined[23]. Anti-CYP 2A6 autoantibodies have also been detected in patients with the autoimmune polyglandular syndrome type 1 (APS-1).

UPDATE ON GENETIC STUDIES OF AIH

AIH and the major histocompatibility complex (MHC)

Approximately 1000 human leucocyte antigens (HLA) have been identified to date. The MHC is encoded on a 4000 kbp portion of chromosome 6p21.3, it is characterized by considerable genetic polymorphism and is divided into three regions: MHC class I and II encode HLA A, B, Cw, Dr, Dq and DP; the MHC class II region encompasses several immune-reactive proteins including the complement proteins C2, C4A, CaB, the heat-shock protein (HSP-70) family, tumour necrosis factor (TNF) α and β, and MHC class I chain-related proteins (MICA, MICB). Interest in HLA association in AIH stems from the fact that the molecular structure and variation of the MHC-peptide alpha-helical region at the floor of the peptide-binding groove determines antigen presentation to the T cell receptor, and that most of the interindividual variability in HLA alleles is relevant to the amino acid sequence of this peptide-binding groove. HLA A1 and B8 association and AIH was reported by Mackay and Morris in 1972[45]. Genetic variability is not limited to the HLA I and II genes but equally affects TNF-α and TNF-β, complement genes, MICA, and MICB. Although the functional changes of TNF gene promoter polymorphisms are not clear the role of the molecule in inflammation, cell death, apoptosis and the up-regulation of MHC expression makes it an interesting candidate gene.

Association of HLA A1, Cw7, B8 and DR3 – which is inherited as a haplotype – as well as DR4 with AIH and other autoimmune diseases, has been conclusively demonstrated in a number of studies[46,47]. In turn an association with HLA B, Cw, and TNF-α have been found not to be the major factor. Studies from Europe and the USA have identified DRB1*0301 and DRB1*0401 as susceptibility alleles, and DRB1*1501 as a resistance allele[48,49]. However, immunogenetic findings appear not to apply universally, and it was noted that significant geographic differences exist. In Japan DR2 (DRB1*1501) is a weak susceptibility rather than a resistance allele[50] and in South American children DRB1*1301 is a strong susceptibility allele[51] not found in any of the other studies. Molecular comparisons of amino acid residues at the alpha helical binding groove region of the HLA molecule have suggested that in Japanese patients histidine at position 13; in US and European as well as Japanese, Mexican and Argentinian patients lysine at position 71; and for South American children valine at position 86 appear to confer AIH risk. These data illustrate that genetic association varies in study populations. A number of explanations may account for this finding. An exogenous factor present in a distinct population may be necessary (molecular footprint) which is in line with the current hypothesis of environmental trigger and genetic susceptibility, but the model may just be too simple altogether in its assumption of the relevance of single amino acid residue differences. AIH is not likely to be a monogenetic or oligogenetic disease. It is obvious that a polygenetic profile of factors which remains to be elucidated will define predisposition for this disease.

Non-MHC genes and AIH

The CD152 (cytotoxic lymphocyte antigen-4, CTLA-4) molecule on immune regulatory (CD25-positive) T cells interacts with CD80 and CD86 on the antigen-presenting cell with up to 50-fold higher affinity than CD28. Co-recognition of CD152 results in a reduction of the immune response. The CTLA-4 gene has been shown to exhibit more than 16 single nucleotide polymorphisms, and the CTLA-4 A+49G allele was found to be associated with diabetes, primary biliary cirrhosis and autoimmune thyroiditis. An association with AIH makes biological sense, and is indeed the strongest non-MHC association yet[52]. The study of interleukin 1 and 10 did not reveal associations. Variants of the vitamin D receptor, associated with a number of autoimmune diseases as well as with the tyrosine phosphatase CD45, showed a week association with AIH[53,54].

AIH in the autoimmune polyendocrine syndrome (APS) type 1: a model disease?

The APS-1 syndrome is characterized by a number of autoimmune disorders involving endocrine and non-endocrine organs including mucocutaneous candidiasis, hypoparathyroidism and adrenal insufficiency (establishing the diagnosis when two of the latter are present)[55]. In 10% of patients AIH is present. APS-1 has greatly increased our understanding of autoimmune diseases since it has a monogenic association with mutations in the autoimmune regulator (*AIRE*) gene. AIRE is expressed in medullary epithelial cells of the thymus accounting for less than 01.% of thymic cells[56]. The transcription factor encoded by the *AIRE* gene regulates the expression of a multitude of antigens required for the negative selection of autoreactive T cells in the thymus. In AIRE-deficient mice less autoantigen is expressed in thymic medullary epithelial cells, resulting in a higher number of higher reactive T cells in the periphery, which contributes to the establishment of autoimmune disease[57]. AIH in APS-1 syndrome leads to the formation of autoantibodies against CYP1A2 and CYP2A6. AIH can be the first clinically apparent component of this syndrome, in particular in children. However, the retrospective analysis of adult patients with AIH has not detected an increased frequency of variant *AIRE* alleles[58].

UPDATE ON THE CLINICAL PRESENTATION AND TREATMENT OF AIH

AIH is a remarkably well-treatable chronic liver disease[1]. In view of a potentially excellent prognosis with the induction of disease remission, which is more favourable than that of any other chronic inflammatory liver disease, the timely diagnosis and an efficient therapy represent a continuous challenge for clinical hepatologists. Untreated, the prognosis of active AIH is dismal with 5- and 10-year survival rates between 50% and 10% and a well-recognized therapeutic effect exemplified by the last placebo-controlled treatment trial reported in 1980[59,60]. In comparison to chronic hepatitis C (HCV) and B (HBV) AIH is frequently reported to represent a rare disease. However, based on limited epidemiological data, it is believed to account for between 50 and 200 cases per million in Western countries, and is diagnosed in as much as 20% of all cases of chronic hepatitis. In contrast to chronic HCV and HBV infection the diagnosis of AIH is reached by exclusion rather than by a single sensitive and specific diagnostic parameter, which considerably confounds the possibility of an accurate assessment of prevalence. A serological clue pointing to AIH are autoantibodies and hypergammaglobulinaemia (i.e. immunoglobulin G). However, serum autoantibodies in AIH are characterized by heterogeneity but do not carry a significant relevance for differences in therapeutic strategy[2]. With the exception of SLA/LP none of these autoantibodies is very specific for AIH. They occur in other conditions including rheumatological diseases[61], drug-allergic reactions[32], virus-associated autoimmunity[13] and genetic autoimmune disease[55]. However, from the point of view of clinical presentation, serum autoantibodies are also indicative of heterogeneity in AIH[2]. LKM-positive and also anti-liver cytosol type 1 (LC1)-positive[62] AIH is more prevalent among younger patients and is characterized by higher inflammatory activity, higher bilirubin and aminotransferase levels, a higher rate of acute presentation and an increased prevalence of cirrhosis at diagnosis[63,64]. In contrast, ANA-positive AIH has been designated by some authors a disease of elderly patients[65] although it is undoubtedly also prevalent in children[63,64]. Viewed from a historical perspective AIH was first described in young jaundiced women by Waldenstrom et al.[66] and Kunkel et al.[67], and subsequently defined as lupoid hepatitis and thereby coined as a separate disease entity by Mackay et al.[68]. This clinical description of a disease of young women was later expanded to older (male and female) patients and described with a bimodal occurrence[69], which acknowledges patients presenting with AIH not only between 10 and 30 years of age but also later than 40–50 years of age.

This descriptive view of AIH leaves a number of questions open. One question is the issue surrounding the presentation of AIH. The presentation of acute hepatitis, clinical symptoms of jaundice, abdominal pain and malaise has a high likelihood of attracting medical attention and subsequently leading to the diagnosis of AIH[70]. More subtle courses of AIH may not lead to clinically relevant signs and may develop unnoticed other than by routine work-up for other problems, or by screening programmes. The question of disease onset in terms of an initiation of the immune-mediated liver disease versus the clinical consequences that become noticeable after an unknown period of disease progression cannot be easily resolved. Thus, 'late-onset' AIH may just simply reflect a less severe course of the disease with a slower progression to cirrhosis. While LKM-positive patients display a tendency towards an earlier presentation, both acute and subtle (earlier and late presentation) variants appear to exist in ANA-positive AIH. In practice the diagnostic dilemma is that AIH is still perceived by many as a disease of younger individuals, and that therefore this differential diagnosis is less frequently considered in elderly patients with 'cryptogenic' hepatitis or cirrhosis. Another relevant question resulting from these considerations is the issue of treatment. Standard therapy in AIH consists of steroids alone or in combination with azathioprine. In maintenance therapy azathioprine monotherapy can also be implemented, but induction with azathioprine alone is not effective.

One way to reduce steroid side-effects is the administration of budenoside, which shows a high first-pass metabolism in the liver and is therefore likely to reduce systemic steroid side-effects. An initial study performed with difficult-to-treat patients, budesonide given in conjuction with prior immunosuppression in 10 patients, was reviewed pessimistically[71]. In a later study with 12 patients monotherapy was administered in naive AIH patients and showed 58% complete remissions[72], indicating a role beyond maintenance therapy. The authors noted that portosystemic shunts increased the likelihood of side-effects and also demonstrated higher AUC values in patients with higher grades of inflammation. This is in line with the expected pharmacokinetic profile of the drug. At present, a large multicentre trial with over 200 patients is ongoing that is likely to clarify the future role of budesonide therapy in AIH.

UPDATE ON ELDERLY PATIENTS AND AIH

From a general standpoint most internists will use caution when administering steroids to elderly patients, especially in women in whom osteopenia or diabetes may be present. Recommendations for the treatment of AIH suggest that the steroid side-effects be weighed against the potential benefit of therapy, and that consequently not all patients with AIH are good candidates for steroid treatment[1]. However, controversy exists surrounding the benefit of therapy in this group of elderly patients. One series reported 12 patients aged over 65 years out of 54 AIH patients. Cirrhosis had developed after follow-up in 26% irrespective of age, although the histological grade of AIH activity was more severe in the elderly group. Forty-two per cent of the patients over 65 did not receive therapy, yet deaths were reported only in the younger group[73]. In a different series of 20 patients aged >65 years, a longer time to the establishment of the diagnosis (8.5 vs 3.5 months) was reported; patients presented mainly with jaundice and acute-onset AIH and showed a response rate to immunosuppression comparable to that of younger patients[74]. The authors also noted that the prevalence of the HLA-A1-B8 allotype was less frequent in older patients, suggesting a role for immunogenetics. This point was further elaborated by a recent report analysing 47 patients with ANA-positive AIH aged 60 years and older, as well as 31 patients of 30 years and younger in whom $DR4^{+}/DR3^{-}$ prevalence was 47% (older) versus 13% (younger) patients[75]. In older patients steroid responsiveness was better, which is in line with previous findings in the same collective[76]. Cirrhosis and extrahepatic immune-mediated syndromes, including thyroid and rheumatological disease (47% vs 26%), were more prevalent in older AIH patients. However, although more treatment failures were observed in younger patients (24% vs 5%) the rate of remission, sustained remission and relapse was similar. Interestingly, an assessment of age-stratified prevalence showed an increase after the age of 40 from 15% to over 20%. From all of these data AIH in elderly patients appears to be characterized by a distinct clinical feature, a distinct immunogenetic profile, favourable response rates and higher rates of cirrhosis present at diagnosis, which contributes to the heterogeneity of AIH.

Recently, a study presented data that contribute to the further elucidation of the aforementioned issues surrounding AIH in elderly patients[77]. In a cohort of 164 patients from the UK, including 43 individuals 60 years and older, AIH characterized by ANA, SMA and LKM is analysed. Not surprisingly, LKM-positive AIH was found only in the younger (group 2) patients (7%), data on SLA/LP-positive patients were not reported or available. At first glance the groups show no significant differences regarding serum biochemistry, autoantibody titres, time to establishment of diagnosis, or mode of presentation. The authors provide a substratification of patients below and above 40 years of age, and find that older patients had a higher median histological stage and a comparable median grade, but younger patients had more median relapse episodes and a higher median stage at follow-up biopsy. The most distinguishing clinical sign was a higher prevalence of ascites in the older group. However, rates of complete, partial and failed response were similar, and the median number of relapses higher in younger patients, which nevertheless did not lead to differences in liver-related deaths in both groups (12% vs 15%). In comparison to the study of ANA-positive AIH patients from the USA[75] the differing findings regarding HLA association are notewothy. In the UK study there was no differential distribution of HLA DR3 and DR4, and this questions the suggested hypothesis of a primary influence of immunogenetics on the observed clinical distinctions. The presence of LKM-positive AIH is an unlikely explanation because of the low number of patients. Referral bias in two tertiary-care centres may have had an impact. However, both studies agree with respect to most of the clinical findings, including the differential prevalence of extrahepatic immune-mediated disease. One can speculate about the reasons for the clinical differences of AIH in older and younger patients. The authors discuss differences in hepatic blood flow, but alterations involving the regulation of cellular immunity during ageing are also a likely pathophysiological reason[78,79].

UPDATE ON HEPATOCELLULAR CARCINOMA AND AIH

Two additional points require consideration: one is that of hepatocellular carcinoma, which was diagnosed in 10.4% of cirrhotic patients and in 5% of the overall cohort, contrasting previous reports from the same institution[77]. Since viral disease was excluded from the analysis this leaves room for speculation regarding the effects of azathioprine or other external factors that may have escaped the medical work-up. The other point is that of fibrosis progression. Despite a rate of complete response of over 90% in both groups there was a clear increase of histological grade in both groups under treatment. Although sampling error may play a role this does suggest that therapeutic monitoring in AIH, irrespective of age, is a critical issue, which appears to be only incompletely represented by biochemistry and clinical presentation.

What are the practical clinical lessons regarding older patients? Older patients present with AIH with an increased prevalence after 40 years of age and require treatment which has a favourable outcome. Whether this represents *de-novo* initiation of AIH or clinical presentation of previously ongoing asymptomatic AIH is scientifically interesting and remains unresolved, but should not influence decision-making to initiate therapy[80]. Clinical assessment should consider that older patients present with more extrahepatic immune-mediated disease components, which may be related to altered regulation of cellular immunity during ageing and possibly to immunogenetics. However, the main point is a wary clinical work-up with an early diagnosis and the prospect of an effective therapeutic intervention even in older AIH patients.

References

1. Manns MP, Strassburg CP. Autoimmune hepatitis: clinical challenges. Gastroenterology. 2001;120:1502–17.
2. Strassburg CP, Manns MP. Autoantibodies and autoantigens in autoimmune hepatitis. Semin Liver Dis. 2002;22:339–52.
3. Donaldson PT, Czaja AJ. Genetic effects on susceptibility, clinical expression, and treatment outcome of type 1 autoimmune hepatitis. Clin Liver Dis. 2002;6:419–37.

4. Strassburg CP, Obermayer-Straub P, Manns MP. Autoimmunity in liver diseases. Clin Rev Allergy Immunol. 2000;18:127–39.
5. Strassburg CP, Manns MP. Autoimmune tests in primary biliary cirrhosis. Baillieres Best Pract Res Clin Gastroenterol. 2000;14:585–99.
6. Tan MT. Autoantibodies in pathology and cell biology. Cell. 1991;67:841–2.
7. Toh BH. Smooth muscles autoantibodies and autoantigens. Clin Exp Immunol. 1979;38: 621–8.
8. Czaja AJ, Cassani F, Cataleta M, Valentini P, Bianchi FB. Frequency and significance of antibodies to actin in type 1 autoimmune hepatitis. Hepatology. 1996;24:1068–73.
9. Obermayer-Straub P, Strassburg CP, Manns MP. Target proteins in human autoimmunity: cytochromes P450 and UDP-glucuronosyltransferases. Can J Gastroenterol. 2000;14:429–39.
10. Tukey RH, Strassburg CP. Human UDP-glucuronosyltransferases: metabolism, expression, and disease. Annu Rev Pharmacol Toxicol. 2000;40:581–616.
11. Zimmerman HJ, Lewis JH, Ishak KG, Maddrey W. Ticrynafen-associated hepatic injury: analysis of 340 cases. Hepatology. 1984;4:315–23.
12. Homberg JC, Abuaf N, Helmy-Khalil S et al. Drug induced hepatitis associated with anticytoplasmic organelle autoantibodies. Hepatology. 1985;5:722–5.
13. Strassburg CP, Obermayer-Straub P, Alex B et al. Autoantibodies against glucuronosyltransferases differ between viral hepatitis and autoimmune hepatitis. Gastroenterology. 1996;111:1576–86.
14. Czaja AJ, Manns MP, Homburger HA. Frequency and significance of antibodies to liver/kidney microsome type 1 in adults with chronic active hepatitis. Gastroenterology. 1992; 103:1290–5.
15. Homberg JC, Abuaf N, Bernard O et al. Chronic active hepatitis associated with anti liver/kidney microsome type 1: a second type of 'autoimmune' hepatitis. Hepatology. 1987;7: 1333–9.
16. Gelpi C SE, Rodriguez-Sanchez JL. Autoantibodies against a serine tRNA-protein complex implicated in cotranslational selenocysteine insertion. Proc Natl Acad Sci USA. 1992;89:9739–43.
17. Manns M, Gerken G, Kyriatsoulis A, Staritz M, Meyer zum Buschenfelde KH. Characterisation of a new subgroup of autoimmune chronic active hepatitis by autoantibodies against a soluble liver antigen. Lancet. 1987;1:292–4.
18. Volkmann M, Martin L, Baurle A et al. Soluble liver antigen: isolation of a 35-kd recombinant protein (SLA-p35) specifically recognizing sera from patients with autoimmune hepatitis. Hepatology. 2001;33:591–6.
19. Wies I, Brunner S, Henninger J et al. Identification of target antigen for SLA/LP autoantibodies in autoimmune hepatitis [see comments]. Lancet. 2000;355:1510–15.
20. Czaja AJ, Donaldson PT, Lohse AW. Antibodies to soluble liver antigen/liver pancreas and HLA risk factors for type 1 autoimmune hepatitis. Am J Gastroenterol. 2002;97:413–19.
21. Ma Y, Okamoto M, Thomas MG et al. Antibodies to conformational epitopes of soluble liver antigen define a severe form of autoimmune liver disease. Hepatology. 2002;35:658–64.
22. Clemente MG, Meloni A, Obermayer-Straub P, Frau F, Manns MP, DeVirgiliis S. Two cytochromes P450 are major hepatocellular autoantigens in autoimmune polyglandular syndrome type 1. Gastroenterology. 1998;114:324–8.
23. Dalekos GN, Obermayer-Straub P, Maeda T, Tsianos EV, Manns MP. Antibodies against cytochrome P4502A6 (CYP2A6) in patients with chronic viral hepatitis are mainly linked to hepatitis C virus infection. Digestion. 1998;59:S36.
24. Bourdi M, Larrey D, Nataf J et al. Anti-liver endoplasmic reticulum antibodies are directed against human cytochrome P-450IA2. A specific marker of dihydralazine hepatitis. J Clin Invest. 1990;85:1967–73.
25. Manns MP, Griffin KJ, Quattrochi L et al. Identification of cytochrome P450 IA2 as a human autoantigen. Arch Biochem Biophys. 1990;280:229–32.
26. Sacher M, Blümel P, Thaler H, Manns M. Chronic active hepatitis associated with vitiligo, nail dystrophy, alopecia and a new variant of LKM antibodies. J Hepatol. 1990;10:364–9.
27. Bourdi M, Gautier JC, Mircheva J et al. Anti-liver microsomes autoantibodies and dihydralazine induced hepatitis: specificity of autoantibodies and inductive capacity of the drug. Mol Pharmacol. 1992;42:280–5.

28. Van Pelt F, Straub P, Manns M. Molecular basis of drug-induced immunological liver injury. Semin Liver Dis. 1995;15:283–300.
29. Crivelli O, Lavarini C, Chiaberge E et al. Microsomal autoantibodies in chronic infection with HBsAg associated delta (delta) agent. Clin Exp Immunol. 1983;54:232–8.
30. Kenna JG, Neuberger JM. Immunopathology and treatment of halothane hepatitis. Clin Immunother. 1995;3:108–24.
31. Pessayre D. Toxic and immune mechanisms leading to acute and subacute drug induced liver injury. In: Miguet JP, Dhumeaux D, editors. Progress in Hepatology. Paris: John Libbey Eurotext, 1993:23–39.
32. Beaune P, Dansette PM, Mansuy D et al. Human anti-endoplasmic reticulum autoantibodies appearing in a drug-induced hepatitis are directed against a human liver cytochrome P-450 that hydroxylates the drug. Proc Natl Acad Sci USA. 1987;84:551–5.
33. Eliasson E, Kenna G. Cytochrome P450 2E1 is a cell surface autoantigen in halothane hepatitis. Mol Pharmacol. 1996;50:573–82.
34. Leeder JS, Riley RJ, Cook VA, Spielberg SP. Human anti-cytochrome P450 antibodies in aromatic anticonvulsant-induced hypersensitivity reactions. J Pharmacol Exp Ther. 1992; 263:360–7.
35. Clot P, Albano E, Eliasson E et al. Cytochrome P450 2E1 hydroxyethyl radical adducts as the major antigen in autoantibody formation among alcoholics. Gastroenterology. 1996; 111:206–16.
36. Strassburg CP, Obermayer-Straub P, Manns MP. Autoimmunity in hepatitis C and D virus infection. J Viral Hepatol. 1996;3:49–59.
37. Yamamoto AM, Cresteil D, Homberg JC, Alvarez F. Characterization of anti-liver–kidney microsome antibody (anti-LKM1) from hepatitis C virus-positive and -negative sera. Gastroenterology. 1993;104:1762–7.
38. Ma Y, Peakman M, Lobo-Yeo A et al. Differences in immune recognition of cytochrome P4502D6 by liver kidney microsomal (LKM) antibody in autoimmune hepatitis and chronic hepatitis C virus infection. Clin Exp Immunol. 1994;97:94–9.
39. Muratori L, Lenzi M, Ma Y et al. Heterogeneity of liver/kidney microsomal antibody type 1 in autoimmune hepatitis and hepatitis C virus related liver disease. Gut. 1995;37:406–12.
40. Yamamoto AM, Johanet C, Duclos-Vallee JC et al. A new approach to cytochrome CYP2D6 antibody detection in autoimmune hepatitis type-2 (AIH-2) and chronic hepatitis C virus (HCV) infection: a sensitive and quantitative radioligand assay. Clin Exp Immunol. 1997;108:396–400.
41. Durazzo M, Philipp T, Van Pelt FN et al. Heterogeneity of liver–kidney microsomal autoantibodies in chronic hepatitis C and D virus infection. Gastroenterology. 1995;108: 455–62.
42. Sugimura T, Obermayer-Straub P, Kayser A et al. A major CYP2D6 autoepitope in autoimmune hepatitis type 2 and chronic hepatitis C is a three-dimensional structure homologous to other cytochrome P450 autoantigens. Autoimmunity. 2002;35:501–13.
43. Todros L, Touscoz G, D'Urso N et al. Hepatitis C virus-related chronic liver disease with autoantibodies to liver-kidney microsomes (LKM). Clinical characterization from idiopathic LKM-positive disorders. J Hepatol. 1991;13:128–31.
44. Dalekos GN, Wedemeyer H, Obermayer-Straub P et al. Epitope mapping of cytochrome P4502D6 autoantigen in patients with chronic hepatitis C during alpha-interferon treatment. J Hepatol. 1999;30:366–75.
45. Mackay IR, Morris PJ. Association of autoimmune chronic hepatitis with HLA-A1-B8. Lancet. 1972;2:793–5.
46. Donaldson PT, Doherty DG, Hayllar KM, McFarlane IG, Johnson PJ, Williams R. Susceptibility to autoimmune chronic active hepatitis: human leukocyte antigens DR 4 and A1-B8-DR-3 are independent risk factors. Hepatology. 1991;13:701–6.
47. Mackay IR, Tait BD. HLA association with autoimmune type chronic active hepatitis: identification of B8-DRw3 haplotype by family studies. Gastroenterology. 1980;79:95–8.
48. Doherty DG, Donaldson PT, Underhill JA et al. Allelic sequence variation in the HLA class II genes and proteins on patients with autoimmune hepatitis. Hepatology. 1994;19:609–15.
49. Strettell MD, Thomson LJ, Donaldson PT, Bunce M, O'Neill CM, Williams RW. HLA-C genes and susceptibility to type 1 autoimmune hepatitis. Hepatology. 1997;26:1023–6.

50. Ota M, Seki T, Kiyoshawa K et al. A possible association between basic amino acids of position 13 of DRB1 chains and autoimmune hepatitis. Immunogenetics. 1992;36:40–55.
51. Pando M, Larriba J, Fernandez GC et al. Pediatric and adult forms of type I autoimmune hepatitis in Argentina: evidence for differential genetic predisposition. Hepatology. 1999;30:1374–80.
52. Agarwal K, Czaja AJ, Jones DE, Donaldson PT. Cytotoxic T lymphocyteantigen-4 (CTLA-4) gene polymorphisms and susceptibility to type 1 autoimmune hepatitis. Hepatology. 2000;31:49–53.
53. Vogel A, Strassburg CP, Manns MP. Genetic association of vitamin D receptor polymorphisms with primary biliary cirrhosis and autoimmune hepatitis. Hepatology. 2002;35:126–31.
54. Vogel A, Strassburg CP, Manns MP. 77 C/G mutation in the tyrosine phosphatase CD45 and autoimmune hepatitis: evidence for a genetic link. Genes Immun. 2003;4:79–81.
55. Obermayer-Straub P, Perheentupa J, Braun S et al. Hepatic autoantigens in patients with autoimmune polyendocrinopathy–candidiasis–ectodermal dystrophy. Gastroenterology. 2001;121:668–77.
56. Pitkanen J, Vahamurto P, Krohn K, Peterson P. Subcellular localization of the autoimmune regulator protein. Characterization of nuclear targeting and transcriptional activation domain. J Biol Chem. 2001;276:19597–602.
57. Ramsey C, Winqvist O, Puhakka L et al. AIRE-deficient mice develop multiple features of APECED phenotype and show altered immune response. Hum Mol Genet. 2002;11:397–409.
58. Vogel A, Liermann H, Harms A, Strassburg CP, Manns MP, Obermayer-Straub P. Autoimmune regulator AIRE: evidence for genetic differences between autoimmune hepatitis and hepatitis as part of the autoimmune polyglandular syndrome type 1. Hepatology. 2001;33:1047–52.
59. Kirk AP, Jain S, Pocock S, Thomas HC, Sherlock S. Late results of the Royal Free Hospital prospective controlled trial of prednisolone therapy in hepatitis B surface antigen negative chronic active hepatitis. Gut. 1980;21:78–93.
60. Soloway RD, Summerskill WH, Baggenstoss AH et al. Clinical, biochemical, and histological remission of severe chronic active liver disease: a controlled study of treatments and early prognosis. Gastroenterology. 1972;63:820–33.
61. Tan EM, Chan EKL, Sullivan KF, Rubin RL. Antinuclear antibodies (ANAs): diagnostically specific immune markers and clues toward the understanding of systemic autoimmunity. Clin Immunol Immunopathol. 1988;47:121–41.
62. Bridoux-Henno L, Maggiore G, Johanet C et al. Features and outcome of autoimmune hepatitis type 2 presenting with isolated positivity for anti-liver cytosol antibody. Clin Gastroenterol Hepatol. 2004;2:825–30.
63. Gregorio GV, Portmann B, Reid F et al. Autoimmune hepatitis in childhood: a 20-year experience. Hepatology. 1997;25:541–7.
64. Oettinger R, Brunnberg A, Gerner P, Wintermeyer P, Jenke A, Wirth S. Clinical features and biochemical data of Caucasian children at diagnosis of autoimmune hepatitis. J Autoimmun 2005;24:79–84.
65. Parker DR, Kingham JG. Type I autoimmune hepatitis is primarily a disease of later life. Q J Med. 1997;90:289–96.
66. Waldenström J. Leber, Blutproteine und Nahrungseiweisse. Dtsch Gesellsch Verd Stoffw. 1950;15:113–19.
67. Kunkel HG, Ahrens EHJ, Eisenmenger WJ. Extreme hypergammaglobulinemia in young women with liver disease of unknown etiology. J Clin Invest. 1951;30:654.
68. Mackay I, R, Taft LI, Cowling DC. Lupoid hepatitis. Lancet. 1956;2:1323–6.
69. McFarlane IG. The relationship between autoimmune markers and different clinical syndromes in autoimmune hepatitis. Gut. 1998;42:599–602.
70. Nikias GA, Batts KP, Czaja AJ. The nature and prognostic implications of autoimmune hepatitis with acute presentation. J Hepatol. 1994;21:866–71.
71. Czaja AJ, Lindor KD. Failure of budesonide in a pilot study of treatment-dependent autoimmune hepatitis. Gastroenterology. 2000;119:1312–16.
72. Wiegand J, Schuler A, Kanzler S et al. Budesonide in previously untreated autoimmune hepatitis. Liver Int. 2005;25:927–34.

73. Newton JL, Burt AD, Park JB, Mathew J, Bassendine MF, James OF. Autoimmune hepatitis in older patients. Age Ageing. 1997;26:441–4.
74. Schramm C, Kanzler S, zum Buschenfelde KH, Galle PR, Lohse AW. Autoimmune hepatitis in the elderly. Am J Gastroenterol. 2001;96:1587–91.
75. Czaja AJ, Carpenter HA. Distinctive clinical phenotype and treatment outcome of type 1 autoimmune hepatitis in the elderly. Hepatology. 2006;43:532–8.
76. Czaja AJ, Carpenter HA, Santrach PJ, Moore SB. Significance of HLA DR4 in type 1 autoimmune hepatitis. Gastroenterology. 1993;105:1502–7.
77. Al-Chalabi T, Boccato S, Portmann BC, McFarlane IG, Heneghan MA. Autoimmune hepatitis (AIH) in the elderly: a systematic retrospective analysis of a large group of consecutive patients with definite AIH followed at a tertiary referral centre. J Hepatol. 2006;45:575–83.
78. Prelog M. Aging of the immune system: a risk factor for autoimmunity? Autoimmun Rev. 2006;5:136–9.
79. Talor E, Rose NR. Hypothesis: the aging paradox and autoimmune disease. Autoimmunity. 1991;8:245–9.
80. Strassburg CP, Manns MP. Autoimmune hepatitis in the elderly: what is the difference? J Hepatol. 2006;45:480–2.

12
Overlap syndromes

U. LEUSCHNER

PROBLEMS OF DEFINITION

At present there is no unequivocal definition of what we call overlap syndrome (OLS). An OLS could be defined as the coexistence of two different autoimmune liver diseases in one single patient. The two diseases might be detected simultaneously or successively during the course of the disease. Since OLS are rare, nobody has ample experience with respect to characterization, diagnosis and treatment. This means that the following data are based on investigations performed in small patient numbers only, and the picture may change when more information becomes available.

We know three well-defined autoimmune liver diseases, namely autoimmune hepatitis (AIH), primary biliary cirrhosis (PBC) and primary sclerosing cholangitis (PSC), although according to recent investigations there are some doubts as to whether PSC is an autoimmune liver disease at all[1]. For example, PSC does not have organ-specific autoantibodies, there is no female preponderance and it does not respond to corticosteroid therapy.

When we consider all data available for autoimmune liver disease we can stratify AIH into classic AIH, classic AIH with peculiar features (e.g. acute onset, spontaneous periods of remissions, relapses after therapy and others), and variant forms. Variant forms are AMA-positive AIH, AIH with cholestasis or AIH with bile duct injury. A similar classification has been proposed for PBC, e.g. absence of typical histological features or existence of piecemeal necrosis, and PSC, e.g. small duct cholangitis, and finally we have two rather well-defined overlap syndromes[2] (Table 1), the AIH/PBC- and the AIH/PSC-OLS.

This means that we have three autoimmune liver diseases, AIH, PSC and PBC, but 18 peculiar and variant forms. This great variety of data, unspecific and not even typical for each of the three liver diseases, raises the question of what OLS means? Does this mixture of data represent different entities, worthwhile of having an extra name, or is it nothing else but a mixture of symptoms, histological, biochemical and serological findings which can be diagnosed in great abundance due to better and more sensitive methods? In order to make the definition of AIH easier, and to discriminate AIH from OLS

Table 1 Autoimmune liver diseases

Classic form of AIH	
Peculiar features of classic AIH (acute, insidious, fluctuating, relapsing, etc.)	
Variant forms of AIA	
AIH + AMA	AIH + bile duct injury
AIH + cholestasis	AIH without ANA, SMA, LKM
AIH + ulcerative colitis	AIH resistant to immunosuppressants
Classic form of PBC	
Peculiar features of classic PBC (missing histological features, piecemeal necrosis, etc.)	
Variant forms of PBC	
PBC–AMA negative	PBC without biochemical cholestasis
Classic and variant forms of PSC (different antibody spectrum, small duct PSC)	
AIH/PBC-OLS	
AIH/PSC-OLS	
PBC/PSC-OLS (very rare condition)	

(particularly from AIH/PBC and AIH/PSC), the International Autoimmune Hepatitis Group (IAIHG) suggested a scoring system which can help to differentiate AIH from other autoimmune liver conditions. The IAIHG scoring system is suitable for scientific questions but unsuitable for clinical routine[3].

OVERLAP SYNDROMES

Although there is still no generally agreed nomenclature of liver diseases called OLS, according to data available in the year 2006 we are able to describe two conditions presently named OLS: the most commonly reported AIH-PBC-OLS and the AIH-PSC-OLS. Furthermore, we have two conditions which cannot be classified, namely autoimmune cholangitis (AIC) and cryptogenic chronic hepatitis and, to make things even more complicated, in some patients diagnosis can change during the course of the disease (Table 2).

How often do we find so-called OLS or atypical manifestations of AIH, another term for OLS? The AIH-PBC-OLS has been observed in about 8% and the AIH-PSC-OLS in about 6%; the two unclassifiable conditions AIC and cryptogenic hepatitis in about 10% and 13%. This means that OLS are rare conditions, but we have to establish the correct diagnosis, because different treatment regimens will probably be necessary[4,5].

Which tools do we have to characterize and diagnose an OLS? What do we know about the aetiology; can genetics be helpful; which biochemical, serological and histological features help to characterize an OLS; do different OLS respond differently to treatment?

The aetiology of the OLS is unclear. Pathogens or toxins, for example, could induce two different ALD, which simply overlap; or do pathogens or toxins directly induce the development of the OLS as a specific entity? This question cannot be answered at present, because we have no clear picture of the aetiology of any of the autoimmune liver diseases (Figure 1).

Table 2 Atypical manifestations of autoimmune hepatitis

Overlap syndromes (two diseases in one liver)
AIH-PBC
AIH-PSC
AIH-PBC = AIC
AIH/HepC, B?
Outlier syndromes (cannot be classified)
AIC (AMA-negative PBC?)
Cryptogenic chronic hepatitis
Changing diagnosis (switch over time)
PBC → AIH
AIH → PSC

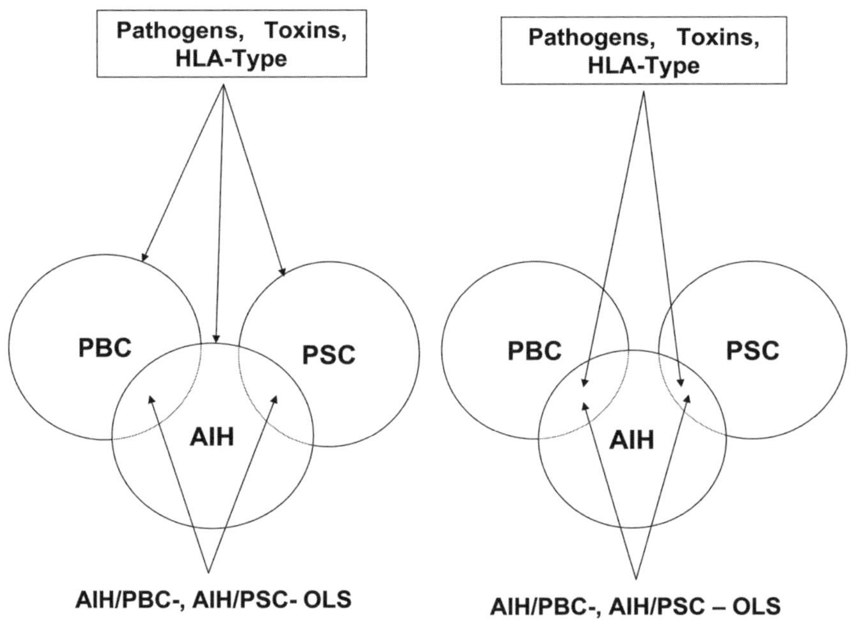

Figure 1 Aetiology of OLS. Left: pathogens or toxins induce two different diseases, which overlap. Right: pathogens or toxins directly induce OLS

Table 3 Genetics of autoimmune liver disease

Disease	*HLA association*	*Characteristics*
AIH type 1 f>m	DR3	Early onset, severe disease
	DR4	Older-onset, extrahepatic manifestations, good response to therapy, protective in children?
	DR2	Alleles in Japan
	DR4	See above
AIH type 2	–	–
PBC	DR8	Stage III, IV 24%, stage I, II 5%, marginally significant
AIC	DR3, DR4	Has to be confirmed
PSC	B8-DR3, DR2, DR6	Strong with B8-DR3, negative with DR4
PSC/IBD		Little genetic overlap between genes

Could immunogenetic investigations help to identify and to classify an OLS? It is well known that AIH is associated with the two susceptibility alleles HLA DR3 and DR4[6]; DR3 is responsible for early onset, more likely to relapse and a severe course of the disease, DR4 for later onset and good response to immunosuppressive therapy[7–9]. For the chronic cholestatic autoimmune diseases it has been shown that PBC is associated with HLA DR8, but only in the late stages III and IV, and even this association is only marginally significant[10–12]. PSC seems to be strongly associated with B8-DR3, to a lesser extent with DR2 and DR6 and negatively with DR4[13,14] (Table 3).

This means, as suggested in 1999, that a genetic susceptibility to develop an autoimmune liver disease or an OLS seems to exist, at least for the AIH-PBC-OLS. If DR3 or DR4 predominates, a more hepatitic picture of the OLS with an interface hepatitis will develop, otherwise a more cholestatic course will be observed[15]. In conclusion, immunogenetics could be helpful, but are not suitable for daily routine.

AIH-PBC-OLS

How do routine laboratory parameters and histology help to characterize and diagnose an OLS?

The AIH-PBC-OLS is characterized by biochemical and histological features of both AIH and PBC. In these patients we find AMA type 2[16,17], directed against PDC-E2 (IFT, ELISA, Western blotting), located on the inner mitochondrial membrane, IgG is rather low, IgM high; furthermore we find ANA and SMA, although in lower concentrations[5]. Histologically the OLS is characterized by the same alterations as seen in PBC and AIH: destructive bile

Table 4 AIH-PBC-OLS

Histology: criteria of PBC+AIH
Serology: AMA-M_2 (PDC-E_2)
Proposed definition[16]
AIH: ALT 4 times, IgG 2 times, SMA, histology
PBC: AP 2 times, or GGT 5 times, AMA, histology
IgG rather low, IgM higher concentrations
ANA, SMA lower titres than in AIH
HLA: DR3, DR4

duct injury, typical for PBC, infiltrates in the portal tract and the parenchyma, interface hepatitis and granulomas as seen in AIH. In about 30% of patients these characteristic markers may be absent, and the pathologist describes alterations ‘compatible with’ AIH and PBC. It has to be stated that the histological findings are not specific for either AIH or PBC, because some of them are also found in non-alcoholic steatohepatitis, alcoholic or drug hepatitis, chronic viral hepatitis and other liver diseases. Immunogenetically the AIH-PBC-OLS seems to be predominantly associated with DR3 and DR4, as is AIH (Table 4), while DR8, which has been found in stage III and IV of PBC, is absent. Nevertheless, today some authors believe that the AIH-PBC-OLS should rather be regarded as part of primary biliary cirrhosis than of AIH[18,19].

AUTOIMMUNE CHOLANGITIS

There is a second variant of this OLS; this variant is AMA-negative with histological features of PBC only, and not of AIH[18]. This means there is no histological over-lapping. It is not quite clear what these findings signify; therefore this condition together with cryptogenic chronic hepatitis is classified as an outlier syndrome. This condition is called autoimmune cholangitis (AIC) by some authors; others prefer the term AMA-negative PBC, others call it the second variant of an AIH-PBC-OLS[20,21]. At present AMA-negative PBC is the term most often used because there are no histological differences between the AMA-negative and positive variant, and because, as in the classic PBC, PDC-E2 is expressed on bile duct cells, the course of the disease is identical to that of classic PBC and both respond to UDCA therapy[22]. Some authors call AIC a separate disease, because of rather high titres of ANA and SMA, because the PBC-specific antibodies to the 2-oxoglutarate-dehydrogenase, the branched-chain 2-oxoacid dehydrogenase- and the PDC-E2-complex are absent and concentrations of the routine laboratory data differ from those of classic PBC. AST and IgM concentrations seem to be lower than in AMA-positive cases. Furthermore it has been shown that immunogenetically this second variant of the OLS is associated with DR3 or DR4 as seen in AIH, but not DR8, characteristic for PBC[23].

CRYPTOGENIC CHRONIC HEPATITIS

In 15–20% of patients with chronic liver disease AIH-typical autoantibodies such as ANA, SMA and LKM are missing[24]. Some 15–20% of patients express SLA/LP, LC1 or pANCA autoantibodies instead[25]. This condition is called cryptogenic chronic hepatitis, but because of similarities concerning sex, age distribution, HLA alleles and associated extrahepatic immune diseases cryptogenic hepatitis is a variant form of AIH and not an OLS.

AIH-PSC-OLS

The second and better-defined OLS is the AIH-PSC-OLS. Morphologically we have the picture of PSC (histology, ERC, MRC), IgG, ANA and SMA suggest AIH, and like PSC this OLS is associated with a clinical or subclinical course of inflammatory bowel disease, mainly ulcerative colitis[26–28]. There are three reports on the AIH-PSC-OLS in adults, having applied the old and new IAIHG scoring system[26,29,30], but children with either AIH or PSC have a 50% greater chance of developing an AIH-PSC-OLS[27,28]. Figure 2 shows the data of a 39-year-old man; in the ERC we find the typical picture of PSC, corroborated by liver biopsy. In the upper right corner interface hepatitis is present, in the lower part of the picture lobular infiltration, and in the middle (second biopsy) bile duct obstruction can be seen.

In children AIH as well as PSC are rare conditions, but when they exist they present with the picture of an AIH-PSC-OLS, which in paediatrics is also called autoimmune sclerosing cholangitis (ASC). ASC responds well to immunosuppressive therapy[27], but although liver function tests improve, the histological process may progress and progression of bile duct abnormalities has been observed in about 50% of ASC patients by retrograde cholangiograms[31].

AIH/HEP C-, B-, D-OLS

Coming back to the question what an OLS is, I would like to briefly discuss the question of whether an AIH/Chron Hep C-, B- or D-OLS exists. It is possible that, during the coincidental appearance of AIH and chronic viral hepatitis, AIH predominates over HCV or HBV infection. On the other hand, viral hepatitis may predominate over AIH. Is that an OLS, an unspecific stimulation of the immune system or rather comorbidity?

Does an AIH/Hep C-OLS exist? In patients with HCV infection we have several markers of autoimmunity[32–34], and in patients with AIH we have findings typical for viral hepatitis. Is that non-specific? There is no female preponderance and only a weak association with HLA DR3 and DR4 (Table 5). Therefore most investigators believe that an AIH/Hep C-OLS does not exist[35]. But why, for example, in patients with AIH where markers of chronic hepatitis C but not of chronic hepatitis B are so frequent, is there a difference between the two chronic viral infections with respect to overlapping? Since

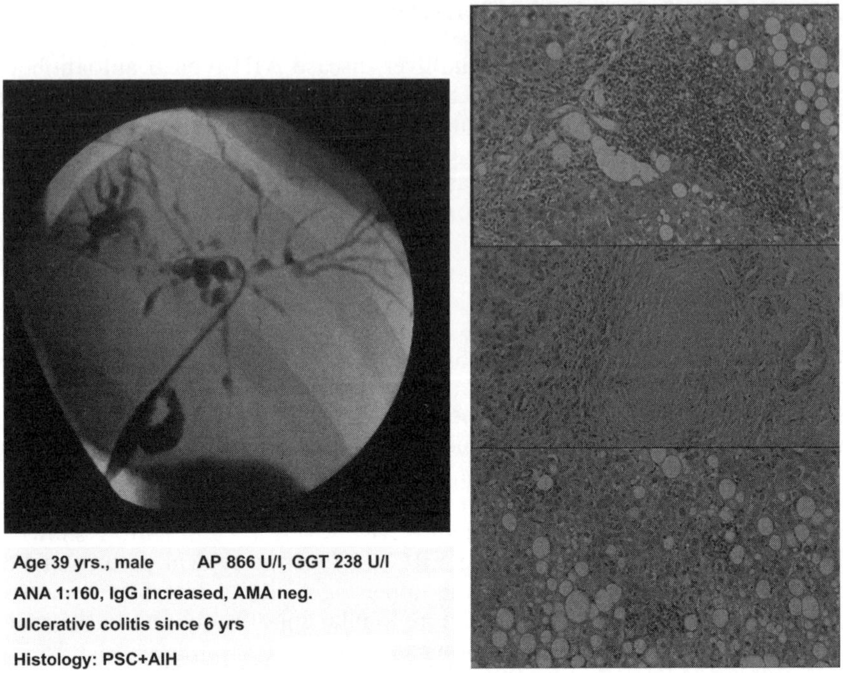

Figure 2 Findings of a PSC/AIH-OLS in a 39-year-old male patient (liver histology see text)

Table 5 Hepatitis C/B/D – AIH-OS: does it exist?

In patients with chronic HCV infection					
ANA:	9–38%	p-ANCA:	10%		
ASMA:	5–91%	AAA:	8%	HMG1,2 AB:	11%
LKM1,3:	0–10% (different epitopes from AIH)				
Hypergammaglobulinaemia					
Moderate/severe interface hepatitis, lobular infiltrates, granulomatous					
In patients with AIH					
HCV-AB:	44–77% (EIA)				
	12–35% (RIBA)				
HCV RNA:	19%				
Histological characteristics of HCV hepatitis					
But: No female preponderance, weak association with DR3, DR4					

there is no exact definition of what an OLS is, and the aetiology of autoimmune liver diseases is unknown, we cannot exclude that many more OLS may exist than we presently believe.

At present we are treating a 40-year-old female patient who has suffered from chronic hepatitis B since 1999. She was HBs-, HBe- and anti-HBc-positive and had a rather high virus load. The diagnosis of chronic hepatitis B had been confirmed by liver biopsy. She was AMA-negative but had ANA of the multiple nuclear dot-type in high concentrations, typical for PBC. She was severely symptomatic. A second opinion given by two different pathologists was chronic hepatitis B and PBC. Therefore we started treatment with UDCA. Eleven months later she had serum conversion from HBe to anti-HBe, the virus load had dropped very significantly, the liver function tests normalized completely and symptoms disappeared. Is this a PBC/Hep B-OLS or comorbidity? Why did the virus marker respond to UDCA therapy? Was it spontaneous serum conversion? This case report again raises the question of whether OLS may occur between viral hepatitis and autoimmune liver diseases, in our case it was PBC.

TREATMENT OF OLS

What do we know about treatment of OLS? There are only limited studies available, but from the data one can foresee how treatment should be performed and that the response to a given therapy is not helpful to establish the diagnosis of an OLS.

AIC, or the so-called AMA-negative PBC, should be treated like classic PBC with UDCA in a dosage of 13–15 mg/kg daily. This is a common guideline, because AIC is believed not to be an OLS but a variant of classic PBC[22]. Is this correct? For 4 years we have been treating a 45-year-old female patient with histologically and biochemically proven AMA-negative but ANA-positive PBC. She had severe pruritus and xanthelasmas. She responded insufficiently to UDCA; therefore we added, one after another prednisolone, the non-steroidal anti-inflammatory drug sulindac and mycophenolate mofetil. Pruritus disappeared, but when we reduced the dose of one of the mentioned drugs liver biochemistries deteriorated and pruritus recurred. Is this really an AMA-negative PBC or is it an OLS which needs additional immunosuppressive therapy?

A small study, which has been corroborated by others[36,37], showed that out of 130 patients 12 had an AIH-PBC-OLS[16]. The response to UDCA alone, or to monotherapy with steroids, was incomplete in most patients, and the combination of the two substances was superior; therefore the treatment of choice for the AIH-PBC-OLS seems to be the combination of UDCA and glucocorticoids in usual doses.

There are no convincing data on the treatment of the AIH-PSC-OLS. Since a small treatment study with steroids had a poor outcome in 78%, and 33% of the patients died or had to be transplanted, steroid therapy is not recommended[5]. Patients should be treated with UDCA alone; when competent bile duct strictures develop drug treatment should be performed in combination with endoscopic bile duct dilation. Whether the treatment with bile salts in combination with azathioprine is superior to monotherapy has not been investigated sufficiently.

Assuming that an AIH/Chron Hep C-OLS exists and the SMA and/or ANA titres exceeded 1:360, if in addition hypergammaglobulinaemia and interface hepatitis can be found, immunosuppressive therapy is recommended[33,35,38]. When the titres are lower, antiviral treatment seems to be superior[33,38,39]. The same recommendation is given for comorbidity of chronic viral hepatitis and AIH or an unspecific stimulation of the immune system. In all three cases treatment has to be checked repeatedly, because therapy is more or less experimental.

In patients with OLS the treating physician has to monitor the course of the disease very carefully, and first to select, and later to adapt, treatment. Probably it will not be possible to give strict treatment recommendations since chronic liver diseases present with a variety of genetic, biochemical, serological and histological peculiarities which may change during the course of the disease.

SWITCH OF DIAGNOSIS

As already mentioned it has been shown that, in a few patients, diagnosis can change during the course of the disease. This has been observed for patients with PBC, who later developed AIH[16,40], and mainly for children with AIH, who developed PSC[41,42]. A rather high percentage of these children were male, and presented with inflammatory bowel disease (Figure 3).

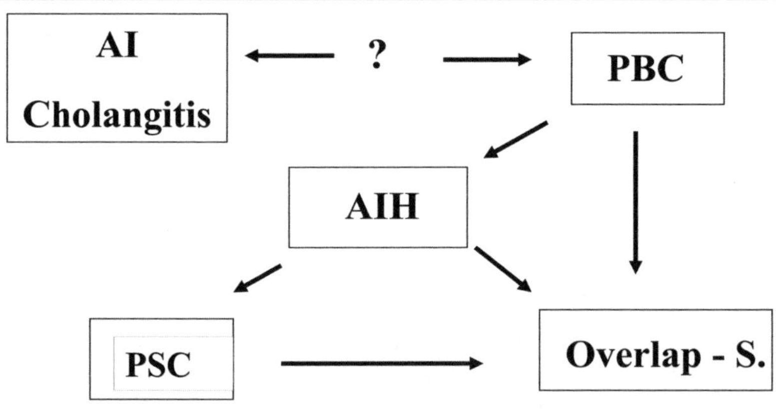

Figure 3 Changing diagnosis and OLS in chronic autoimmune liver diseases

Therefore again the question rises, what are and how should we define OLS? Three well-known autoimmune diseases of the liver, 18 peculiarities and variant forms, and finally changing diagnoses which influence both variations and peculiarities!

When I started to study clinical medicine in 1965, I learned from the legendary Ferdinand Hoff, Professor of Internal Medicine of the Frankfurt University in Germany, that in humans never does one system or one organ or one structure alone present with abnormal findings, and that a specific name of a disease, e.g. Wilson's disease, hepatitis, lupus erythematosus, nephritis, or rheumatism, is helpful only to physicians in order to communicate. Therefore, OLS are not recently discovered conditions, but were well known to physicians of past decades.

SUMMARY AND CONCLUSIONS

Overlap syndromes represent old concepts, which have been rediscovered due to more sensitive investigational methods. They are difficult to define and to diagnose. The correct diagnosis of an OLS is important, because they probably require new therapy concepts.

References

1. Chapman RW. Primary sclerosing cholangitis as an autoimmune disease: pros and cons. In: Immunology and Liver, Manns MP, Paumgartner G, Leuschner U, editors. Dordrecht: Kluwer, 2000:279–87.
2. Poupon R. Autoimmune overlapping syndromes. In: Heathcote EJ, editor. Clinics in Liver Disease, vol. 7, no. 4. Philadelphia: WB Saunders, 2003:865–78.
3. Alvarez F, Berg PA, Bianchi FB et al. International Autoimmune Hepatitis Group report: review of criteria for diagnosis of autoimmune hepatitis. J Hepatol. 1999;31:929–38.
4. Czaja AJ. The variant forms of autoimmune hepatitis. Ann Intern Med. 1996;125:588–98.
5. Czaja AJ. Frequency and nature of the variant syndromes of autoimmune liver disease. Hepatology. 1998;28:360–5.
6. Doherty DG, Donaldson PT, Underhill JA et al. Allelic sequence variation in the HLA class II genes and proteins in patients with autoimmune hepatitis. Hepatology 1994;19:609–15.
7. Donaldson PT, Doherty DG, Hayllar KM et al. Susceptibility to autoimmune chronic active hepatitis: human leucocyte antigens DR4 and A1-B8-DR3 are independent risk factors. Hepatology. 1991;13:701–6.
8. Czaja AJ, Strettell MDJ, Thomson LJ et al. Association between alleles of the major histocompatibility complex type 1 autoimmune hepatitis. Hepatology. 1997;25:317–23.
9. Czaja AJ, Carpenter HA, Santrach PJ et al. Significance of HLA DR4 in type 1 autoimmune hepatitis. Gastroenterology. 1993;195:1502–7.
10. Underhill JA, Donaldson PT, Bray G et al. Susceptibility to primary biliary cirrhosis is associated with the HLADR8-DQB1*0402 haplotype. Hepatology. 1992;16:1404–8.
11. Gregory W, Mehal W, Dunn AN et al. Primary biliary cirrhosis: contribution of HLA class II allele DR8. Q J Med. 1993;86:393–9.
12. Begovich AB, Klitz W, Moonsamy PV et al. Genes within the HLA Class II region confer both predisposition and resistance to primary biliary cirrhosis. Tissue Antigens. 1994; 43:71–7.
13. Donaldson PT, Farrant JM, Wilkinson ML et al. Dual association of HLA DR2 and DR3 with primary sclerosing cholangitis. Hepatology. 1991;13:129–33.
14. Mehal W, Lo Y, Wordsworth B et al. HLA DR4 is a marker for rapid disease progression in primary sclerosing cholangitis. Gastroenterology. 1994;106:160–7.

15. Lohse AW, Meyer zum Büschenfelde KH, Franz B et al. Characterization of the overlap syndrome of primary biliary cirrhosis (PBC) and autoimmune hepatitis: evidence for it being a hepatic form of PBC in genetically susceptible individuals. Hepatology. 1999;29: 1078–84.
16. Chazouillères O, Wendum D, Serfaty L et al. Primary biliary cirrhosis–autoimmune hepatitis overlap syndrome: clinical features and response to therapy. Hepatology. 1998; 28:296–301.
17. Talwalkar JA, Keach JC, Angulo P et al. Overlap of autoimmune hepatitis and primary biliary cirrhosis: an evaluation of a modified scoring system. Am J Gastroenterol. 2002;97: 1191–7.
18. Invernizzi P, Crosignani A, Battezzati PM et al. Comparison of the clinical features and clinical course of antimitochondrial antibody-positive and -negative primary biliary cirrhosis. Hepatology. 1997;25:1090–5.
19. Heathcote J. Overlap syndromes and changing diagnoses. In: Manns PM, Paumgartner G, Leuschner U, editors. Immunology and Liver. Dordrecht: Kluwer, 2000:288–95.
20. Michieletti P, Wanless IR, Katz A et al. Antimitochondrial antibody negative primary biliary cirrhosis: a distinct syndrome of autoimmune cholangitis. Gut. 1994;35:260–5.
21. Taylor SL, Dean PJ, Riely CA. Primary autoimmune cholangitis. An alternative to antimitochondrial antibody-negative primary biliary cirrhosis. Am J Surg Pathol. 1994;18: 91–9.
22. Kim WR, Poterucha JJ, Jorgensen RA et al. Does antimitochondrial antibody status affect response to treatment in patients with primary biliary cirrhosis? J Hepatol. 2002;36:8–13.
23. Stone J, Wade JA, Couch-Dudek K et al. Human leucocyte antigen class II associations in serum antimitochondrial antibodies (AMA)-positive and AMA-negative primary biliary cirrhosis. J Hepatol. 2002;36:8–13.
24. Czaja AJ, Carpenter HA, Santrach PJ et al. The nature and prognosis of severe cryptogenic chronic active hepatitis. Gastroenterology. 1993;104:1755–61.
25. Manns M, Gerken G, Kyriatsoulis A et al. Characterization of a new subgroup of autoimmune chronic active hepatitis by autoantibodies against soluble liver antigen. Lancet. 1987;1:292–4.
26. Boberg KM, Fausa O, Haaland T et al. Features of autoimmune hepatitis in primary sclerosing cholangitis: an evaluation of 114 primary sclerosing cholangitis patients according to a scoring system for the diagnosis of autoimmune hepatitis. Hepatology. 1996;23:1369–76.
27. Gregorio GV, Portmann B, Karani J et al. Autoimmune hepatitis/sclerosing cholangitis overlap syndrome in childhood: a 16-year prospective study. Hepatology. 2001;33:544–53
28. Wilshanski M, Chait P, Wade JA et al. Primary sclerosing cholangitis in 32 children. Clinical laboratory, and radiographic features with survival analysis . Hepatology. 1995;22: 1415–22.
29. Kaya M, Angulo P, Lindor KD. Overlap of autoimmune hepatitis and primary sclerosing cholangitis: an evaluation of a modified scoring system. J Hepatol. 2000;33:537–42.
30. Van Buuren HR, van Hoogstraten HJF, Terkivatan T et al. High prevalence of autoimmune hepatitis among patients with primary sclerosing cholangitis. J Hepatol. 2000;33:543–8.
31. Mieli-Vergani G, Bargiota K, Samyn M et al. Therapeutic aspects of autoimmune liver disease in children. In: Dienes H-P, Leuschner U, Lohse AW, Manns MP, editors. Autoimmune Liver Disease. Berlin: Springer, 2005:278–82.
32. Abuaf N, Lunel F, Giral P et al. Non-organ specific autoantibodies associated with chronic C virus hepatitis. J Hepatol. 1993;18:359–64.
33. Czaja AJ, Carpenter HA, Santrach PJ et al. Evidence against hepatitis viruses as important causes of severe autoimmune hepatitis in the United States. J Hepatol. 1993;18:342–52.
34. Cassani F, Cataleta M, Valentini P et al. Serum autoantibodies in chronic hepatitis C: comparison with autoimmune hepatitis and impact on the disease profile. Hepatology. 1997;26:561–6.
35. Czaja AJ. The autoimmune hepatitis/hepatitis C overlap syndrome: does it exist? In: Leuschner U, Broomé U, Stiehl A, editors. Cholestatic Liver Diseases. Therapeutic Options and Perspectives. Dordrecht: Kluwer, 2004:132–46.
36. Joshi S, Cauch-Dudek K, Walness IR et al. Response to ursodeoxycholic acid in primary biliary cirrhosis patients with additional features of autoimmune hepatitis. Hepatology. 2002;35:409–13.

37. Li C-P, Tong MJ, Hwang S-J et al. Autoimmune cholangitis with featrures of autoimmune hepatitis: successful treatment with immunosuppressive agents and ursodeoxycholic acid. J Gastroentereol Hepatol. 2000;15:95–8.
38. Czaja AJ, Carpenter HA. Histological findings in chronic hepatitis C with autoimmune features. Hepatology. 1997;26:459–66.
39. Clifford BD, Donahue D, Smith L et al. High prevalence of serological markers of autoimmunity with chronic hepatitis C. Hepatology. 1995;21:613–19.
40. Colombato LA, Alvarez F, Cote J et al. Autoimmune cholangiopathy: the result of consecutive primary biliary cirrhosis and autoimmune hepatitis? Gastroenterology 1994; 107:1839–43.
41. Abdo AA, Bain VG, Kichian K et al. Evolution of autoimmune hepatitis to primary sclerosing cholangitis: a sequential syndrome. Hepatology. 2002;36:1393–9.
42. Mieli-Vergani G, Gregorio GV, Portmann BC et al. Evolution from autoimmune hepatitis to sclerosing cholangitis documented by sequential cholangiographic studies. Hepatology. 2000;32:309A.

Laudatio on the Occasion of Professor Gerok's 80th Birthday

H. E. BLUM

On March 27, 2006 Professor Wolfgang Gerok turned 80. This event calls for a celebration also in this forum because Professor Gerok was involved in the organization of the Falk Liver Weeks from the very beginning, i.e. from the II. Falk Liver Week in 1970 in Freiburg with the Falk Symposium No. 5 on 'Bile Acids' until today.

In the following I would like to briefly summarize the major contributions of Prof. Gerok to academic medicine in general and to hepatology in particular.

Wolfgang Gerok studied medicine at the universities of Freiburg and Tübingen from 1945–1950, followed by a postdoctoral training in biochemistry and pathology at the University Tübingen and a 2-year research stay at the University Zürich. This postdoctoral training was followed by a residency in internal medicine and further specialization in gastroenterology and hepatology.

After his clinical training and a period as chief resident at the Universities of Marburg and Mainz he took the chair for internal medicine at the University Hospital in Freiburg in 1968, a position that he filled until his retirement in 1994. Among the many facets of the contributions of Wolfgang Gerok I would like to mention only three:

1. His excellence as an academic teacher
2. His great personal interest in clinical research
3. His continuous support for the organization and funding of clinical research in Germany

ACADEMIC TEACHER

The centre of the clinical and scientific activities of Wolfgang Gerok was the individual patient. He was always available and had a tremendous talent to quickly gain the confidence of his patients and their relatives. For his students and associates, he practiced patient-oriented medicine, limiting diagnostic procedures and therapeutic interventions to the essential aspects, always addressing also the personal needs and anxieties of his patients.

For his students he was a most revered teacher with the ability to explain complex issues in a simple manner, taking into account the pathophysiological and pathobiochemical basis of diseases.

As a clinician and academic teacher Wolfgang Gerok always had a charismatic personality with a natural authority that made him a caring physician for his patients and a role model for his colleagues, associates and students.

RESEARCH

The research interest of Wolfgang Gerok was in general terms the understanding of the physiological, biochemical and molecular basis of diseases. Given this interest, he focused on the protein and amino acid metabolism of the liver in healthy individuals as well as in patients with liver and kidney diseases. Further, he and his associates studied the bile acid metabolism and in particular the transport systems for bile acids and the kinetics of bile acid transport.

Many of these projects were part of the Deutsche Forschungsgemeinschaft (DFG)-funded Centre of Collaborative Research (Sonderforschungsbereich) 'Clinical and Experimental Hepatology' that he organized together with Prof. Decker from 1982–1994. In addition to his interest in basic sciences he also supported the clinical development of new procedures and therapeutic strategies in hepatology. In particular, he and his associates, especially Prof. Rössle brought the transjugular, intrahepatic portosystemic stent (TIPS) procedure to a now worldwide established clinical application.

SUPPORT OF RESEARCH ORGANIZATION AND FUNDING

Apart from his enthusiasm for basic as well as applied clinical research, Wolfgang Gerok was very much involved in the activities of agencies for research funding, such as the DFG. For many years he presided the Committee on 'Internal Medicine' and from 1979–1985 he was Vice-President of the DFG. In addition, he was a member of numerous scientific advisory boards, including the Max-Planck-Gesellschaft, the Thyssen Foundation and many other organizations.

As mentioned already, the major commitment of Wolfgang Gerok was the establishment of basic research in a clinical setting. In this context he very early recognized the problems associated with performing laboratory-based research and at the same time being involved in clinical activities. In 1979 he published a DFG-sponsored memorandum entitled 'State and improvement of clinical research in the Federal Republic of Germany'. This memorandum had a major impact on the research programmes of the DFG in the ensuing years.

Through his untiring personal input, tremendous working discipline and his enthusiasm for patient care as well as teaching and research, he was a role model for many of his associates and greatly contributed to their career.

For his eminent clinical and scientific contributions Prof. Gerok already received many honours and awards. Among others, he became a member of the German Academy of Natural Scientist Leopoldina in Halle, a member of the Ordre pour le Mérite for Sciences and Arts and received numerous medals and prizes from different institutions and societies worldwide, among others the Watson Prize of the University of Minnesota and the Thannhauser Medal of the German Gastroenterology Society.

Given all his activities and talents Wolfgang Gerok certainly was and is one of the most prominent personalities in internal medicine in Germany. For his outstanding personal engagement for medicine, teaching and research as well as his personal continuous support of the funding of clinical research in Germany, he can be certain of the unconditional admiration and appreciation of his students, associates and colleagues. But also from his many patients and their families for whom he was a charismatic and trusted advisor.

Dear Wolf, on behalf of all participants of the XIII. Falk Liver Week, I would like to congratulate you to your birthday and to wish you the very best for the future: health, happiness and time for many of your non-medical hobbies, such as music, hiking, skiing as well as in recent years travelling to far away places.

Happy birthday from all of us: health, good luck and well-being for many years to come.

13
Wolfgang Gerok State-of-the-Art Lecture: Pathogenesis of hepatic encephalopathy

D. HÄUSSINGER, B. GÖRG and F. SCHLIESS

INTRODUCTION

Hepatic encephalopathy (HE) defines a frequent neuropsychiatric manifestation of chronic and acute liver disease with a broad spectrum of neuropsychological and motoric symptoms[1–4]. Even in minimal HE without obvious clinical symptoms neuropsychological and neurophysiological testing uncovers deficits with impact on quality of life and the fitness to drive a car[5–7].

Pathogenetic models on HE in liver cirrhosis have to explain the rapid kinetics of symptom severity, their reversibility and the fact that HE episodes are exacerbated by rather heterogeneous precipitating factors such as ingestion of high protein diets, intestinal bleeding, trauma, diuretics, sedatives, hyponatraemia and infections[2,8–13]. They have also to explain the multiple alteration of neurotransmitter systems and their receptors in the brain of HE patients and HE animal models, as well as the selective alterations in blood–brain barrier permeability. There is consensus that ammonia is a key toxin in HE, which may sensitize the brain to the different precipitating factors[14,15]. Further, there is substantial evidence that astrocytes play an important role in the pathogenesis of HE with consequences for neuronal function[16]. At the neurophysiological level the motor deficits seen in patients with HE were explained by a pathologically altered oscillatory coupling within the central motor system and the cognitive deficits were assigned to pathologically altered oscillatory activity in higher cognitive brain areas[17–19].

Whereas in fulminant hepatic failure the development of a cerebral oedema is a well-known complication[20], patients with liver cirrhosis and HE usually show no clinical signs of an overt cerebral oedema and increased intracranial pressure. In 1994, however, *in vivo* proton magnetic resonance (^{1}H-MRS) studies on the human brain provided first evidence for the development of a low-grade cerebral oedema in cirrhotic patients[21], when it became clear that the myo-inositol peak in ^{1}H-MRS reflects an osmosensitive pool. Its depletion in the brain of cirrhotic patients with HE, together with a consistent elevation of the glutamine/glutamate signal, was therefore indicative for the presence of a low-grade brain (glial) oedema. From this it was hypothesized that the pathogenetic action of ammonia and HE-precipitating factors integrates at the level of astrocyte swelling[21]. The presence of a low-grade cerebral oedema in cirrhotic patients was meanwhile confirmed by means of magnetization transfer ratio measurements[22,23] and the recently introduced TAPIR (T1 mapping with partial inversion recovery) technique, which allows for quantitative water mapping of the brain *in vivo*[24]. Astrocyte swelling primarily affects astrocyte function at multiple levels and in this way impairs neuronal signal transmission, finally leading to the HE-related symptoms[10,13,21]. One consequence of astrocyte swelling is the generation of reactive nitrogen and oxygen species (RNOS), which again promotes astrocyte swelling and thereby contributes to the development of an autoamplificatory loop.

This chapter summarizes evidence that the development of a low-grade cerebral oedema plays a major role for the pathogenesis of HE and critically determines the susceptibility of cirrhotics to HE-precipitating factors.

HE IN LIVER CIRRHOSIS IS A CLINICAL MANIFESTATION OF A LOW-GRADE CEREBRAL OEDEMA

Cerebral ammonia is detoxified by the astrocytes which are the predominant cellular brain compartment expressing glutamine synthetase[25]. Exposure of cultured astrocytes to ammonia leads to astrocyte swelling due to the intracelluar accumulation of glutamine[26]. In order to prevent excessive swelling, astrocytes like most other cell types, make an effort to restore their initial volume by a release of osmolytes, including inorganic ions and organic compounds such as myoinositol, amino acids and ascorbate[27–30].

In vivo ^{1}H-MRS studies on the brain of cirrhotic patients consistently show an increase of glutamine plus glutamate accompanied by a depletion of myoinositol, which is indicative for a partially compensated glial oedema[10,13,21,31–33]. These ^{1}H-MRS changes correlate well to the clinical severity of HE[21–23,32,34–36]. Further, the increase of brain glutamine plus glutamate in cirrhotic patients was correlated to both a decrease in magnetization transfer ratio (which indicates an increase of free water) and neuropsychological deterioration[35]. Brain myoinositol levels may even act as a marker in cirrhotics without overt HE to predict the patient's risk for developing overt HE. Low baseline myoinositol levels were found to predispose for neuropsychological deterioration by an experimental amino acid load[37]. Interestingly, the most relevant HE-precipitating factors such as hyponatraemia[38], tumour necrosis factor alpha (TNF-α)[39], extracellular glutamate[40] and agonists of the peripheral-type benzodiazepine receptor[41] trigger astrocyte swelling and/or increase astrocyte swelling synergistically with ammonia.

These findings support the view that the volume-regulatory capacity of the astrocyte as determined by its osmolyte content is important for the compensation of astrocyte swelling induced by HE-precipitating factors, and represents a major determinant of the neuropsychological outcome. Non-cirrhotics and cirrhotics without HE may tolerate such precipitating factors because astroglial osmolyte pools are not exhausted. However, when the organic osmolytes under hyperammonaemic conditions are largely depleted in order to compensate for swelling, astrocytes are sensitized to swelling induction by HE-precipitating factors. It was hypothesized that the action of HE-precipitating factors largely integrates at the level of astrocyte swelling[10,13,21], which in turn alters astrocyte function. It is well recognized that cell volume changes are registered by 'osmosensing' structures, leading to the activation of signal transduction pathways involved in the control and regulation of metabolism and gene expression[42–44].

FUNCTIONAL CONSEQUENCES OF ASTROCYTE SWELLING

Astrocyte swelling affects glial and neuronal functions in different ways. Hypo-osmotic astrocyte swelling triggers activation of Erk- and p38-type MAP-kinases[45–47] and elevation of the intracellular Ca^{2+} concentration[28,45,48,49] and may increase proliferation[50,51]. Astrocyte swelling stimulates glycogen synthesis[52], which may account for the increased astroglial deposition of glycogen in animal models of chronic HE[14] and astrocyte swelling increases the pH in endocytotic vesicles[53], which could affect receptor/ligand sorting and neurotransmitter processing. Increased expression of the peripheral-type benzodiazepine receptor (PBR) in response to astrocyte swelling augments the synthesis of neurosteroids, which are potent modulators of neuronal $GABA_A$ receptor activity[41]. The interaction between astrocyte swelling, PBR expression and increased neurosteroid synthesis may in part explain the increased GABA-ergic tone found in HE. Further, hypo-osmolarity and ammonia induce the expression of the rat homologue to the 'mouse proline-rich protein expressed in brain' and of clusterin mRNA in cultured rat astrocytes[54]. Whereas the functional impact of the former is unknown, the latter may play a role in cytoprotection. Hypo-osmotically swollen astrocytoma cells display remodelling of the actin cytoskeleton[55], elevated nucleoside trisphosphate levels, stimulation of fatty acid biosynthesis, and a decrease of cytoplasmic pH[56].

Swollen astrocytes predispose to neuronal dysfunction due to impairment of their protective homeostatic functions such as the clearance of excitatory neurotransmitters and maintenance of glial–neuronal communication[57–60]. In the brain *in vivo,* moderate astrocyte swelling decreases the extracellular space, and the possibility was taken into account that astroglial volume changes could be one mechanism to control extracellular fluid volume in the brain[61]. A reduced extracellular space volume fraction could decrease diffusion of molecules and increase the accumulation of ions, excitotoxic transmitters and neurotoxic metabolites, thereby contributing to neuronal injury[62,63].

Of crucial importance for the development of HE appears the production of RNOS by different HE-precipitating toxins.

HE-PRECIPITAING FACTORS AND HYPO-OSMOTIC ASTROCYTE SWELLING PRODUCE OXIDATIVE STRESS

The term 'oxidative stress'[64] defines the cell challenge induced by a high burden of RNOS, which may result from their increased production and/or diminished removal due to a decrease of cellular antioxidative capacity. Although human studies concerning the involvement of oxidative stress in HE are rare up to now[65–68], there is substantial evidence from animal and cell culture studies suggesting that the generation of RNOS may play a major role in the pathogenesis of HE[69,70].

Acute ammonia intoxication of animals increases the cerebral production of superoxide, nitric oxide (NO), and protein tyrosine nitration, which were sensitive to the NMDA receptor antagonist MK-801[20,71–78]. Likewise, chronic hyperammonaemia due to institution of a portacaval shunt in rats elevates brain levels of tyrosine-nitrated proteins[73,79]. Similarly, diazepam and endotoxin, which precipitate human HE, increased cerebral oxidative stress and protein tyrosine nitration in animals *in vivo*[80–83]. Immunohistochemical analysis of brain slices from ammonia-, diazepam-, and endotoxin-treated rats unravelled a considerable colocalization of 3′-nitrotyrosine with the astrocyte markers glutamine synthetase and glial fibrillary acidic protein, indicating protein tyrosine nitration in astrocytes[20,73,81,83]. Nitrotyrosine immunoreactivity was particularly enriched in perivascular astrocytes, which are constituents of the blood–brain barrier[20,73,81,83].

There is evidence that astrocytes represent a major cellular compartment of cerebral RNOS generation under HE-relevant conditions. Ammonia[73,78,84], diazepam[81], and TNF-α[83] increase RNOS and protein tyrosine nitration also in cultured rat astrocytes in a NMDA receptor- and Ca^{2+}-dependent manner. Of interest, swelling of cultured astrocytes as induced by hypo-osmolarity was sufficient to stimulate a NMDA receptor- and Ca^{2+}-mediated increase of RNOS and protein tyrosine nitration[85], suggesting that RNOS production in response to HE-precipitating compounds at least in part depends on astrocyte swelling. This is further supported by the sensitivity of ammonia-induced NO production and protein tyrosine nitration in rat brain *in vivo* to the glutamine synthetase inhibitor methionine sulfoximine (MSO)[75,78], which also prevents astrocyte swelling. Also the ammonia-induced protein tyrosine nitration in cultured astrocytes was sensitive to MSO[73]. On the other hand there is evidence that NMDA receptor activation and RNOS mediate astrocyte swelling by ammonia[86,87], indicating that astrocyte swelling and RNOS production are interdependent.

RNOS production by astrocyte swelling may contribute to the pathogenesis of HE. Whereas astrocytes are relatively resistant to the action of RNOS[88,89], astroglial RNOS affect the respiratory chain of neighboured neurons[90] and may contribute to the compromised brain energy metabolism and neuronal transmission in HE. Further, there is substantial evidence that RNOS and protein tyrosine nitration affect the integrity of the blood–brain barrier. In a mouse model of experimental allergic encephalomyelitis the increase of blood–brain barrier permeability is sensitive to the peroxynitrite scavenger uric acid[91–93]. Protein tyrosine nitration in the perivascular astrocytes, as observed in HE-related animal models[20,73,81,83], may affect transastrocytic substrate transport, which could correspond to the well-known specific alterations of blood–brain barrier permeability in HE.

Among the tyrosine-nitrated proteins in ammonia-treated cultured astrocytes the MAP-kinase Erk-1, the PBR, and the glutamine synthetase were identified[73]. The effect of tyrosine nitration of the PBR and Erk-1 is currently unknown, although nitration of Erks does not activate these MAP-kinases[94] and oxidation of the PBR increased its binding affinity for the PBR ligand PK11195 and accelerates cholesterol transport across the mitochondrial membrane[95], thereby increasing neurosteroid synthesis.

The tyrosine nitration of glutamine synthetase in ammonia-exposed astrocytes is associated with an inactivation of the enzyme[73]. Likewise, tyrosine nitration of hepatic glutamine synthetase in endotoxin-treated rats and of peroxynitrite-treated purified sheep GS is associated with inactivation of the enzyme[96,97]. Mass spectrometric analysis of peptides derived from the peroxynitrite-treated sheep GS identified a nitrated tyrosine residue corresponding to Tyr^{335} of rat glutamine synthetase, which is highly conserved among eukaryotic glutamine synthetases. This tyrosine residue may directly stack with the purine system of GS-bound ADP/ATP, and in addition may contribute to the interface with the neighbouring subunit. Both functions may be impaired by nitration of Tyr^{335} (see refs 78, 96 and 98). Indeed, the extent of peroxynitrite-induced glutamine synthetase nitration *in vitro* inversely correlates with inactivation of the enzyme and treatment of the tyrosine-nitrated glutamine synthetase with protein preparations containing a putative 'denitrase' activity fully recovers its activity[97]. Likewise, the increase of brain glutamine synthetase activity after administration of the NMDA receptor antagonist MK-801 or the NOS inhibitor nitroarginine may result from denitration of the enzyme[99]. A reversible inactivation of glutamine synthetase by tyrosine nitration under hyperammonaemic conditions may protect the brain from excessive glutamine accumulation and astrocyte swelling.

That inhibition of glutamine synthetase and NMDA receptors abolishes ammonia-induced protein tyrosine nitration on the one hand, and on the other hand improves clinical signs of ammonia toxicity in experimental animals[71,76,77,100–104], may argue for a role of protein tyrosine nitration in the development of HE symptoms. However, the possibility remains that protein tyrosine nitration represents only a surrogate marker for other pathogenetically more important alterations. For example, ammonia-induced RNOS production also leads to N-acetylglucosaminidation of proteins and RNA oxidation, respectively.

AMMONIA INDUCES OXIDATION OF RNA IN CULTURED ASTROCYTES AND RAT BRAIN *IN VIVO*

RNOS can hydroxylate guanine to produce 8-oxo-7,8-dihydro-2′-deoxyguanosine (8OHdG) in DNA and 8-oxo-7,8-dihydro-2′-guanosine (8OHG) in RNA that are highly sensitive biomarkers for oxidative stress[105,106]. 8OHdG leads to the transversion of G:C to A:T in the course of

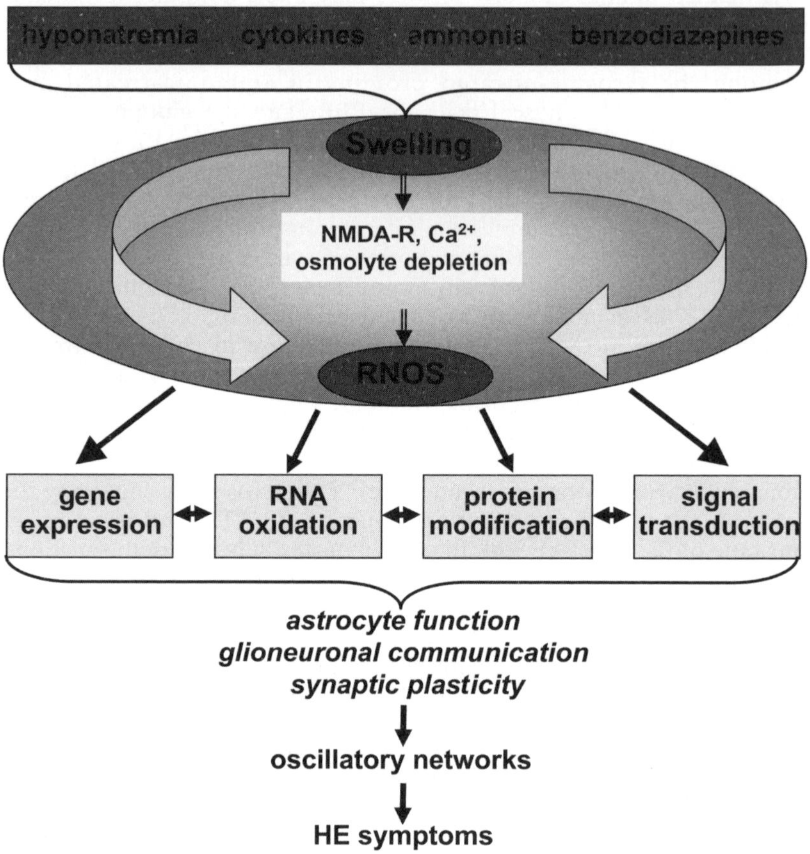

Figure 1 Low-grade cerebral oedema in HE. Ammonia induces astrocyte swelling which is in part counteracted by a volume-regulatory osmolyte depletion but can be aggravated by a heterogeneous set of precipitating factors. Astrocyte swelling involves NMDA receptor activation and the generation of RNOS, which again promotes astrocyte swelling. In this way an autoamplificatory loop is generated which produces signals changing astrocyte function at multiple levels. As a consequence, glioneuronal communication and neuronal signal transmission are impaired, which finally accounts for the HE symptoms. Adapted from ref. 13

DNA replication, which can be avoided by effective repair. 8OHG under oxidative conditions is essentially present in cytosolic RNA. Potential mechanisms repairing oxidized RNA or preventing its translation are less well characterized. Compared to DNA, RNA is probably more sensitive to oxidants because it is single-stranded and not protected by packaging proteins. Oxidized RNA were identified and characterized in post-mortem brain tissue of Alzheimer's disease patients[107].

We recently found that 8OH(d)G immunoreactivity is increased in brains of ammonia-loaded rats, where it localized in part to astrocytes (glutamine synthetase- and GFAP-positive) and neurons (MAP-2-positive). Remarkably, 8OH(d)G immunoreactivity was identified in granular structures within the postsynaptic dendritic compartment, where it associated in part with the RNA binding and transporting proteins NOVA-1 and 2, indicating oxidation of transported RNA. An increase of 8OH(d)G immunoreactivity was also observed in cultured rat astrocytes exposed to ammonia. Of interest, 8OH(d)G immunoreactivity was essentially confined to the cytosol and abolished upon RNase-catalysed digestion, supporting the theory that it defines RNA oxidation rather than oxidation of nuclear or mitochondrial DNA, respectively (Görg, Qvartskhava, Schliess, Häussinger, unpublished data). The implications of RNA oxidation for ammonia neurotoxicity and HE are currently unknown; however, it is conceivable that RNA oxidation compromises translation accuracy. Impairment of axonal protein synthesis is known to affect synaptic plasticity and memory formation[108] and could relate to alterations in memory function and receptor expression in the setting of HE.

A PATHOGENETIC MODEL

A current working hypothesis integrates the potential contribution of both osmotic and oxidative stress to the pathogenesis of HE (Figure 1). HE results from the synergistic action of HE-precipitating factors including ammonia, glutamate, benzodiazepines, hyponatraemia and inflammatory cytokines. Ammonia triggers astrocyte swelling leading to a compensatory depletion of osmolytes, which reduces their volume-regulatory capacity and thus predisposing the astrocytes to induction of swelling by HE-precipitating factors. Astrocyte swelling increases RNOS production which, on the other hand, promotes astrocyte swelling. Mutual amplification of astroglial swelling and RNOS production create an autoamplificatory signalling loop. Apart from the astroglial osmolyte status, $[Ca^{2+}]_i$, the NMDA receptor, and possibly the PBR are players within this loop. Thus, the action of different HE-relevant factors at least in part integrates at the level of astroglial swelling/RNOS production, leading to alterations in covalent protein and nucleic acid modifications, signal transduction, gene expression, and metabolism. Impaired astroglial support of neuronal function may finally cause HE symptoms.

References

1. Häussinger D, Warskulat U, Fischer R, Sinning R, Schliess F. Hepatic encephalopathy. In: Lam SK, Paumgartner G, Wang B, editors. Update on Hepatobiliary Diseases. Dordrecht: Kluwer, 1996:176–88.
2. Hazell AS, Butterworth RF. Hepatic encephalopathy: an update of pathophysiologic mechanisms. Proc Soc Exp Biol Med. 1999;222:99–112.
3. Ferenci P, Lockwood A, Mullen K, Tarter R, Weissenborn K, Blei AT. Hepatic encephalopathy – definition, nomenclature, diagnosis, and quantification: final report of the working party at the Eleventh World Congresses of Gastroenterology, Vienna, 1998. Hepatology. 2002;35:716–21.

4. Häussinger D, Cordoba J, Kircheis G et al. Definition and assessment of low grade hepatic encephalopathy. In: Häussinger D, Kircheis G, Schliess F, editors. Hepatic Encephalopathy and Nitrogen Metabolism. Dordrecht: Springer, 2006:423–32.
5. Groeneweg M, Quero JC, De Bruijn I et al. Subclinical hepatic encephalopathy impairs daily functioning. Hepatology. 1998;28:45–9.
6. Wein C, Koch H, Popp B, Oehler G, Schauder P. Minimal hepatic encephalopathy impairs fitness to drive. Hepatology. 2004;39:739–45.
7. Kircheis G, Wettstein M, Timmermann L, Schnitzler A, Häussinger D. Critical flicker frequency for quantification of low-grade hepatic encephalopathy. Hepatology. 2002;35: 357–66.
8. Blei AT. Infection, inflammation and hepatic encephalopathy: synergism redefined. J Hepatol. 2004;40:327–30.
9. Odeh M. Endotoxin and tumor necrosis factor-alpha in the pathogenesis of hepatic encephalopathy. J Clin Gastroenterol. 1994;19:146–53.
10. Häussinger D, Kircheis G, Fischer R, Schliess F, vom Dahl S. Hepatic encephalopathy in chronic liver desease: a clinical manifestation of astrocyte swelling and low grade cerebral edema. J Hepatol. 2000;32:1035–8.
11. Jones EA. Ammonia, the GABA neurotransmitter system, and hepatic encephalopathy. Metab Brain Dis. 2002;17:275–81.
12. Shawcross D, Jalan R. The pathophysiologic basis of hepatic encephalopathy: central role for ammonia and inflammation. Cell Mol Life Sci. 2005;62:2295–304.
13. Häussinger D. Low grade cerebral edema and the pathogenesis of hepatic encephalopathy in cirrhosis. Hepatology. 2006;43:1187–90.
14. Norenberg MD. Astrocytic–ammonia interactions in hepatic encephalopathy. Semin Liver Dis. 1996;16:245–53.
15. Ong JP, Aggarwal A, Krieger D et al. Correlation between ammonia levels and the severity of hepatic encephalopathy. Am J Med. 2003;114:188–93.
16. Norenberg MD, Neary JT, Bender AS, Dombro RS. Hepatic encephalopathy: a disorder in glial–neuronal communication. Prog Brain Res. 1992;94:261–9.
17. Timmermann L, Gross J, Kircheis G, Häussinger D, Schnitzler A. Cortical origin of mini-asterixis in hepatic encephalopathy. Neurology. 2002;58:295–8.
18. Timmermann L, Butz M, Gross J, Kircheis G, Häussinger D, Schnitzler A. Neural synchronization in hepatic encephalopathy. Metab Brain Dis. 2005;20:337–46.
19. Schnitzler A, Butz M, Gross J, Kircheis G, Häussinger D. Timmermann L. Altered neural oscillations and synchronization: a pathophysiological hallmark of hepatic encephalopathy. In: Häussinger D, Kircheis G, Schliess F, editors. Hepatic Encephalopathy and Nitrogen Metabolism. Dordrecht: Springer, 2006:245–54.
20. Vaquero J, Chung C, Cahill ME, Blei AT. Pathogenesis of hepatic encephalopathy in acute liver failure. Semin Liver Dis. 2003;23:259–69.
21. Häussinger D, Laubenberger J, vom Dahl S et al. Proton magnetic resonance spectroscopy studies on human brain myo-inositol in hypo-osmolarity and hepatic encephalopathy. Gastroenterology. 1994;107:1475–80.
22. Miese F, Kircheis G, Wittsack HJ et al. ^{1}H-MR spectroscopy, magnetization transfer, and diffusion-weighted imaging in alcoholic and nonalcoholic patients with cirrhosis with hepatic encephalopathy. Am J Neuroradiol. 2006;27:1019–26.
23. Minguez B, Garcia-Pagan JC, Bosch J et al. Noncirrhotic portal vein thrombosis exhibits neuropsychological and MR changes consistent with minimal hepatic encephalopathy. Hepatology. 2006;43:707–14.
24. Shah NJ, Neeb H, Zaitsev M et al. Quantitative T1 mapping of hepatic encephalopathy using magnetic resonance imaging. Hepatology. 2003;38:1219–26.
25. Martinez HA, Bell KP, Norenberg MD. Glutamine synthetase: glial localization in brain. Science. 1977;195:1356–8.
26. Norenberg MD, Baker L, Norenberg LO, Blicharska J, Bruce-Gregorios JH, Neary JT. Ammonia-induced astrocyte swelling in primary culture. Neurochem Res. 1991;16:833–6.
27. Chamberlin ME, Strange K. Anisosmotic cell volume regulation: a comparative view. Am J Physiol. 1989;257:C159–73.
28. O'Connor ER, Kimelberg HK. Role of calcium in astrocyte volume regulation and in the release of ions and amino acids. J Neurosci. 1993;13:2638–50.

29. Ordaz B, Tuz K, Ochoa LD, Lezama R, Pena-Segura C, Franco R. Osmolytes and mechanisms involved in regulatory volume decrease under conditions of sudden or gradual osmolarity decrease. Neurochem Res. 2004;29:65–72.
30. Siushansian R, Dixon SJ, Wilson JX. Osmotic swelling stimulates ascorbate efflux from cerebral astrocytes. J Neurochem. 1996;66:1227–33.
31. Morgan MY. Cerebral magnetic resonance imaging in patients with chronic liver disease. Metab Brain Dis. 1998;13:273–90.
32. Ross BD, Jacobson S, Villamil F et al. Subclinical hepatic encephalopathy: proton MR spectroscopic abnormalities. Radiology. 1994;193:457–63.
33. Binesh N, Huda A, Thomas MA et al. Hepatic encephalopathy: a neurochemical, neuroanatomical, and neuropsychological study. J Appl Clin Med Phys. 2006;7:86–96.
34. Laubenberger J, Häussinger D, Bayer S, Gufler H, Hennig J, Langer M. Proton magnetic resonance spectroscopy of the brain in symptomatic and asymptomatic patients with liver cirrhosis. Gastroenterology. 1997;112:1610–16.
35. Cordoba J, Alonso J, Rovira A et al. The development of low-grade cerebral edema in cirrhosis is supported by the evolution of ^{1}H-magnetic resonance abnormalities after liver transplantation. J Hepatol. 2001;35:598–604.
36. Kale RA, Gupta RK, Saraswat VA et al. Demonstration of interstitial cerebral edema with diffusion tensor MR imaging in type C hepatic encephalopathy. Hepatology. 2006;43:698–706.
37. Shawcross DL, Balata S, Olde-Damink SW et al. Low myo-inositol and high glutamine levels in brain are associated with neuropsychological deterioration after induced hyperammonemia. Am J Physiol. 2004;287:G503–9.
38. Cordoba J, Crespin J, Gottstein J, Blei AT. Mild hypothermia modifies ammonia-induced brain edema in rats after portacaval anastomosis. Gastroenterology. 1999;116:686–93.
39. Bender AS, Rivera IV, Norenberg MD. Tumor necrosis factor alpha induces astrocyte swelling. Trans Am Soc Neurochem. 1992;23:113.
40. Bender AS, Schousboe A, Reichelt W, Norenberg MD. Ionic mechanisms in glutamate-induced astrocyte swelling: role of K^+ influx. J Neurosci Res. 1998; 52:307–21.
41. Bender AS, Norenberg MD. Effect of benzodiazepines and neurosteroids on ammonia-induced swelling in cultured astrocytes. J Neurosci Res. 1998;54:673–80.
42. Häussinger D. The role of cellular hydration in the regulation of cell function. Biochem J. 1996;313:697–710.
43. Lang F, Busch GL, Ritter M et al. Functional significance of cell volume regulatory mechanisms. Physiol Rev. 1998;78:247–306.
44. Schliess F, Häussinger D. The cellular hydration state: role in apoptosis and proliferation. Signal Transduction. 2005;6:297–302.
45. Schliess F, Sinning R, Fischer R, Schmalenbach C, Häussinger D. Calcium-dependent activation of Erk-1 and Erk-2 following hypoosmotic astrocyte swelling. Biochem J. 1996; 320:167–71.
46. Sinning R, Schliess F, Kubitz R, Häussinger D. Osmosignalling in C6 glioma cells. FEBS Letters. 1997;400:163–7.
47. Xu D, Wang L, Olson JE, Lu L. Asymmetrical response of p38 kinase activation to volume changes in primary rat astrocytes. Exp Biol Med. 2001;226:927–33.
48. Fischer R, Schliess F, Häussinger D. Characterization of the hypoosmolarity-induced Ca^{2+} response in cultured rat astrocytes. Glia. 1997;20:51–8.
49. Morales MS, Vaca L, Hernandez CA, Pasantes MH. Osmotic swelling-induced changes in cytosolic calcium do not affect regulatory volume decrease in rat cultured suspended cerebellar astrocytes. J Neurochem. 1998;71:2330–8.
50. Reichenbach A. Glia: neuron index: review and hypothesis to account for different values in various mammals. Glia. 1989;2:71–7.
51. Rouzaire-Dubois B, O'Regan S, Dubois JM. Cell size-dependent and independent proliferation of rodent neuroblastoma × glioma cells. J Cell Physiol. 2005;203:243–50.
52. Dombro RS, Bender AS, Norenberg MD. Association between cell swelling and glycogen content in cultured astrocytes. Int J Dev Neurosci. 2000;18:161–9.
53. Busch GL, Wiesinger H, Gulbins E, Wagner HJ, Hamprecht B, Lang F. Effect of astroglial cell swelling on pH of acidic intracellular compartments. Biochim Biophys Acta. 1996; 1285:212–18.

54. Warskulat U, Kreuels S, Müller HW, Häussinger D. Identification of osmosensitive and ammonia-regulated genes in rat astrocytes by Northern blotting and differential display reverse transcriptase-polymerase chain reaction. J Hepatol. 2001;35:358–66.
55. Moran J, Sabanero M, Meza I, Pasantes MH. Changes of actin cytoskeleton during swelling and regulatory volume decrease in cultured astrocytes. Am J Physiol. 1996;271: C1901–7.
56. Flögel U, Niendorf T, Serkowa N, Brand A, Henke J, Leibfritz D. Changes in organic solutes,volume, energy state and metabolism associated with osmotic stress in a glial cell line – a multinuclear NMR study. Neurochem Res. 1995;20:793–802.
57. Norenberg MD. Astrocyte responses to CNS injury. J Neuropathol Exp Neurol. 1994;53: 213–20.
58. Kimelberg HK. Current concepts of brain edema – review of laboratory investigations. J Neurosurg. 1995;83:1051–9.
59. Sykova E, Vargova L, Prokopova S, Simonova Z. Glial swelling and astrogliosis produce diffusion barriers in the rat spinal cord. Glia.1999;25:56–70.
60. Bezzi P, Domercq M, Vesce S, Volterra A. Neuron–astrocyte cross-talk during synaptic transmission: physiological and neuropathological implications. Prog Brain Res. 2001;132: 255–65.
61. Hansson E, Rönnbäck L. Astrocytes in glutamate neurotransmission. FASEB J. 1995;9: 343–50.
62. Sykova E. Glia and volume transmission during physiological and pathological states. J Neural Transm. 2005;112:137-47.
63. Simard M, Nedergaard M. The neurobiology of glia in the context of water and ion homeostasis. Neuroscience. 2004;129:877–96.
64. Sies H, Cadenas E. Oxidative stress: damage to intact cells and organs. Phil Trans R Soc Lond B Biol Sci. 1985;311:617–31.
65. Negru T, Ghiea V, Pasarica D. Oxidative injury and other metabolic disorders in hepatic encephalopathy. Rom J Physiol. 1999;36:29–36.
66. Harrison PM, Wendon JA, Gimson AE, Alexander GJ, Williams R. Improvement by acetylcysteine of hemodynamics and oxygen transport in fulminant hepatic failure. N Engl J Med. 1991;324:1852–7.
67. Wendon JA, Harrison PM, Keays R, Williams R. Cerebral blood flow and metabolism in fulminant liver failure. Hepatology. 1994;19:1407–13.
68. Sushma S, Dasarathy S, Tandon RK, Jain S, Gupta S, Bhist MS. Sodium benzoate in the treatment of acute hepatic encephalopathy: a double-blind randomized trial. Hepatology. 1992;16:138–44.
69. Norenberg MD, Jayakumar AR, Rama-Rao KV. Oxidative stress in the pathogenesis of hepatic encephalopathy. Metab Brain Dis. 2004;19:313–29.
70. Norenberg MD. Oxidative and nitrosative stress in ammonia neurotoxicity. Hepatology. 2003;37:245–8.
71. Kosenko E, Kaminski Y, Lopata O, Muravyov N, Felipo V. Blocking NMDA receptors prevents the oxidative stress induced by acute ammonia intoxication. Free Radic Biol Med. 1999;26:1369–74.
72. Kosenko E, Montoliu C, Giordano G et al. Acute ammonia intoxication induces an NMDA receptor-mediated increase in poly(ADP-ribose) polymerase level and NAD metabolism in nuclei of rat brain cells. J Neurochem. 2004;89:1101–10.
73. Schliess F, Görg B, Fischer R et al. Ammonia induces MK-801-sensitive nitration and phosphorylation of protein tyrosine residues in rat astrocytes. FASEB J. 2002;16:739–41.
74. Larsen FS, Gottstein J, Blei AT. Cerebral hyperemia and nitric oxide synthase in rats with ammonia-induced brain edema. J Hepatol. 2001;34:548–54.
75. Master S, Gottstein J, Blei AT. Cerebral blood flow and the development of ammonia-induced brain edema in rats after portacaval anastomosis. Hepatology. 1999;30:876–80.
76. Hermenegildo C, Marcaida G, Montoliu C, Grisolia S, Minana MD, Felipo V. NMDA receptor antagonists prevent acute ammonia toxicity in mice. Neurochem Res. 1996;21: 1237–44.
77. Hermenegildo C, Monfort P, Felipo V. Activation of *N*-methyl-D-aspartate receptors in rat brain *in vivo* following acute ammonia intoxication: characterization by *in vivo* brain microdialysis. Hepatology. 2000;31:709–15.

78. Schliess F, Görg B, Reinehr RM, Bidmon HJ, Häussinger D. Osmotic and oxidative stress in hepatic encephalopathy. In: Häussinger D, Kircheis G, Schliess F, editors. Hepatic Encephalopathy and Nitrogen Metabolism. Dordrecht: Springer, 2006:20–42.
79. Suarez I, Bodega G, Arilla E, Felipo V, Fernandez B. The expression of nNOS, iNOS and nitrotyrosine is increased in the rat cerebral cortex in experimental hepatic encephalopathy. Neuropathol Appl Neurobiol. 2006;32:594–604.
80. Musavi S, Kakkar P. Diazepam induced early oxidative changes at the subcellular level in rat brain. Mol Cell Biochem. 1998;178:41–6.
81. Görg B, Foster N, Reinehr RM et al. Benzodiazepine-induced protein tyrosine nitration in rat astrocytes. Hepatology. 2003;37:334–42.
82. Possel H, Noack H, Putzke J, Wolf G, Sies H. Selective upregulation of inducible nitric oxide synthase (iNOS) by lipopolysaccharide (LPS) and cytokines in microglia: *in vitro* and *in vivo* studies. Glia. 2000;32:51–9.
83. Görg B, Bidmon HJ, Keitel V et al. Inflammatory cytokines induce protein tyrosine nitration in rat astrocytes. Arch Biochem Biophys. 2006; 449:104–14.
84. Murthy CR, Rama-Rao KV, Bai G, Norenberg MD. Ammonia-induced production of free radicals in primary cultures of rat astrocytes. J Neurosci Res. 2001;66:282–8.
85. Schliess F, Foster N, Görg B, Reinehr R, Häussinger D. Astrocyte swelling increases protein tyrosine nitration in cultured rat astrocytes. Glia. 2004;47:21–9.
86. Zielinska M, Law RO, Albrecht J. Excitotoxic mechanism of cell swelling in rat cerebral cortical slices treated acutely with ammonia. Neurochem Int. 2003;43:299–303.
87. Norenberg MD, Rao KV, Jayakumar AR. Mechanisms of ammonia-induced astrocyte swelling. Metab Brain Dis. 2005;20:303–18.
88. Knorpp T, Robinson SR, Crack PJ, Dringen R. Glutathione peroxidase-1 contributes to the protection of glutamine synthetase in astrocytes during oxidative stress. J Neural Transm. 2006;113:1145–55.
89. Cassina P, Peluffo H, Pehar M et al. Peroxynitrite triggers a phenotypic transformation in spinal cord astrocytes that induces motor neuron apoptosis. J Neurosci Res. 2002;67:21–9.
90. Stewart VC, Sharpe MA, Clark JB, Heales SJ. Astrocyte-derived nitric oxide causes both reversible and irreversible damage to the neuronal mitochondrial respiratory chain. J Neurochem. 2000;75:694–700.
91. Kean RB, Spitsin SV, Mikheeva T, Scott GS, Hooper DC. The peroxynitrite scavenger uric acid prevents inflammatory cell invasion into the central nervous system in experimental allergic encephalomyelitis through maintenance of blood–central nervous system barrier integrity. J Immunol. 2000;165:6511–18.
92. Hooper DC, Scott GS, Zborek A et al. Uric acid, a peroxynitrite scavenger, inhibits CNS inflammation, blood–CNS barrier permeability changes, and tissue damage in a mouse model of multiple sclerosis. FASEB J. 2000;14:691–8.
93. Hooper DC, Kean RB, Scott GS et al. The central nervous system inflammatory response to neurotropic virus infection is peroxynitrite dependent. J Immunol. 2001;167:3470–7.
94. Pinzar E, Wang T, Garrido MR, Xu W, Levy P, Bottari SP. Angiotensin II induces tyrosine nitration and activation of ERK1/2 in vascular smooth muscle cells. FEBS Lett. 2005;579:5100–4.
95. Delavoie F, Li H, Hardwick M et al. *In vivo* and *in vitro* peripheral-type benzodiazepine receptor polymerization: functional significance in drug ligand and cholesterol binding. Biochemistry. 2003;42:4506–19.
96. Görg B, Wettstein M, Metzger S, Schliess F, Häussinger D. Lipopolysaccharide-induced tyrosine nitration and inactivation of hepatic glutamine synthetase in the rat. Hepatology. 2005;41:1065–73.
97. Görg B, Qvartskhava N, Voss P, Grune T, Häussinger D, Schliess F. Reversible inhibition of glutamine synthetase by tyrosine nitration. FEBS Lett. 2007;581:84–91.
98. Görg B, Wettstein M, Metzger S, Schliess F, Häussinger D. LPS-induced tyrosine nitration and inactivation of hepatic glutamine synthetase in the rat. Hepatology. 2005;42:499.
99. Rose C, Felipo V. Limited capacity for ammonia removal by brain in chronic liver failure: potential role of nitric oxide. Metab Brain Dis. 2005;20:275–83.
100. Willard-Mack CL, Koehler RC, Hirata T et al. Inhibition of glutamine synthetase reduces ammonia-induced astrocyte swelling in rat. Neuroscience. 1996;71:589–99.

101. Takahashi H, Koehler RC, Brusilow SW, Traystman RJ. Glutamine synthetase inhibition prevents cerebral oedema during hyperammonemia. Acta Neurochir Suppl Wien. 1990;51: 346–7.
102. Warren KS, Schencker S. Effect of an inhibitor of glutamine synthesis (methionine sulfoximine) on ammonia toxicity and metabolism. J Lab Clin Med. 1964;64:442–9.
103. Kosenko E, Kaminsky Y, Stavroskaya IG, Felipo V. Alteration of mitochondrial calcium homeostasis by ammonia-induced activation of NMDA receptors in rat brain *in vivo*. Brain Res. 2000;880:139–46.
104. Vogels BA, Maas MA, Daalhuisen J, Quack G, Chamuleau RA. Memantine, a noncompetitive NMDA receptor antagonist improves hyperammonemia-induced encephalopathy and acute hepatic encephalopathy in rats. Hepatology. 1997;25:820–7.
105. Kasai H, Chung MH, Jones DS et al. 8-Hydroxyguanine, a DNA adduct formed by oxygen radicals: its implication on oxygen radical-involved mutagenesis/carcinogenesis. J Toxicol Sci. 1991;16:95–105.
106. Nunomura A, Honda K, Takeda A et al. Oxidative damage to RNA in neurodegenerative diseases. J Biomed Biotechnol. 2006;3:1–6.
107. Shan X, Lin CL. Quantification of oxidized RNAs in Alzheimer's disease. Neurobiol Aging. 2006;27:657–62.
108. Costa-Mattioli M, Gobert D, Harding H et al. Translational control of hippocampal synaptic plasticity and memory by the eIF2alpha kinase GCN2. Nature. 2005;436:1166–73.

Section VI
Sequelae of chronic hepatitis (Part 1)

Chair: KH MEYER ZUM BÜSCHENFELDE and J RODÉS

Presentation of the International Dame Sheila Sherlock Award 2006

H. E. BLUM

The International Dame Sheila Sherlock Award, inaugurated in 2003 and generously sponsored by the Falk Family and Falk Foundation, is given for outstanding contributions to clinical hepatology, including advances in the diagnosis, therapy and prevention of acute and chronic liver diseases and their sequelae. The prize committee including Drs Hodgson, Manns, Omata, Powell, Rizzetto, Thomas and myself unanimously elected Dr. Elizabeth Jane Lindsay Heathcote as this year's recipient of the International Dame Sheila Sherlock Award.

Jenny Heathcote is certainly known to most of you as an internationally highly visible clinical scientist in hepatology and in particular viral hepatitis.

Dr. Heathcote obtained her professional qualifications at the Royal College of Physicians and the University of London. After her undergraduate training she started her medical training in 1968 at the University of London. Funded by a Medical Research Council Research Fellowship in Hepatology, she trained at the Royal Free Hospital from 1971 until 1974 and was lecturer in medicine at the same institution from 1974 until 1976. In 1976 she joined the Department of Gastroenterology and Immunology at Stanford University Medical Center in Palo Alto as an NIH research fellow and later on as a clinical fellow. Three years later, in 1979, Dr. Heathcote joined the Department of Medicine at the University of Toronto, first as Assistant Professor and since 1995 as Full Professor. In this academic position Dr. Heathcote held and holds a number of clinical positions with several hospital affiliations, among others as consultant in gastroenterology in the Department of Medicine in the Hospital for Sick Children in the Toronto Western Hospital, in the Institute of Medical Sciences of University of Toronto and the Toronto Western Research Institute.

Dating back to her early days with Dame Sheila Sherlock, she maintained her interest in different aspects of primary biliary cirrhosis and is at present the principal investigator of primary biliary cirrhosis family studies. Her major research interests, however, became and are clinical studies on the treatment or re-treatment of patients with chronic hepatitis B or C. She has made major contributions to numerous clinical trials and is at present principal investigator of trials, among others, of telbivudine and tenofovir, respectively, for patients with chronic hepatitis B, the treatment of patients with chronic hepatitis C with albuferon and the evaluation of an antifibrotic agent.

Her tremendous research activities are documented in close to 200 peer-reviewed publications, starting in 1972 with publications on HBV, later on the treatment of primary biliary cirrhosis with azathioprine and on cholestasis in childhood together with, among others, Dame Sheila Sherlock. These publications appeared in the most prestigious journals, including the *Lancet*, the *New England Journal of Medicine* and *Gastroenterology*, a publication record that continues until today.

Apart from her clinical and research activities, she has served on numerous committees of the University of Toronto, as Associate Editor or Member of the Editorial Board of many journals and in several other academic functions.

Given her outstanding contributions to hepatology, she has received already many honours and awards, among others the Distinguished Service Award of the Canadian Liver Foundation, the Queen's Jubilee Medal of the Government of Canada, the Golden Medal Award of the Canadian Association for the Study of the Liver and the Canadian Liver Foundation, the May Cohen Award for Women Mentors by the Canadian Medical Association, and the Distinguished Achievement Award of the American Association for the Study of Liver Diseases.

Today we have the pleasure to present to you the International Dame Sheila Sherlock Award, named after your first and certainly most important teacher, mentor and role model in academic medicine.

On behalf of the Prize Committee I would like to congratulate you on this highly deserved recognition of your outstanding contributions to clinical hepatology that carry on the legacy of the late Dame Sheila Sherlock.

14
Award Lecture: Past, present and future of hepatology

E. J. L. HEATHCOTE

INTRODUCTION

I had the good fortune to first meet the late Professor Dame Sheila Sherlock when I was her intern in 1968; thus I was exposed to a wide variety of liver disease at a very early stage in my medical career. I believe close to 40 years as a practising hepatologist is a good basis for reflecting on the changes that have taken place since my somewhat 'nerve wracking' experience as a young intern. In those days management of chronic liver disease was nearly always for an acute change in someone with a previously unrecognized, often silent, chronic condition.

PRESENTATION, DIAGNOSIS AND MANAGEMENT OF LIVER DISEASE IN 1968

Acute hepatic decompensation secondary to a sudden gastrointestinal haemorrhage was one of the most common causes of hospital admission in our patients, and sadly often the cause of their rapid demise. In 1968 there was no endoscopy, and the only agent we had available to treat a probable variceal haemorrhage was intravenous vasopressin. We had no intensive-care unit, and I well remember one evening when five patients in the ward simultaneously had an upper gastrointestinal haemorrhage! Little wonder that the mortality rate for variceal haemorrhage at that time was $>50\%$, in marked contrast to the $<16\%$ reported in the 21st century[1]. Ascites, either spontaneous or precipitated by our overzealous resuscitative manoeuvres for haemorrhage, was another common condition in the liver ward. It was recognized that those with fever and/or abdominal tenderness had a poor prognosis but we did not appreciate clinically silent spontaneous bacterial peritonitis (SBP) as both diagnostic and therapeutic paracentesis were considered too risky a procedure! Subsequently Conn[2] showed not only that this procedure was safe, but that 10% of those with ascites had silent SBP. Chronic hepatic encephalopathy was also common as the only satisfactory way to control variceal bleeding was achieved by performing a portacaval shunt.

Obstructive jaundice could be recognized only by the symptom of pruritus and compatible liver biochemical tests, but the aetiology was hard to determine except for patients who tested positive for antimitochondrial antibodies (AMA). The identification of AMA as a serum marker for primary biliary cirrhosis (PBC) by the late Professor Deborah Doniach and her colleagues[3] was a major breakthrough in hepatology, as it saved many individuals from undergoing unnecessary laparotomy. With no ultrasound, endoscopic retrograde cholangiopancreatography (ERCP) or magnetic resonance cholangiography (MRC) it was impossible to determine in AMA-negative yet cholestatic individuals whether they had a biliary stricture, a common duct stone or a tumour!

Professor Sherlock was also responsible for paediatric hepatology, and I well remember the unfortunate children with an unknown cholestatic syndrome manifested often by severe pruritus and almost generalized xanthomatosis, frequently with a xanthomatous neuropathy and sometimes even with green teeth!

Fulminant liver failure caused the same alarm and concern as it does now. As an intern I participated in several hours of exchange transfusion in young persons with acute liver failure, all to no avail because of the lack of liver transplantation. We rarely knew the cause of their liver failure and certainly did not fully appreciate that we were often handling large volumes of infectious material and taking grossly inadequate precautions!

Persistent (chronic) hepatitis of uncertain aetiology was what we observed most often in the ambulatory-care setting. Neither hepatitis B nor hepatitis C had been recognized, but we could recognize autoimmune hepatitis from the autoantibody pattern and presence of hyperglobulinaemia. We also noted that certain medications were associated with chronic hepatitis, e.g. alpha-methyldopa[4].

The diagnostic tools available in the late 1960s were very limited compared to those currently available, so we appropriately concentrated on a detailed history and physical examination. An array of biochemical blood tests was available, but there were few immunological tests. As interns 'on call' we were able to order only a very limited choice of tests: a complete blood count, urea and electrolytes, perhaps a serum alanine aminotransferase (ALT) and calcium if this was really deemed necessary!

The most noticeable changes since 1968 have been technological. As an intern we could order a flat plate of the abdomen and a chest X-ray. We also had access to venography, angiography, and cholangiography. The technetium-labelled colloid scan was the best non-invasive test we had for cirrhosis – marked uptake of the isotope, seen predominantly in the spleen and vertebral column, was what we were looking for to support this diagnosis.

It was the late Professor Dame Sheila Sherlock who pioneered the use of percutaneous liver biopsy in the 1950s. I learned very early that the most important factor determining the success of a liver biopsy was the interpretation by the hepatopathologist. In my case that person was the late Professor Peter Scheuer. I greatly admired his detailed analysis of every cut made. I also learned at this early age the necessity of an adequately sized specimen. Biochemical analyses of liver tissue were well developed and these analyses played an important role in our ability to be able to diagnose several metabolic diseases of childhood.

MAJOR INFLUENCES IN HEPATOLOGY 1968–2007

These have been successively in the fields of immunology followed by virology and more recently by genetics/genomics – part of the 'explosion' in molecular biology. Concurrent with these advances in the basic medical sciences have been those in technology. First to revolutionize our investigation of patients with liver disease was ultrasonography, which entered the field in the early 1970s. To be able to determine who with cholestatic jaundice had dilated bile ducts, without making a risky attempt at percutaneous cholangiography or laparotomy, provided a tremendous advance. Now we can even ask patients to swallow the ultrasound probe, allowing even better delineation of the pancreas and biliary tree! Upper panendoscopy, initially using a fairly rigid endoscope, was followed by the more flexible fibrescope, which has fortunately been replaced by video equipment. Now we can not only visualize the cause of most upper gastrointestinal haemorrhage but we can institute immediate therapy in most instances. This technology has in part been responsible for the dramatic fall in mortality from variceal haemorrhage. The introduction of the computed tomography (CT) scan and accompanying angiography allowed us to delineate hepatic tumours, but it is because of the ever-improving resolution of ultrasound that we are able to safely screen for hepatocellular carcinoma (HCC) and identify them much earlier, in many instances at a size when cure is possible. This was unheard of when I was training; then a diagnosis of HCC was a universal death sentence. The arrival of ERCP dramatically changed the face of gastroenterology and opened up yet another subspecialty, that of therapeutic endoscopy. However, when ERCP was first introduced in the early 1970s, therapeutic manoeuvres were yet to be introduced. Nevertheless, this was the first time we could make a confident diagnosis of primary sclerosing cholangitis without a laparotomy or autopsy! We had to wait a while before non-surgical removal of common bile duct stones was possible. Of course, in the past 10 years magnetic resonance imaging (MRI) technology has for the most part supplanted ERC when only diagnosis of the cause of large duct biliary tract disease is required. Today there is rarely any indication for an ERCP in an anicteric patient with possible cholestatic liver disease.

The downside of the easy availability of abdominal ultrasound is that many individuals with often benign liver lesions, who would otherwise never have known that they had a haemangioma, a cyst, or a benign tumour, are identified, often creating a great deal of unnecessary anxiety. Undue concern about these benign liver lesions is nevertheless a small price to pay for the tremendous advantage of this totally safe and effective non-invasive test.

The recognition of the extent of silent liver disease has definitely changed the face of hepatology. The availability of more advanced investigative tools, the introduction of routine screening and the recognition by physicians that an individual does not have to be an alcoholic to have liver disease, are all responsible for the surge in reporting of 'asymptomatic' liver disease. The first of many journals dedicated only to the liver, i.e. *Hepatology*, was started in 1981. There is now a plethora of journals dedicated to the liver.

AUTOIMMUNE LIVER DISEASE

Autoimmune hepatitis (AIH) was well described in the 1950s and its manifestations were elaborated over the next decade[5]. The three 'classic' trials of immunosuppressive therapy for AIH were all designed and conducted during the late 1960s although publication did not take place until the early 1970s[6–8]. Now we recognize that certain individuals thought to have AIH were in fact suffering from chronic viral hepatitis, some even from another autoimmune disease such as primary sclerosing cholangitis (PSC) or PBC. The arrival of ERC allowed us to recognize the overlap of AIH and PSC, particularly in children[9]. The introduction of including liver biochemical tests at the time of routine checkup, or when applying for some form of insurance coverage, has been another factor responsible for the increase in recognition of asymptomatic liver disease. It would have been unheard of for there to be a case of asymptomatic AIH when the three classic randomized controlled trials of therapy were designed; hence it is erroneous to translate the results of these trials to patients with asymptomatic AIH.

The term 'primary biliary cirrhosis' was first coined in 1950[10]. In the early 1990s it became recognized that some individuals had all the features of PBC but repeatedly tested negative for antimicrobial antibody (AMA), hence the term 'AMA-negative PBC'[11]. Much more sophisticated ELISA incorporating the three inner mitochondrial membrane antigenic determinants of AMA has found some of these cases to test AMA-positive, but not all[12]. Asymptomatic PBC now accounts for 60% or more of affected individuals diagnosed with this form of chronic liver disease. Our current dilemma is how to predict who with early PBC will remain stable, perhaps for ever, and who will progress.

Although ERC and MRC have been the most useful tools to determine who has PSC, we have yet to determine how to diagnose this disease prior to the development of irreversible stricturing of the biliary tree. The identification of small-duct PSC has identified another form of this disease, but long-term follow-up studies do not suggest that this is an earlier manifestation of classical PSC, i.e. with large-duct disease[13]. The introduction of MRC has facilitated recognition of secondary causes of stricturing and beading of the biliary tree, sometimes reversible, such as autoimmune pancreatitis cholangitis[14].

VIRAL HEPATITIS AND ITS ASSOCIATIONS

I returned to the liver unit at the Royal Free Hospital in 1971, just when the first serum test for hepatitis B became available. The topic of my thesis work was to identify risk factors for hepatitis B transmission and thereby the sexual transmission of hepatitis B was first reported[15]. The 1970s and early 1980s was a very exciting time. David Shafritz and colleagues[16] demonstrated the association of primary hepatocellular carcinoma in carriers of hepatitis B, and Mario Rizzetto et al.[17] first described co-infection with delta virus and hepatitis B. I still remember that my very first oral presentation at the Royal College of Physicians in 1972[18] was to describe the role of hepatitis B in patients with polyarteritis nodosa. That same year I made my first visit to the National Institute of Health (NIH) and met with Dr Stephen Feinstone, who asked me to take some slides of what he thought was hepatitis A virus in stool, back to the UK to the then Chief of Virology at the Burroughs Welcome Laboratory, Dr June Almeida, for her opinion. Dr Feinstone published his identification of hepatitis A virus 2 years later in 1974[19].

At about that time it became clear there was another form of viral hepatitis associated with blood transfusion which was caused neither by hepatitis A nor B (non-A, non-B hepatitis). It took more than a decade, an incalculable number of hours of hard work and enormous sums of money before Michael Houghton and his team at Chiron published in 1989 how they had identified hepatitis C[20]. It is now close to 40 years since the epidemic of hepatitis C started in the Western world[21] as a result of the popularity of illicit injection drugs. Hepatitis C is now the most common risk factor for HCC in the West[22]. The non-hepatic complications of chronic infection with hepatitis C have brought this disease into view for all internists. Symptomatic cryoglobulinaemia is predominantly caused by infection with hepatitis C[23] and B cell non-Hodgkin's lymphoma may be a complication of the same[24]. The identification of the human immunodeficiency virus (HIV) has with good reason dominated the field of infectious diseases since its discovery in the early 1980s but only more recently have the dangers of co-infection in those with HIV with either hepatitis B and/ or C become appreciated. The recognition that more severe liver disease occurs in co-infected individuals is being realized only now, when HIV infection can be adequately suppressed in the long term[25].

ADVANCES ACHIEVED USING MOLECULAR BIOLOGICAL TECHNIQUES

Although it was already recognized that several chronic liver diseases appeared to be genetic, e.g. Wilson disease, haemochromatosis, alpha-1 antitripsin deficiency, it was only when more sophisticated molecular biological techniques became available that the specific gene mutation(s) thought responsible for these metabolic disorders were identified. There are now more than 24 different mutations in individuals with Wilson disease, and the full spectrum of gene mutations responsible for iron overload disorders has yet to be identified.

The cloning of the genes for many of the transporters located predominantly on the canalicular membrane of the hepatocyte revolutionized our understanding of the progressive familial intrahepatic cholestatic syndromes 1, 2 and 3[26]. The cause of benign recurrent intrahepatic cholestasis, that mysterious non-progressive rare disease that intermittently makes the life of affected individuals so miserable by causing profound pruritus, has been shown to be due to a different mutation in the same gene as is affected in PFIC type 1[27]. The late Professor Alagille described a syndrome that now bears his name, and has as one of its manifestations intrahepatic bile duct hypoplasia[28]. I later described a series of children with similar manifestations that we thought may be attributed to an intrauterine viral infection[29]. Perhaps they would now fall into the so-called non-syndromatic intrahepatic duct hypoplastic syndrome. The gene defect responsible for Alagille syndrome has been described – a mutation in the jagged 1 gene[30]. Other different gene mutations associated with various other causes of childhood cirrhosis[31,32] have been described in children living in isolated settlements for North American Indians, particularly in Quebec. We are now in the era of identifying the complex disease traits using SNIPS technology for several different forms of chronic liver disease. These studies, along with the development of several animal models[33,34], will allow us to better understand many forms of chronic liver disease, particularly those thought to have an autoimmune basis.

With the advent of so many new medications being brought onto the market, and a society encouraged by advertising to take a 'pill' for every ailment, it is little wonder that there has been a greater appreciation of the frequency of drug-induced liver disease and its diverse manifestations. Perhaps these untoward events may still be avoided in the near future by genetic testing[35].

Perhaps the most shocking revelation in terms of recognizing the aetiology of a specific liver disease has been our slow and hesitant acknowledgement that obesity, particularly central obesity, may lead to significant liver disease! Only now are we beginning to appreciate why only some, and not all, who are overweight develop the metabolic syndrome[36].

CHANGES IN THE PRACTICE OF HEPATOLOGY

There have been two major changes in the practice of hepatology in the past 40 years. One is the recognition of asymptomatic chronic liver disease and the other is the introduction of the life-saving manoeuvre of liver transplantation for end-stage liver disease. Eventually it will be the introduction of effective primary and secondary preventive strategies, developed to either prevent liver disease or to prevent the progression of already-present liver disease that will have the greatest beneficial effect on overall mortality from liver diseases.

ASYMPTOMATIC LIVER DISEASE.

In 1970 a test was developed to determine who was infected with hepatitis B, but as no treatment was then available it was inappropriate to screen for this infection unless the individual was a blood donor. It subsequently became clear that many individuals were chronically infected with hepatitis B – the prevalence of this chronic infection varied across the world, being much more common in the Far East and South Africa than in the West. Now that therapies are available which both reduce the likelihood of progression to cirrhosis and probably reduce the risk of HCC we need to consider the appropriateness/cost-effectiveness of universal screening for viral hepatitis B and C, both infections being asymptomatic in the majority. It is equally important to combine such screening with implementation of universal vaccination against hepatitis B[37]. It has been predicted that, by 2010, the rate of cirrhosis, decompensated liver disease and a need for transplantation due to chronic hepatitis C infection will have escalated since this disease was first recognized in the early 1990s[38]. Still less than half of those infected with viral hepatitis C living in North America have been tested, and new infections of both hepatitis B and C continue to silently take place, mostly in those who are exposed to contaminated needles or equipment. This risk is not limited to the use of illicit injection drugs, rather for many in the developing world infection results from iatrogenic reuse of needles or medical equipment[39]! Hepatitis C is one of the few chronic viral infections that can be cured – it used to be unthinkable that liver disease could be cured!

Asymptomatic autoimmune liver disease poses a different problem. At least half of those individuals given a diagnosis of PBC have no symptoms and the dilemma we now face is that some who have been given this diagnosis will develop progressive liver disease in their 'natural' lifetime and some will not[40]. Risk factors for disease progression in PBC need to be established before we can introduce clinical trials of new, more potent, therapies. Up to 20% of cases of AIH are now identified when asymptomatic[41]. Such individuals were never included in the classic trials of therapy evaluating prednisone ± azathioprine; thus there are no criteria to define who with asymptomatic AIH may benefit from immunosuppressive therapy. We face similar issues in individuals identified with fatty liver; no non-invasive markers that predict disease progression are known in this condition.

LIFE-SAVING PREVENTIVE STRATEGIES IN PATIENTS WITH LIVER DISEASE

Liver transplantation has indeed been little short of a miracle. Babies with biliary atresia, children with progressive familial intrahepatic cholestasis may now live into adulthood and countless adults are given a new lease of life. Less 'eye-catching', but nevertheless no less dramatic, has been the introduction of the hepatitis B vaccine. In Taiwan, where universal vaccination of all newborn babies was introduced in 1986, the hepatitis B chronic infection rate has been reduced 10-fold and a reduction in HCC already reported[37]. The introduction of chelation therapy prevents death in those with Wilson disease[42]. Corticoid steroid therapy reduced the 3-year mortality of symptomatic AIH from 60% to less than 20%[6–8]. A much more common genetic disease, namely haemochromatosis, when recognized and treated with regular phlebotomy, improves overall survival and prevents subsequent HCC in the non-cirrhotic[43]. Early recognition of both Wilson disease and haemochromatosis identified by screening siblings of index cases has allowed prevention of significant liver disease in affected siblings. The use of *N*-acetylcysteine within 48 h of taking an overdose of Tylenol, either with suicidal intent or inadvertently, has saved the lives of many young persons[44].

LIFE-PROLONGING STRATEGIES IN LIVER DISEASE

The introduction of primary prophylaxis with non-selective beta-blockers in patients with cirrhosis and large oesophageal varices has meant fewer cases of life-threatening haemorrhage[45]. Long-term oral antibiotics in patients following their first episode of spontaneous bacterial peritonitis have enhanced the survival of such individuals, so that now there is sufficient time to allow a work-up and subsequent liver transplantation[46]. This same group of investigators has also shown that the onset of the hepatorenal syndrome which may complicate the course of spontaneous bacterial peritonitis (SBP) can be significantly reduced by giving albumin infusions with the initial antibiotic therapy at the time of presentation with SBP[47]. Recently, studies have shown that both type I and II hepatorenal syndrome can be controlled so as to delay death sufficiently to allow liver transplant to occur[48].

Despite these effective strategies in place there appears to have been no overall improvement in mortality of individuals with cirrhosis from 1968 to 1998[49]. The introduction of routine screening of all cirrhotics for small HCC using ultrasound may significantly reduce mortality from HCC[50]. Nucleoside analogue therapy given to individuals with advanced hepatic fibrosis and ongoing viral replication due to hepatitis B has indicated that, within the first 36 months of treatment, a significant reduction in liver failure and HCC can be achieved[51]. Now the challenge is to identify all the individuals who are at risk and introduce treatment appropriately. Eradiction of hepatitis C leads to a significant reduction in HCC as long as the treatment is given prior to the development of cirrhosis[52].

HEPATOLOGY BEYOND 2006

There remain many unsolved mysteries in hepatology. There is an urgent need for a reliable, non-invasive, diagnostic test for cholangiocarcinoma. Epigenetics will probably become more relevant, particularly in diseases such as PSC, PBC, fatty liver, and viral hepatitis. In all branches of medicine long-term compliance with medications shown to be effective either in preventing or treating disease is a problem, particularly in youngsters. Perhaps we need to think of new ways to deliver drugs to combat compliance issues. All of us recognize the tremendous value which organ transplant plays in those with end-stage liver disease, but there surely has to be an alternative to this procedure which does not depend on organ donation from another. We all hope that a strategy using stem cell biological techniques will become the answer. Simultaneously much more attention needs to be paid to effective disease prevention strategies. It is certainly time to refocus on effective public health measures, particularly in our globe-trotting society where a spread of infection may occur so rapidly.

WHERE SHOULD WE BE GOING IN HEPATOLOGY BEYOND 2006?

The introduction of molecular genetics/genomics, regenerative medicine/stem cell biology, proteomics and pharmacogenetics will surely lead to enhanced understanding of the pathophysiology of liver disease. But these 'high-tech' methodologies must not blind us to the fact that more effective public health measures will save many more lives than delaying death in those with end-stage liver disease! The need to attend to social issues such as illicit drug use, inadequate housing and alcohol abuse, *and* the provision of universal vaccination against hepatitis B and A, will be achieved only by directing appropriate funding. But this needs to happen in concert with the recognition by the public of their need to take greater responsibility for their own general health. The introduction of health education programmes, certainly in terms of appropriate nutrition and exercise, are now being introduced in many countries.

References

1. Yan BM, Lee SS. Emergency management of bleeding esophageal varices: drugs, bands or sleep? Can J Gastroenterol. 2006;20:165–70.
2. Conn HO. Bacterial peritonitis: spontaneous or paracentetic? Gastroenterology. 1979;77: 1145–6.
3. Walker JG, Doniach D, Roitt IM et al. Serological tests in diagnosis of primary biliary cirrhosis. Lancet. 1965;1:827–31.
4. Goldstein GB, Lam KC, Mistilis SP. Drug-induced active chronic hepatitis. Am J Dig Dis. 1973;18:177–84.
5. Read AE, Sherlock S, Harrison CV. Active 'juvenile' cirrhosis considered as part of a systemic disease and the effect of corticosteroid therapy. Gut. 1963;4:378–93.
6. Cook GC, Mulligan R, Sherlock S. Controlled prospective trial of corticosteroid therapy in active chronic hepatitis. Q J Med. 1971;40:159–85.
7. Murray-Lyon IM, Stern RB, Williams R. Controlled trial of prednisone and azathioprine in active chronic hepatitis. Lancet. 1973;1:735–7.
8. Soloway RD, Summerskill WH, Baggenstoss AH et al. Clinical, biochemical, and histological remission of severe chronic active liver disease: a controlled study of treatments and early prognosis. Gastroenterology. 1972;63:820–33.
9. Gregorio GV, Portmann B, Karani J et al. Autoimmune hepatitis/sclerosing cholangitis overlap syndrome in childhood: a 16-year prospective study. Hepatology. 2001;33:544–53.
10. Ahrens JR EH, Payne MA, Kunkel HG, Eisenmenger WJ, Blondheim SH. Primary biliary cirrhosis, Medicine. 1950;29:299–364.
11. Michieletti P, Wanless IR, Bassendine MF et al. Antimitochondrial anibody negative primary biliary cirrhosis: a distinct syndrome of autoimmune cholangitis. Gut. 1994;35: 260–5.
12. Miyakawa H, Tanaka A, Kikuchi K et al. Detection of antimitochondrial autoantibodies in immunofluorescent AMA-negative patients with primary biliary cirrhosis using recombinant autoantigens. Hepatology. 2001;34:243–8.
13. Broome U, Glaumann H, Lindstom E et al. Natural history and outcome in 32 Swedish patients with small duct primary sclerosing cholangitis (PSC). J Hepatol. 2002;36:586–9.
14. Erkelens GW, Vleggaar FP, Lesterhuis W, van Buuren HR, van der Werf SD. Sclerosing pancreato-cholangitis responsive to steroid therapy. Lancet. 1999;354:43–4.
15. Heathcote J, Sherlock S. Spread of acute type B hepatitis in London. Lancet. 1973;1:1468–70.
16. Shafritz DA, Shouval D, Sherman HI, Hadziyannis SJ, Kew MC. Integration of hepatitis B virus DNA into the genome of liver cells in chronic liver disease and hepatocellular carcinoma. Studies in percutaneous liver biopsies and post-mortem tissue specimens. N Engl J Med. 1981;305:1067–73.

17. Smedile A, Dentico P, Zanetti A et al. Infection with the delta agent in chronic HBsAg carriers. Gastroenterology. 1981;81:992–7.
18. Heathcote EJL, Dudley FJ, Sherlock S. Association of polyarteritis and Australia antigen. Gut. 1972;13:319.
19. Feinstone SM, Kapikian AZ, Gerin JL, Purcell RH. Buoyant density of the hepatitis A virus-like particle in cesium chloride. J Virol. 1974;13:1412–14.
20. van der Poel CL, Reesink HW, Lelie PN et al. Anti-hepatitis C antibodies and non-A, non-B post-transfusion hepatitis in the Netherlands. Lancet. 1989;2:297–8.
21. Tanaka Y, Hanada K, Mizokami M et al. Inaugural article: A comparison of the molecular clock of hepatitis C virus in the United States and Japan predicts that hepatocellular carcinoma incidence in the United States will increase over the next two decades. Proc Natl Acad Sci USA. 2002;99:15584–9.
22. Davila JA, Morgan RO, Shaib Y, McGlynn KA, El-Serag HB Hepatitis C infection and the increasing incidence of hepatocellular carcinoma: a population-based study. Gastroenterology. 2004;127:1372–80.
23. Zignego AL, Ferri C, Pileri SA, Caini P, Bianchi FB; for the Italian Association of the Study of Liver (AISF) Commission on Extrahepatic Manifestations of HCV Infection. Extrahepatic manifestations of hepatitis C virus infection: a general overview and guidelines for a clinical approach. Dig Liver Dis. 2007;39:2–17.
24. Nieters A, Kallinowski B, Brennan P et al. Hepatitis C and risk of lymphoma: results of the European multicenter case–control study EPILYMPH. Gastroenterology. 2006;131:1879–86.
25. Rosenthal E, Pialoux G, Bernard N et al. for the Germivic Joint Study Group. Liver-related mortality in human-imunodeficiency-virus-infected patients between 1995 and 2003 in the French germivic Joint Study Group network (MORTAVIC 2003 Study). J Viral Hepatol. 2007;14:183–8.
26. Balistreri WF, Bezerra JA, Jansen P, Karpen SJ, Shneider BL, Suchy FJ. Intrahepatic cholestasis: summary of an American Association for the Study of Liver Diseases single-topic conference. Hepatology. 2005;42:222–35
27. van Mil SW, van der Woerd WL, van der Brugge G et al. Benign recurrent intrahepatic cholestasis type 2 is caused by mutations in ABCB11. Gastroenterology. 2004;127:379–84.
28. Alagille D, Estrada A, Hadchouel M, Gautier M, Odievre M, Dommergues JP. Syndromic paucity of interlobular bile ducts (Alagille syndrome or arteriohepatic dysplasia): review of 80 cases. J Pediatr. 1987;110:195–200.
29. Heathcote J, Deodhar KP, Scheuer P, Sherlock S. Intrahepatic cholestasis in childhood. N Engl J Med. 1976;295:801–5.
30. Oda T, Elkahloun AG, Pike BL et al. Mutations in the human Jagged1 gene are responsible for Alagille syndrome. Nat Genet. 1997;16:235–42.
31. Chagnon P, Michaud J, Mitchell G et al. A missense mutation (R565W) in cirhin (FLJ14728) in North American Indian childhood cirrhosis. Am J Hum Genet. 2002;71: 1443–9.
32. Phillips MJ, Azuma T, Meredith SL et al. Abnormalities in villin gene expression and canalicular microvillus structure in progressive cholestatic liver disease of childhood. Lancet. 2003; 362:1112–19.
33. Fickert P, Fuchsbichler A, Wagner M et al. Regurgitation of bile acids from leaky bile ducts causes sclerosing cholangitis in Mdr2 (Abcb4) knockout mice. Gastroenterology. 2004;127: 261–74.
34. Irie J, Wu Y, Wicker LS et al. NOD.c3c4 congenic mice develop autoimmune biliary disease that serologically and pathogenetically models human primary biliary cirrhosis. J Exp Med. 2006;203:1209–19.
35. Frist, William H. Health Care in the 21st Century. N Engl J Med. 2005;352:267–72.
36. Lam TK, Gutierrez-Juarez R, Pocai A et al. Brain glucose metabolism controls the hepatic secretion of triglyceride-rich lipoproteins. Nat Med. 2007;13:171–80.
37. Chang MH, Chen CJ, Lai MS et al. Universal hepatitis B vaccination in Taiwan and the incidence of hepatocellular carcinoma in children. Taiwan Childhood Hepatoma Study Group. N Engl J Med. 1997;336:1855–9.
38. Kim WR. The burden of hepatitis C in the United States. Hepatology. 2002;36(Suppl. 1): S30–4.

39. Hutin YJ, Duclos P, Hogerzeil H, Ball A, Carr R. Unsterile injections and emergence of human pathogens. Lancet. 2002;359:2280.
40. Pares A, Caballeria L, Rodes J. Excellent long-term survival in patients with primary biliary cirrhosis and biochemical response to ursodeoxycholic acid. Gastroenterology. 2006;130:715–20.
41. Feld JJ, Dinh H, Arenovich T, Marcus VA, Wanless IR, Heathcote EJ. Autoimmune hepatitis: effect of symptoms and cirrhosis on natural history and outcome. Hepatology. 2005;42:53–62.
42. Walshe JM, Yealland M. Chelation treatment of neurological Wilson's disease. Q J Med. 1993;86:197–204.
43. Niederau C, Fischer R, Sonnenberg A, Stremmel W, Trampisch HJ, Strohmeyer G. Survival and causes of death in cirrhotic and in noncirrhotic patients with primary hemochromatosis. N Engl J Med. 1985;313:1256–62.
44. Scalley RD, Conner CS. Acetaminophen poisoning: a case report of the use of acetylcysteine. Am J Hosp Pharm. 1978;35:964-7
45. Lebrec D, Poynard T, Hillon P, Benhamou JP. Propranolol for prevention of recurrent gastrointestinal bleeding in patients with cirrhosis: a controlled study. N Engl J Med. 1981; 305:1371–4.
46. Gines P, Rimola A, Planas R et al. Norfloxacin prevents spontaneous bacterial peritonitis recurrence in cirrhosis: results of a double-blind, placebo-controlled trial. Hepatology. 1990;12:716–24.
47. Sort P, Navasa M, Arroyo V et al. Effect of intravenous albumin on renal impairment and mortality in patients with cirrhosis and spontaneous bacterial peritonitis. N Engl J Med. 1999;341:403–9.
48. Ortega R, Gines P, Uriz J et al. Terlipressin therapy with and without albumin for patients with hepatorenal syndrome: results of a prospective, nonrandomized study. Hepatology. 2002;36:941–8.
49. Roberts SE, Goldacre MJ, Yeates D. Trends in mortality after hospital admission for liver cirrhosis in an English population from 1968 to 1999. Gut. 2005;54:1615–21.
50. Bruix J, Sherman M, Llovet JM et al.; EASL Panel of Experts on HCC. Clinical management of hepatocellular carcinoma. Conclusions of the Barcelona-2000 EASL conference. European Association for the Study of the Liver. J Hepatol. 2001;35:421–30.
51. Liaw YF, Sung JJ, Chow WC et al.; Cirrhosis Asian Lamivudine Multicentre Study Group. Lamivudine for patients with chronic hepatitis B and advanced liver disease. N Engl J Med. 2004;351:1521–31.
52. Yoshida H, Shiratori Y, Moriyama M et al. Interferon therapy reduces the risk for hepatocellular carcinoma: a national surveillance program of cirrhotic and noncirrhotic patients with chronic hepatitis C in Japan. IHIT Study Group. Inhibition of hepatocarcinogenesis by interferon therapy. Ann Inter Med. 1999;131:174.

15
Progression of chronic hepatitis: from inflammation to fibrosis

M. PINZANI

INTRODUCTION

This chapter is divided into six sections, each of which is directed at delivering key concepts, and an update on various other key concepts in this area of hepatology.

HEPATIC FIBROSIS DEVELOPS WITH VARYING MORPHOLOGICAL AND SPATIAL PATTERNS

Although cirrhosis is the common result of progressive fibrogenesis there are distinct patterns of fibrotic development related to the underlying disorders causing the fibrosis[1]. Biliary fibrosis, due to the co-proliferation of reactive bile ductules and periductular (myo)fibroblast-like cells at the portal–parenchymal interface, tends to follow a portal to portal direction. This leads to the formation of portal–portal septa surrounding liver nodules, where the central vein and its connections with the portal tract are preserved until late stages. In contrast, the chronic viral hepatitis pattern of fibrosis is considered the result of portal–central (vein) bridging necrosis, thus originating portal–central septa. In addition, this form of fibrogenic evolution is characterized by the presence of 'interface' hepatitis and development of portal–portal septa and septa ending blind in the parenchyma, and by rapid derangement of the vascular connections with the portal system (earlier portal hypertension). The so-called central to central (vein) form of fibrogenic evolution is in general secondary to venous outflow problems (e.g. chronic heart failure) and is characterized by the development of central to central septa and 'reversed lobulation'. Finally, a peculiar type of fibrosis development (pericellular/sinusoidal) is observed in alcoholic and metabolic liver diseases (e.g. NASH), in which the deposition of fibrillar matrix is concentrated around the sinusoids (capillarization) and around groups of hepatocytes (chicken-wire pattern). These different patterns of fibrogenic evolution are related to different factors, and particularly: (1) the topographic localization of tissue damage, (2) the relative concentration of profibrogenic factors, and (3) the prevalent profibrogenic mechanism(s). In addition, these patterns imply the participation of different cellular effectors of the fibrogenic process.

DIFFERENT PROFIBROGENIC CELLS CONTRIBUTE TO FIBROGENESIS

As the liver becomes fibrotic there are both quantitative and qualitative changes in the composition of the hepatic extracellular matrix (ECM). The total content of collagenous and non-collagenous components increases 3–5-fold, accompanied by a shift in the type of ECM in the subendothelial space from the normal low-density basement membrane-like matrix to interstitial-type matrix containing fibril-forming collagens. Most of our present knowledge on the cell and molecular biology of hepatic fibrosis derives from *in vitro* studies employing culture-activated hepatic stellate cells (HSC) isolated from rat, mouse or human liver, mostly because of the possibility of isolating them from liver tissue with a relatively high purity. However, it is now evident that distinct ECM-producing cells, each with a distinct localization and a characteristic immunohistochemical and/or electron microscopic phenotype, are likely to contribute to liver fibrosis[1]. These include: fibroblasts and myofibroblasts of the portal tract, smooth muscle cells localized in vessel walls, and myofibroblasts localized around the centrilobular vein. The relative participation of these various cell types is dependent on the development of distinct patterns of fibrosis. Major efforts are currently being made in order to identify and characterize the origin of the cell types responsible for the fibrogenic process. By employing cell identification markers[1], three populations of ECM-producing cells can be identified: (1) septal myofibroblasts, present in the inner part of fibrous septa, expressing a panel of cell markers identical to portal myofibroblasts; (2) activated HSC, located in capillarized sinusoids adjacent to expanded portal tracts; and (3) interface myofibroblasts, located at the edge of fibrous septa. This last population presents with an expression profile intermediate between portal myofibroblasts and activated HSC. It is conceivable that interface myofibroblasts derive from activated HSC recruited at the site of active fibrogenesis, where the occurrence of cell damage, ECM degradation by gelatinases and inflammatory infiltration are greatest. In addition, consistent evidence obtained both in humans and on animal models of fibrogenesis suggest that bone marrow-derived stromal cells recruited at the site of liver injury could contribute to this population of fibrogenic cells[2,3].

It is likely that all these ECM-producing cell types undergo a process of activation in conditions of chronic liver damage or, anyhow, in conditions in which the physiological homeostasis of the tissue is chronically altered. For reasons previously mentioned, the process of HSC activation has been the object of several studies and a consistent bulk of information is presently available[4]. Following prolonged culture on plastic, HSC undergo a process of activation from the quiescent 'storing' phenotype to the highly proliferative 'myofibroblast-like' phenotype. This process is still regarded as similar to that occurring in liver tissue following chronic damage, although this assumption probably represents an oversimplification. The activated phenotype is characterized by a dramatically increased synthesis of collagen types I and III, that appears predominant over the synthesis of collagen type IV (I > III ≫ IV) and other ECM components. Studies performed in recent years have emphasized some important aspects potentially related to the initiation of HSC activation. A first important element concerns disruption of the normal ECM pattern that follows liver tissue injury and acute inflammation. A perturbation in the composition of the normal hepatic ECM and/or of the cell–cell relationship between epithelial and mesenchymal cells present in liver tissue, typical of some cholestatic disorders (i.e. those characterized by bile duct proliferation and lobular invasion) could also be considered a potent stimulus for the activation and proliferation of HSC, as well as other ECM-producing cells. Indeed, loss of adhesion with the various elements constituting the basal membrane-like ECM of the space of Disse is likely to determine a marked increase in the proliferative and synthetic properties of HSC. This issue is becoming more and more important with the demonstration that the movement, shape and proliferation of cells is greatly influenced by the cooperation of ECM components and cell adhesion molecules.

PROFIBROGENIC MECHANISMS

The mechanisms responsible for the fibrogenic evolution of chronic liver diseases (CLD) can be summarized in three main groups: chronic activation of the wound-healing reaction, oxidative stress-related molecular mechanisms, and the derangement of the so-called 'epithelial–mesenchymal' interaction.

Chronic activation of the wound-healing reaction

This process, which is highly efficient in the presence of a single acute tissue insult, leads to progressive scarring when tissue damage is chronic. In other terms, deposition of fibrillar matrix rather then organized tissue regeneration becomes the best option in order to maintain tissue continuity. Hepatic fibrogenesis due to the chronic activation of the wound-healing reaction is characterized by the following key features: (a) the persistence of hepatocellular/cholangiocellular damage with variable degrees of necrosis and apoptosis; (b) a complex inflammatory infiltrate including mononuclear cells and immunocompetent cells; (c) the activation of different types of ECM-producing cells (HSC, portal myofibroblasts, etc.) with marked proliferative, synthetic and contractile features; (d) marked changes in the quality and quantity of the hepatic ECM associated with very limited or absent possibilities of remodelling and regeneration[5].

Oxidative stress

When chronic liver injury is not clearly associated with an abundant inflammatory infiltrate, other soluble agents may sustain the activation of HSC through pathways that are specific for a particular type of damage. Evidence of oxidative stress has been detected in almost all the clinical and experimental conditions of CLD with different aetiology and fibrosis progression rate, often in association with decreased antioxidant defences. As already proposed for atherosclerosis and chronic degenerative diseases of the central nervous system (CNS), oxidative stress-related molecules such as reactive oxygen intermediates (ROI) and reactive aldehydes, may act as mediators able to modulate tissue and cellular events responsible for the progression of liver fibrosis[6]. In alcoholic liver injury, for example, acetaldehyde, the main metabolite of ethanol, is able to increase gene transcription and synthesis of different ECM components in activated HSC[7,8]. In addition to acetaldehyde, products of lipid peroxidation generated by exposure to ethanol or the production of iron overload may also perpetuate HSC activation[9]. Along these lines, stimulation of lipid peroxidation or exposure to 4-hydroxynonenal (4-HNE), a highly reactive aldehydic end-product of lipid peroxidation, increases procollagen I gene expression in activated human HSC[10,11].

In the increasingly relevant area of non-alcoholic fatty liver diseases (NAFLD), a 'two-hit theory' best describes the progression from simple steatosis to non-alcoholic steatohepatitis (NASH), fibrosis, or cirrhosis. The first hit consists of the accumulation of excessive hepatic fat primarily owing to insulin resistance. The persistence of this metabolic derangement leads to the second hit, characterized by a predominant proinflammatory and profibrogenic role of oxidative stress products[12].

Derangement of the epithelial–mesenchymal interaction

The spectrum of cholangiopathies ranges from conditions in which a normal epithelium is damaged by autoimmunity, infectious agents, toxic compounds or ischaemia, to genetically determined diseases arising from an abnormal bile duct biology, such as cystic fibrosis or biliary atresia. Probably because of the well-established heterogeneity in cholangiocyte function, a different portion of the biliary tree appears to be preferentially affected in specific cholangiopathies[13].

A feature common to all cholangiopathies is the coexistence of cholestasis with cholangiocyte loss (by apoptotic or lytic cell death), cholangiocyte proliferation and various degrees of portal and periportal inflammation and fibrosis. Both bile duct proliferation and ductular metaplasia are associated with profound changes in the surrounding mesenchymal cells and ECM. As already mentioned, it is likely that, at least in the early phases, ECM-producing cells other than HSC are primarily involved, whereas HSC become subsequently involved when proliferating bile ducts tend to invade lobular areas. It is still unclear whether the changing epithelial phenotype directly induces an alteration in portal mesenchymal cells and ECM, or whether the epithelial cell changes are induced by modifications in ECM; however, the intense crosstalk between epithelial and mesenchymal cells is consistent with Desmet's hypothesis that 'ductular reaction is the pace-maker of portal fibrosis'[14,15]. Cytokines and proinflammatory mediators, released in the portal spaces, probably contribute to these processes by activating fibrogenesis, stimulating apoptotic and proliferative responses, damaging the peribiliary circulation, increasing the expression of histocompatibility antigens in cholangiocytes and by altering the transport functions of the epithelium. An emerging concept is that bile duct epithelial cells are active participants in inflammatory diseases and, in pathological conditions, secrete proinflammatory and chemotactic cytokines, such as interleukin-6 (IL-6), tumour necrosis factor alpha (TNF-α), IL-8, and monocyte chemotactic protein-1 (MCP-1)[16–19], together with growth factors able to activate mesenchymal cells and matrix production (endothelin 1 (ET-1), platelet-derived growth factor-BB (PDGF-BB), transforming growth factor beta (TGF-β_2), connective tissue growth factor (CTGF)[20–25].

THE REGRESSION OF FIBROSIS AND HEPATIC STELLATE CELL APOPTOSIS

Regression of liver fibrosis and, possibly, cirrhosis has also been reported in patients with chronic liver disease[26–29], once cessation of the causative agent is removed. Although the possibility for regression of complete cirrhosis is uncertain[30], specific induction of apoptosis in human HSC/MF may represent a realistic objective for cell-targeted therapy of liver fibrosis. However, despite these encouraging observations, information concerning the susceptibility to apoptosis of human HSC[31], immortalized HSC[32] or human hepatic myofibroblasts[33] is still scarce. Recent work by Novo and co-workers has shown that activated human HSC, displaying a fully activated myofibroblast-like phenotype, did not undergo spontaneous apoptosis and survived to prolonged serum deprivation, Fas activation, or exposure to NGF, TNF-α, oxidative stress mediators, doxorubicin and etoposide [34]. Induction of classical, mitochondria-driven apoptosis in HSC was observed only when protein synthesis or transcription were inhibited. Activated human HSC express significantly higher levels of Bcl-2, Akt and phospho-Akt, but lower levels of Bax, than freshly isolated, quiescent cells. However, Bcl-2 overexpression was lost in cells made senescent by prolonged serial culture on plastic. Accordingly, senescent cells as well as Bcl-2-silenced HSC by RNA interference strategy became susceptible to TNF-α-induced apoptosis, indicating the key role of Bcl-2 in sustaining survival to pro-apoptotic stimuli. Immunohistochemical data provided in the study by Novo and co-workers clearly indicate that myofibroblast-like cells at the interface between septa and parenchyma significantly overexpress Bcl-2, confirming the possible '*in vivo*' relevance of '*in vitro*' data. In summary, the marked resistance to apoptosis of human HSC/MF may also suggest that the observed reversal of fibrosis in the human liver may not be entirely dependent on HSC/MF apoptosis, and explain the longer time frame necessary to observe a significant reduction in the amount of ECM.

FIBROGENESIS AND ANGIOGENESIS

Angiogenesis is a hypoxia-stimulated and growth-factor-dependent process consisting in the formation of new vascular structures from pre-existing blood vessels. Formation of new vessels is known to occur in several organs and to be critical for both growth and repair of tissues in several pathophysiological conditions[35–40]. However, it is increasingly clear that angiogenesis occurring during chronic wound-healing and fibrogenesis provides a key contribution to disease progression[41]. Hepatic angiogenesis differs from homologous processes in other tissues for a number of reasons, including the existence of two different types of microvascular structures in the liver (i.e. large vessels lined by a continuous endothelium vs sinusoids lined by a phenestrated endothelium)[42], the apparent production of a liver-specific angiogenic factor AN-GPTL3[43] and the unique but not homogeneous phenotypic profile and functional role of HSC[44–47]. A possible interplay and/or association between fibrogenesis and angiogenesis in CLD is suggested and supported by several findings: (a) angiogenesis and up-regulation of vascular endothelial growth factor (VEGF) expression has been documented in various models of acute and chronic liver injury[41,48–50], as well as in specimens from human fibrotic/cirrhotic liver and hepatocellular carcinoma[41,51–53]; (b) in HSC, hypoxia has been shown to up-regulate expression of VEGF[49,54–56], VEGF receptor type I (fms-like tyrosine kinase receptor or Flt-1)[49,54] and collagen type I[49]; (c) VEGF has been proposed to directly stimulate proliferation and expression of α_1(I)-procollagen mRNA in activated rat HSC[50]; (d) paracrine expression of VEGF by rat HSC as well as by hepatocytes has been shown to regulate the phenotype (i.e. fenestration and CD-31 expression) of liver sinusoidal endothelial cells[57], a feature of possible relevance in CLD. Recent data from our group outline a direct link between exposure to hypoxic conditions of HSC and release of pro-angiogenic cytokines able to stimulate their migration and chemotaxis. In this context VEGF seems to play major role by interacting with Flk-1, and that, following their activation and phenotypical modulation, HSC tend to acquire a generic 'pro-angiogenic' phenotype (Novo E et al., unpublished observations). In aggregate the available experimental evidence suggest that HSC may represent a cellular crossroad connecting neo-angiogenesis to inflammation and fibrogenesis. Indeed, these cells represent a target for the multiple action of VEGF and Ang-1, including stimulation of proliferation, collagen type I synthesis and recruitment of HSC[48,49]. At the same time HSC/MF are a significant source of these angiogenic cytokines under conditions of hypoxia[40,54,56], acute and chronic liver injury[40,48–50,53], possibly through the contribution of a number of growth factors, proinflammatory cytokines and conditions of altered metabolic control, as recently suggested by data indicating that leptin is able to up-regulate VEGF[57].

References

1. Cassiman D, Roskams T. Beauty is in the eye of the beholder: emerging concepts and pitfalls in hepatic stellate cell research. J Hepatol. 2002;37:527–35.
2. Forbes SJ, Russo F, Burra P, Rugge M, Wright N, Alison M. A significant proportion of myofibroblasts are of bone marrow origin in human liver fibrosis. Gastroenterology. 2004;126:955–63.
3. Russo FP, Alison MR, Bigger BW et al. The bone marrow functionally contributes to liver fibrosis. Gastroenterology. 2006;130:1807–21.
4. Friedman SL. Molecular regulation of hepatic fibrosis, an integrated cellular response to tissue injury. J Biol Chem. 2000;275:2247–50.
5. Pinzani M. Liver fibrosis. Springer Semin Immunopathol. 2000;21:475–90.
6. Parola M, Robino G. Oxidative stress-related molecules and liver fibrosis J Hepatol. 2001; 35:297–306.
7. Casini A, Cunningham M, Rojkind M, Lieber CS. Acetaldehyde increases procollagen type I and fibronectin gene transcription in cultured rat fat-storing cells through a protein synthesis-dependent mechanism. Hepatology. 1991;13:758–65.
8. Casini A, Ceni E, Salzano R et al. Neutrophil-derived superoxide anion induces lipid peroxidation and stimulates collagen synthesis in human hepatic stellate cells. Role of nitric oxide. Hepatology. 1997;25:361–7.
9. Pietrangelo A. Metals, oxidative stress, and hepatic fibrogenesis. Semin Liver Dis. 1996;16: 13–30.
10. Parola M, Pinzani M, Casini A et al. Induction of procollagen type I expression and synthesis in human hepatic stellate cells by 4-hydroxy-2,3-nonenal and the other 4-hydroxy-2,3-alkenals is related to their molecular structure. Biochem Biophys Res Commun. 1996;222:261–4.
11. Parola M, Robino G, Marra F et al. HNE interacts directly with JNK isoforms in human hepatic stellate cells. J Clin Invest. 1998;102:1942–50.
12. Clark JM, Diehl AM. Nonalcoholic fatty liver disease: an underrecognized cause of cryptogenic cirrhosis. J Am Med Assoc. 2003;289:3000–4.
13. Alpini G, McGill JM, Larusso NF. The pathobiology of biliary epithelia. Hepatology. 2002; 35:1256–68.
14. Desmet V, Roskams T, Van Eyken P. Ductular reaction in the liver. Pathol Res Pract. 1995; 191:513–24.
15. Roskams T, Desmet V. Ductular reaction and its diagnostic significance. Semin Diagn Pathol. 1998;15:259–69.
16. Spirli C, Nathanson MH, Fiorotto R et al. Proinflammatory cytokines inhibit secretion in rat bile duct epithelium. Gastroenterology. 2001;121:156–69.
17. Morland CM, Fear J, McNab G, Joplin R, Adams DH. Promotion of leukocyte transendothelial cell migration by chemokines derived from human biliary epithelial cells *in vitro*. Proc Assoc Am Phys. 1997;109:372–82.
18. Marra F, DeFranco R, Grappone C et al. Increased expression of monocyte chemotactic protein-1 during active hepatic fibrogenesis: correlation with monocyte infiltration. Am J Pathol. 1998;152:423–30.
19. Hsieh CS, Huang CC, Wu JJ et al. Ascending cholangitis provokes IL-8 and MCP-1 expression and promotes inflammatory cell infiltration in the cholestatic rat liver. J Pediatr Surg. 2001;36:1623–8.
20. Caligiuri A, Glaser S, Rodgers RE et al. Endothelin-1 inhibits secretin-stimulated ductal secretion by interacting with ETA receptors on large cholangiocytes. Am J Physiol. 1998; 275:G835–46.
21. Grappone C, Pinzani M, Parola M et al. Cholangiocytes contribute to cholestatic fibrosis in the rat through platelet-derived growth factor. J Hepatol. 1999;31:100–9.
22. Kinnman N, Hultcrantz R, Barbu V et al. PDGF-mediated chemoattraction of hepatic stellate cells by bile duct segments in cholestatic liver injury. Lab Invest. 2000;80:697–707.
23. Kinnman N, Goria O, Wendum D et al. Hepatic stellate cell proliferation is an early platelet-derived growth factor-mediated cellular event in rat cholestatic liver injury. Lab Invest. 2001;81:1709–16.
24. Milani S, Herbst H, Shuppan D, Stein H and Surrenti C. Transforming growth factor-β1 and -β2 are differentially expressed in fibrotic liver disease. Am J Pathol. 1991;139:1221–9.

25. Sedlaczek N, Jia JD, Bauer M et al. Proliferating bile duct epithelial cells are a major source of connective tissue growth factor in rat biliary fibrosis. Am J Pathol. 2001;158:1239–44.
26. Dufour JF, DeLellis R, Kaplan MM. Reversibility of hepatic fibrosis in autoimmune hepatitis. Ann Intern Med. 1997;127:981–5.
27. Wanless IR, Nakashima E, Sherman M. Regression of human cirrhosis: morphologic features and the genesis of incomplete septal cirrhosis. Arch Pathol Lab Med. 2000;124: 1599–607.
28. Hammel P, Couvelard A, O'Toole D et al. Regression of liver fibrosis after biliary drainage in patients with chronic pancreatitis and stenosis of the common bile duct. N Engl J Med. 2001;344:418–23.
29. Poynard T, McHutchison J, Manns M et al. Impact of pegylated interferon alfa-2b and ribavirin on liver fibrosis in patients with chronic hepatitis C. Gastroenterology. 2002;123: 1061–9.
30. Desmet VJ, Roskams T. Cirrhosis reversal: a duel between dogma and myth. J Hepatol. 2004;40:860–7.
31. Zamara E, Novo E, Marra F et al. 4-Hydroxynonenal as a selective pro-fibrogenic stimulus for activated human hepatic stellate cells. J Hepatol. 2004;40:60–8.
32. Kweon YO, Paik YH, Schnabl B, Qian T, Lemasters JJ, Brenner DA. Gliotoxin-mediated apoptosis of activated human hepatic stellate cells. J Hepatol. 2003;39:38–46.
33. Li L, Tao J, Davaille J et al. 15-Deoxy-$\Delta^{12,14}$-prostaglandin J2 induces apoptosis of human hepatic myofibroblasts. J Biol Chem. 2001;276:38152–8.
34. Novo E, Marra F, Zamara E et al. Overexpression of Bcl-2 by activated human hepatic stellate cells: resistance to apoptosis as a mechanism of progressive hepatic fibrogenesis in humans. Gut. 2006;55:1174–82.
35. Yancopoulos GD, Davis S, Gale N, Rudge J, Wiegand S, Holash J. Vascular specific growth factors and blood vessel formation. Nature. 2000;407:242–8.
36. Carmeliet P. Mechanisms of angiogenesis and arteriogenesis. Nat Med. 2000;6:389–95.
37. Ferrara N, Gerber H, LeCouter J. The biology of VEGF and its receptors. Nat Med. 2003; 9:669–76.
38. Pugh CW, Ratcliffe P. Regulation of angiogenesis by hypoxia: role of the HIF system. Nat Med. 2003;9:677–84.
39. Jain RK. Molecular regulation of vessel maturation. Nat Med. 2003;9:685–93.
40. Carmeliet P. Angiogenesis in health and disease. Nat Med. 2003;9:653–60.
41. Medina J, Arroyo AG, Sanchez-Madrid F, Moreno-Otero R. Angiogenesis in chronic inflammatory liver disease. Hepatology. 2004;39:1185–95.
42. McCuskey RS, Reilly F. Hepatic microvasculature: dynamic structure and its regulation. Semin Liver Dis. 1993;13:1–12.
43. Camenisch G, Pisabarro M, Sherman D et al. ANGPTL3 stimulates endothelial cell adhesion and migration via integrin alpha vbeta 3 and induces blood vessel formation *in vivo*. J Biol Chem. 2002;227:17281–90.
44. Friedman SL. Liver fibrosis: from bench to bedside. J Hepatol. 2003;38:S38–53.
45. Friedman SL. A moving target in hepatic fibrogenesis. Hepatology. 2004;40:1041–3.
46. Pinzani M, Marra F. Cytokine receptors and signaling in hepatic stellate cells. Semin Liver Dis. 2001;21:397–416.
47. Cassiman D, Libbrecht L, Desmet V, Denef C, Roskams T. Hepatic stellate cell/ myofibroblast subpopulations in fibrotic human and rat livers. J Hepatol. 2002;36:200–9.
48. Ankoma-Sey V, Matli M, Chang KB et al. Coordinated induction of VEGF receptors in mesenchymal cell types during rat hepatic wound healing. Oncogene. 1998;17:115–21.
49. Corpechot C, Barbu V, Wendum D et al. Hypoxia-induced VEGF and collagen I expressions are associated with angiogenesis and fibrogenesis in experimental cirrhosis. Hepatology. 2002;35:1010–21.
50. Yoshiji H, Kuriyama S, Yoshii J et al. Vascular endothelial growth factor and receptor interaction is a prerequisite for murine hepatic fibrogenesis. Gut. 2003;52:1347–54.
51. Medina J, Caveda L, Sanz-Cameno P et al. Hepatocyte growth factor activates endothelial pro-angiogenic mechanisms relevant in chronic hepatitis C-associated neoangiogenesis. J Hepatol. 2003;38:660–7.
52. Shimoda K, Mori M, Shibuta K, Banner B, Barnard G. Vascular endothelial growth factor/ vascular permeability factor mRNA expression in patients with chronic hepatitis C and hepatocellular carcinoma. Int J Oncol. 1999;14:353–9.

53. Medina J, Sanz-Cameno P, Garcia-Buey L, Martin-Vilchez S, Lopez-Cabrera M, Moreno-Otero R. Evidence of angiogenesis in primary biliary cirrhosis: an immunohistochemical descriptive study. J Hepatol. 2005;42:124–31.
54. Ankoma-Sey V, Wang Y, Dai Z. Hypoxic stimulation of vascular endothelial growth factor expression in activated rat hepatic stellate cells. Hepatology. 2000;31:141–8.
55. Wang YQ, Luk JM, Ikeda K et al. Regulatory role of vHL/HIF-1α in hypoxia-induced VEGF production in hepatic stellate cells. Biochem Biophys Res Commun. 2004;317:358–62.
56. Aleffi S, Petrai I, Bertolani C et al. Upregulation of proinflammatory and proangiogenic cytookines by leptin in human hepatic stellate cells. Hepatology. 2005;42:1339–48.
57. DeLeve LD, Wang X, Hu L, McCuskey MK, McCuskey RS. Rat liver sinusoidal endothelial cell phenotype is maintained by paracrine and autocrine regulation. Am J Physiol Gastrointest Liver Physiol. 2004;287:G757–63.

16
Treatment of portal hypertension and variceal bleeding

T. SAUERBRUCH and J. HELLER

Portal hypertension is defined as a portal pressure that surpasses 5 mmHg. This pressure can be easily determined by measuring the gradient between the wedged hepatic venous pressure and the free hepatic venous pressure (HVPG). This gradient reflects rather adequately the portal pressure due to the structural changes of the liver. By contrast, direct measurement of the portal pressure requires an invasive percutaneous transhepatic approach or intraoperative puncture of the portal vein. Measurement of the portal flow is not an adequate substitute for HVPG.

If the HVPG surpasses the threshold of 10 mmHg oesophageal varices may develop[1]. There is a further important threshold of the HVPG, namely 12 mmHg, which marks the risk of bleeding. It is extremely rare that patients with values below this pressure bleed from their varices[2,3]. The average portal pressure in patients who have bled from their oesophageal collaterals is much higher (20 ± 5 mmHg). The average HVPG in patients who have not bled from their varices is somewhat lower[2] (17 ± 3 mmHg). However, it has been questioned whether HVPG is a good tool to discriminate between bleeders and non-bleeders[4,5]. A HVPG above 20 mmHg is associated with a bad outcome in patients bleeding from oesophageal varices[6].

Thus, it is a major goal in prophylaxis of primary or recurrent bleeding to keep the pressure as low as possible; yet it has also been shown that reduction of the portal pressure by more than 20% is already rather sufficient to shelter the patient from oesophageal bleeding[7–9]. Such a reduction of portal pressure can be achieved by drugs or by portal venous shunts. A further method to reduce the bleeding risk from oesophageal variceal bleeding is eradication of those vessels which bear the highest risk of haemorrhage, namely varices in the distal part of the oesophagus, by using endoscopic ligation.

PREPRIMARY PROPHYLAXIS OF VARICEAL BLEEDING

It remains an open question whether we can prevent the development of varices, variceal bleeding or other complications of liver cirrhosis, such as ascites or encephalopathy in patients with established compensated liver cirrhosis. One trial used the non-selective β-blocker timolol for an average period of 4 years[10]. Unfortunately there was no reduction of complications at all as compared to the randomized controls. The trial points to the open question of whether there is a drug that can reverse fibrosis or stop the ongoing process towards complications. Obviously non-selective β-blockers cannot prevent the dynamics of hepatic fibrosis, inflammation and its sequelae; yet there may be other drugs that should be tested in the forthcoming decade, such as statins, AT 1-receptor antagonists or interferons[11,12]. All these drugs may have beneficial effects; for example it has been shown in animal models that statins increase intrahepatic nitric oxide formation in liver cirrhosis and decrease fibrosis if applied early[13]. Similarly, AT 1-receptor blockers have been shown to have an – albeit small – portal pressure-lowering[14,15] and antifibrotic effect[15,16]. Interferons might also reduce fibrosis independent from the antiviral activity[12].

PRIMARY PROPHYLAXIS OF VARICEAL BLEEDING

As early as the 1970s it had been shown that unselective β-blockers, mainly propranolol, are able to reduce the risk of first bleeding from large oesophageal varices by 50% within 2 years after initiation of prophylactic treatment[18,19]. The effect of prophylactic sclerotherapy was regarded as less clear, mainly because the controlled trials were heterogeneous, showing a rather high complication rate in some studies[20]. By contrast, trials initiated in the 1990s using endoscopic ligation, known to harbour less complications than sclerotherapy as an endoscopic procedure, showed unequivocally that ligation can also be used as a prophylactic therapy to prevent first bleeding[21–25]. According to different meta-analyses it at least equals the preventive effect of non-selective beta-adrenoceptor antagonists[23]. Current expert opinion suggests using propranolol first, and ligation only in those patients who cannot tolerate the drug for various reasons[24]. This advice is mainly based on the fact that ligation may induce early lethal complications from ulcer bleeding, which has not been observed for propranolol[25].

TREATMENT OF ACUTE BLEEDING

Treatment of acute bleeding includes blood volume restitution, administration of antibiotics immediately at admission[26], endoscopy as soon as possible (within 12 h) together with vasoactive drugs for 2–5 days. Ligation is the endoscopic treatment of choice for acute variceal haemorrhage, although sclerotherapy may also be used in the acute situation if necessary[27–29].

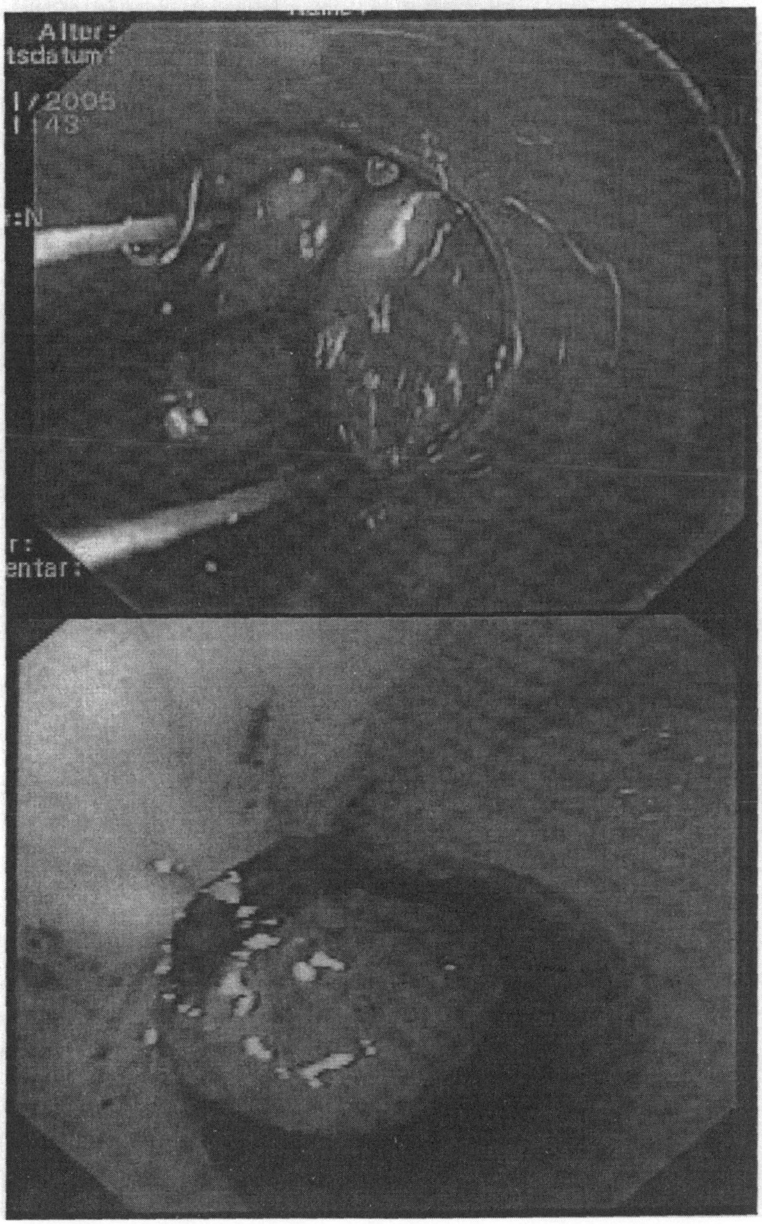

Figure 1 Endoscopic variceal ligation

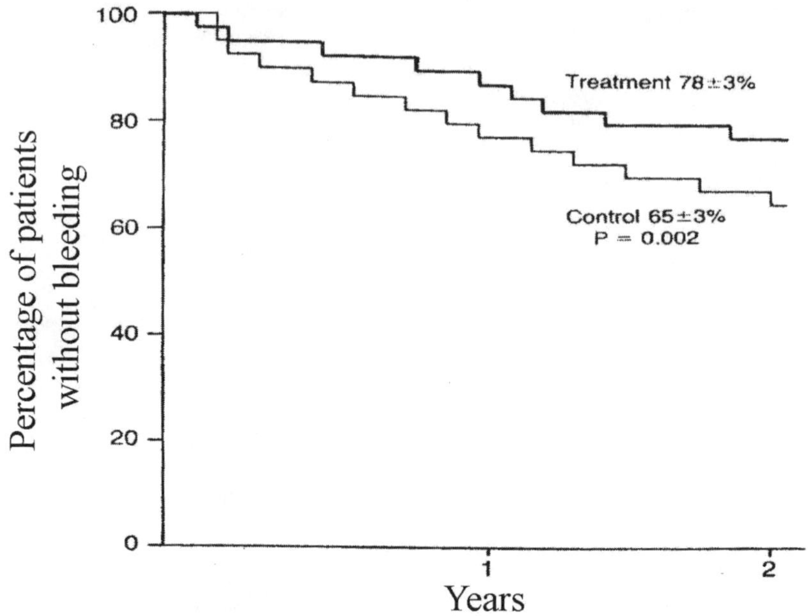

A

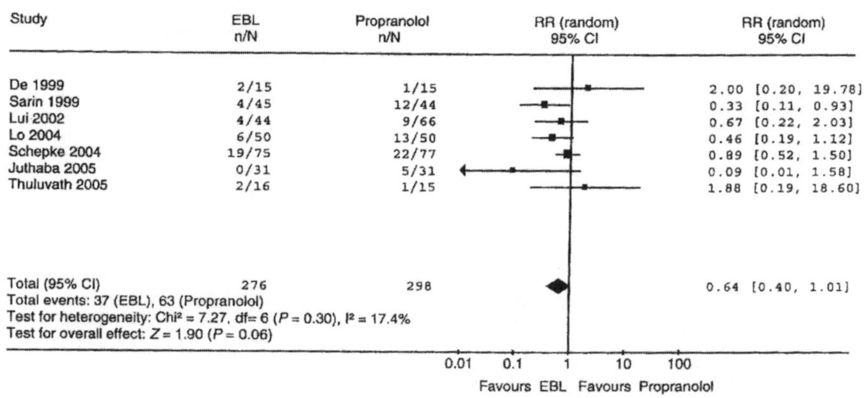

B

Figure 2 A: Beta-adrenergic antagonist drugs in the prevention of gastrointestinal bleeding in patients with cirrhosis and oesophageal varices. Compared to placebo, propranolol reduces the risk of first variceal bleeding from 35% to 22%[18]. **B**: Endoscopic band ligation at least equals the preventive effect of non-selective β-adrenoceptor antagonists in the prevention of first variceal haemorrage[24]

If initial treatment (antibiotics, endoscopic haemostasis, vasoactive drugs) fails, a second attempt at endoscopic haemostasis may be tried, otherwise the patient should receive emergency TIPS[27]. Recently, the use of self-expanding stents has been introduced in the treatment of acute variceal bleeding. The effectiveness of this challenging technique, however, needs to be studied in further trials[30]

PROPHYLAXIS OF REBLEEDING

Since patients who survived an acute bleeding episode have a high risk of rebleeding (60–70%), rebleeding prophylaxis should be started as soon as possible from day 6 of the index variceal bleeding episode[31,32]. Adding β-blockers to ligation is superior to ligation alone for the prevention of rebleeding[33]. On the other hand it has recently been demonstrated that adding ligation to medical rebleeding prophylaxis with nadolol plus isosorbite mononitrate has no significant advantage with respect to overall rebleeding, although it might have some positive impact on variceal rebleeding[34].

In patients who fail endoscopic and pharmacological treatment for prevention of rebleeding TIPS or surgical shunts are effective for patients with compensated or moderately decompensated cirrhosis, and should be used[32]. Of course transplantation should be considered in these patients.

In patients who had bled from isolated gastric varices type 1 or gastro-oesophageal varices type 2 glue (*N*-butyl-cyanoacrylate), TIPS or β-blockers are recommended[32], although there are no large controlled trials covering this situation.

References

1. LeMoine O, Hadengue A, Moreau R et al. Relationship between portal pressure, esophageal varices, and variceal bleeding on the basis of the stage and cause of cirrhosis. Scand J Gastroenterol. 1997;32:731–5.
2. Garcia-Tsao G, Groszmann RJ, Fisher RL et al. Portal pressure, presence of gastroeophageal varices and variceal bleeding. Hepatology. 1985;5:419–24.
3. Viallet A, Marleau D, Huet M et al. Hemodynamic evaluation of patients with intrahepatic portal hypertension. Relationship between bleeding varices and the portohepatic gradient. Gastroenterology. 1975;69:1297–300.
4. Lebrec D, de Fleury P, Rueff B et al. Portal hypertension, size of esophageal varices, and risk of gastrointestinal bleeding in alcoholic cirrhosis. Gastroenterology. 1980:79:1139–44.
5. Armonis A, Patch D, Burroughs A. Hepatic venous pressure measurement: an old test as a new prognostic marker in cirrhosis? Hepatology. 1997;25:245–8.
6. Moitinho E, Escorell A, Bandi JC et al. Prognostic value of early measurements of portal pressure in acute variceal bleeding. Gastroenterology. 1999;117:626–31.
7. Vorobioff J, Groszmann RJ, Picabea E et al. Prognostic value of hepatic venous pressure gradient measurement in alcoholic cirrhosis: a 10-year prospective study. Gastroenterology. 1996;111:701–9.
8. Merkel C, Bolognesi M, Sacerdoti D et al. The hemodynamic response to medical treatment of hypertension as a predictor of clinical effectiveness in the primary prophylaxis of variceal bleeding. Hepatology. 2000;32:930–4.
9. Turnes J, Garcia-Pagan JC, Abraldes JG et al. Pharmacological reduction of portal pressure and long term risk of first variceal bleeding in patients with cirrhosis. Hepatology. 2003;38:62A.

10. Groszmann RJ, Gascia-Tsao G, Bosch J et al. Beta-blockers to prevent gastroesophageal varices in patients with cirrhosis. N Engl J Med. 2005;353:2254–61.
11. Rincon D, Ripoll C, Catalina C et al. Endoscopic treatment versus endoscopic plus pharmacologic treatment for acute variceal bleeding: a meta-analysis. J Antimicrob Chemother. 2006;58:7–12.
12. Zafra C, Albraldes JG, Turnes J et al. Simvastatin enhances hepatic nitric oxide production and decreases the hepatic vascular tone in patients with cirrhosis. Gastroenterology. 2004; 126:749–55.
13. Trebicka J, Hennenberg M, Biecker E et al. Atorvastatin lowers portal pressure in cirrhotic rats via inhibition of RhoA/Rho-kinase and activation of eNOS. Hepatology. 2007 (In press).
14. Schepke M, Werner E, Biecker E et al. Hemodynamic effects of the angiotensin II receptor antagonist irbesartan in patients with cirrhosis and portal hypertension. Gastroenterology. 2001:121:389–95.
15. Croquet V, Moal F, Veal N et al. Hemodynamic and antifibrotic effects of losartan in rats with liver fibrosis and/or portal hypertension. J Hepatol. 2002;37:773–80.
16. Sookoian S, Fernandez MA, Castano G. Effects of six months losartan administration on liver fibrosis in chronic hepatitis C patients: a pilot study. World J Gastroenterol. 2005;11:7560–3.
17. Rincon D, Ripoll C, Lo lacono O et al. Antiviral therapy decreases hepatic venous pressure gradient in patients with chronic hepatitis C and advanced fibrosis. Am J Gastroenterol. 2006;101:2269–74.
18. Poynard T, Gales P, Pasta L et al. Beta-adrenergic antagonist drugs in the prevention of gastrointestinal bleeding in patients with cirrhosis and esophageal varices. An analysis of data and prognostic factors in 589 patients from four randomized clinical trials. Franco-Italian Multicenter Study Group. N Engl J Med. 1991;30:324:1532–8.
19. Garcia-Tsao G. Current management of the complications of cirrhosis and portal hypertension: variceal hemorrhage, ascites, and spontaneous bacterial peritonitis. Gastroenterology. 2001;120:726–48.
20. Pagliaro L, D'Amico G, Sorensen Tl et al. Prevention of first bleeding in cirrhosis. A meta-analysis of randomized trials of nonsurgical treatment. Ann Intern Med. 1992;117:59–70.
21. Imperiale T, Chalasani N. A meta-analysis of endoscopic variceal ligation for primary prophylaxis of esophageal variceal bleeding. Hepatology. 2001;33:802–7.
22. Lay CS, Tsai Yt, Teg CY et al. Endoscopic variceal ligation in propylaxis of first variceal bleeding in cirrhotic patients with high-risk esophageal varices. Hepatology. 1997;25:1346–50.
23. Khuroo MS, Khuroo NS, Farahat KL et al. Meta-analysis: endoscopic variceal ligation for primary prophylaxis of oesophageal variceal bleeding. Aliment Pharmacol Ther. 2005;21:347–61.
24. Grace ND, Garcia-Pagan JC, Albillos A et al. Baveno IV consensus statements: prevention of the first bleeding episode. In: de Franchis. editor. Baveno IV. Proceedings of the fourth Baveno International Consensus Workshop on Methodology of Diagnosis and Treatment. Blackwell, 2006.
25. Schepke M, Kleber G, Nurnberg D et al. Ligation versus propranolol for the primary prophylaxis of variceal bleeding in cirrhosis. Hepatology. 2004;40:65–72.
26. Bernard B, Grange JD, Khac EN et al. Antibiotic prophylaxis for the prevention of bacterial infections in cirrhotic patients with gastrointestinal bleeding: a meta-analysis. Hepatology. 1999;29:1655–61.
27. Bosch J, Laine L, Banares R et al. Baveno IV consensus statements: treatment of the acute bleeding episode. In: de Franchis, editor. Baveno IV. Proceedings of the fourth Baveno International Consensus Workshop on Methodology of Diagnosis and Treatment. Blackwell, 2006.
28. Gross M, Schiemann U, Muhlhofer A et al. Meta-analysis: efficacy of therapeutic regimens in ongoing variceal bleeding. Endoscopy. 2001;33;737–46.
29. Banares R, Albillos A, Rincon D et al. Endoscopic treatment versus endoscopic plus pharmacologic treatment for acute variceal bleeding: a meta-analysis. Hepatology. 2002;32:605–15.
30. Hubmann R, Bodlaj G, Czompo M et al. The use of self-expanding metal stents to treat acute esophageal variceal bleeding. Endoscopy. 2006;38:896–901.

31. Bernard B, Lebrec D, Mathurin P et al. Beta-adrenergic antagonists in the prevention of gastrointestinal rebleeding in patients with cirrhosis. A meta-analysis. Hepatology. 1997;25:63–70.
32. Lebrec D, Sauerbruch T, Bernard-Chabert B et al. Baveno IV consensus statements: Prevention of rebleeding. In: de Franchis, editor. Baveno IV. Proceedings of the fourth Baveno International Consensus Workshop on Methodology of Diagnosis and Treatment. Blackwell, 2006.
33. De La Pena J, Sanchez-Hernandez E, Rivero M et al. Variceal ligation plus nadolol compared with ligation for prophylaxis of variceal rebleeding: a multicenter trail. Hepatology. 2005; 41:572–8.
34. Study Group Spanish Cooperative Variceal Rebleeding. Multicenter RCT comparing drug therapy vs the combination of drug therapy + endoscopic variceal ligation in the prevention of rebleeding in patients with cirrhosis. Hepatology. 2006;44:202A.

17
Management of ascites and spontaneous bacterial peritonitis

A. CÁRDENAS, J. FERNÁNDEZ and P. GINÈS

INTRODUCTION

Ascites is defined as the accumulation of free fluid in the peritoneal cavity. This entity was first described by the ancient Egyptians and Greeks. Erasitratus of Cappadoccia, in 300 BC, noted the relationship between hardness of the liver and ascites[1].

The most common cause of ascites is portal hypertension secondary to cirrhosis, which accounts for over 80% of patients with ascites. In the natural history of cirrhosis, patients develop abnormalities in renal function manifested by impaired sodium and solute-free water excretion. These are responsible for fluid accumulation in the form of ascites and oedema. Ascites is the most common complication of cirrhosis and nearly 60% of patients with compensated cirrhosis develop ascites within 10 years of the disease[2]. Once patients develop ascites they have a poor prognosis; approximately half of these patients will die in approximately 3 years without liver transplantation[3]. Therefore patients with ascites should be evaluated for liver transplantation[4]. Patients with cirrhosis and ascites are also at risk of developing infections, particularly infection of the ascitic fluid, which is known as spontaneous bacterial peritonitis (SBP).

PATHOPHYSIOLOGY OF ASCITES IN CIRRHOSIS

Most patients with advanced cirrhosis have an inability to maintain extracellular fluid volume within normal limits, which results in increased total extracellular fluid volume and subsequent accumulation of fluid in the peritoneal and/or pleural cavities and interstitial tissue[5–9]. The main factor responsible for this increase in extracellular fluid volume is an abnormal increase in renal sodium reabsorption[6]. Although the exact mechanism of abnormal fluid regulation in cirrhosis is unknown, a large body of evidence indicates that it is secondary to arterial splanchnic vasodilation with a subsequent decrease in effective arterial blood volume (the volume sensed by arterial and cardiopulmonary receptors)[5–9]. The accumulation of fluid and the abnormalities in renal function are the consequence of the homeostatic activation of vasoconstrictor and antinatriuretic factors triggered to compensate for the relative arterial underfilling. The theory that best explains the pathogenesis of sodium retention and ascites formation in cirrhosis is the arterial vasodilation theory (Figure 1)[5]. According to this hypothesis, sodium retention is the consequence of a homeostatic response raised by an underfilling of the arterial circulation secondary to arterial vasodilation in the splanchnic vascular bed. The underfilling of the arterial circulation is sensed by arterial and cardiopulmonary receptors and results in hypervolaemia by activating antinatriuretic factors. The retained fluid initially compensates for the disturbance in the arterial circulation and suppresses the activation of sodium-retaining mechanisms. However, as the vasodilation in the splanchnic circulation causes more marked arterial underfilling, the retained fluid does not adequately fill the intravascular compartment, mainly because fluid is leaking continuously into the peritoneal cavity and the sodium-retaining mechanisms become permanently activated. A detailed review of the pathogenesis of ascites is beyond the scope of this review and can be found elsewhere[5–7].

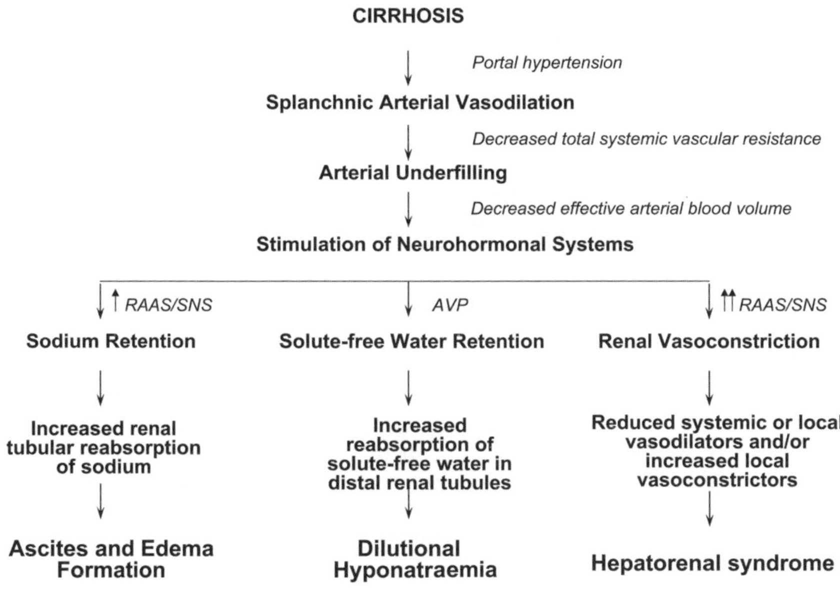

Figure 1 The pathogenesis of ascites formation, dilutional hyponatraemia and hepatorenal syndrome in patients with cirrhosis and ascites. The neurohormonal systems are represented by the renin–angiotensin–aldosterone system (RAAS), sympathetic nervous system (SNS) and arginine vasopressin (AVP). The neurohumoral effects of these systems on the systemic circulation and renal function in cirrhosis with ascites are responsible for sodium and solute-free water retention as well as hepatorenal syndrome

DIAGNOSIS AND EVALUATION OF PATIENTS WITH CIRRHOSIS AND ASCITES

Clinical features

The main clinical symptom of patients with ascites and oedema is an increase in abdominal girth, often accompanied by lower extremity oedema. In some cases the accumulation of fluid is so important that respiratory function and physical activity may be impaired. In most cases the first episode of ascites develops insidiously over the course of several weeks or a few months. Patients must have approximately 1500 ml of fluid for ascites to be detected reliably by physical examination. Increased intra-abdominal pressure may favour the development of abdominal hernias (mainly umbilical) in patients with cirrhosis and long-standing ascites[10].

The current classification of ascites defined by the International Ascites Club divides patients into three groups[8]. Patients with grade 1 ascites are those in whom ascites is detected only by ultrasonography. Patients with grade 2 ascites (moderate ascites) are those in which ascites causes moderate symmetrical distension of the abdomen. Patients with grade 3 ascites (large ascites) have increased amounts or tense ascites causing marked abdominal distension. Patients with refractory ascites are those who do not respond to high doses of diuretics or who develop side-effects that preclude their use (see later).

Evaluation

The evaluation of a cirrhotic patient with ascites should include standard haematology, electrolyte, renal, coagulation (prothombin time or international normalized ratio), and liver tests (aminotransferases, bilirubin, albumin, total protein, alkaline phosphatase, serum alpha-fetoprotein). An abdominal ultrasonography to rule out hepatocellular carcinoma and evaluate patency of the portal venous system must always be done. In addition an upper gastrointestinal endoscopy to assess the presence and characteristics of oesophageal and gastric varices or portal hypertensive gastropathy should be performed. In patients with previously unknown liver disease the diagnosis of cirrhosis should be confirmed either histologically or by a combination of findings on the physical examination, ultrasonographic (diffuse parenchymal heterogeneity, nodular liver edge, or signs of portal hypertension), and endoscopic findings (presence of gastro-oesophageal varices).

Renal and circulatory function should be evaluated in patients with cirrhosis and ascites as this helps with prognosis and tailoring of therapy (see later[11–14]). Evaluation of renal function must be performed with low sodium intake (approximately 90 mEq/day) and at least 5 days off diuretics. Parameters to be measured include 24-h urine volume and sodium excretion, serum electrolytes, and serum creatinine. A strong prognostic marker of cirrhotic patients with ascites is the ability to handle water by means of a water load test. In this test a patient receives 20 ml/kg of intravenous 5% dextrose and following this the renal excretion of water is measured. An impaired ability to excrete the water load (urine volume of <8 ml/min) is associated with a poor prognosis[14]. In patients with renal failure (serum creatinine greater than 1.5 mg/dl), urine sediment and 24-h urine protein should be assessed, and the kidneys should be examined by ultrasonography. Evaluation of circulatory function should include measurement of arterial pressure.

Ascitic fluid analysis

Diagnostic paracentesis (approximately 30 ml of fluid) is required in all patients presenting with their first episode of ascites, those requiring hospitalization for any reason, and those with any evidence of clinical deterioration such as fever, abdominal pain, gastrointestinal bleeding, hepatic encephalopathy, hypotension or renal failure to rule out SBP. Necessary tests in the ascitic fluid include cell count, culture in blood culture bottles (10 ml of fluid injected at the bedside), albumin, and total protein[15]. Patients with a protein concentration in ascitic fluid lower than 15 g/l seem to have a greater risk of SBP than do patients with higher ascitic fluid protein levels[16–18]. Patients with low ascitis fluid protein should be evaluated for prophylactic administration of quinolones to reduce the risk of SBP and hepatorenal syndrome (HRS) (see later).

In most cases the ascitic fluid white blood cell count is less than 500/mm^3 with a predominance of mononuclear cells (>75%) and a very low number of polymorphonuclear (PMN) cells. An increased number of white blood cells with predominance of PMN indicates peritoneal infection. The diagnosis of SBP is made when the fluid sample has a PMN count greater than 250/mm^3 (see refs 19 and 20). Use of the so-called urine dipsticks (reagent strips) has been proposed for the rapid diagnosis of SBP[21]. A low proportion of patients may have a positive ascitic fluid culture without increased PMN cell count. This condition is known as bacterascites[19,22]. Conversely, as indicated above, many patients with high PMN count suggestive of peritoneal infection may have a negative ascitic fluid culture. This condition is known as culture-negative SBP and should be managed as culture-positive SBP[19,23].

TREATMENT OF ASCITES

Therapeutic methods

Liver transplantation

The most important aspect of the management of all patients with cirrhosis and ascites is an evaluation for liver transplantation. Probably the most reliable factors indicating a poor prognosis in cirrhotics with ascites are those related to renal and circulatory function. These factors include dilutional hyponatraemia (serum sodium <130 mEq/L), low arterial blood pressure, serum creatinine >1.2 mg/dl, and intense sodium retention (urine sodium less than 10 mEq/day)[3,12]. Interestingly, in patients with ascites, parameters of renal function and systemic haemodynamics are better predictors of prognosis than are liver tests, such as albumin, bilirubin, and prothrombin time. Allocation for liver transplantation in the United States and other countries is currently based on the Model for End-Stage Liver disease (MELD) score that includes serum bilirubin, serum creatinine and the INR as variables[24]. This scoring system is objective, includes an important parameter of renal function, and predicts survival in cirrhotics. Unpublished data from our unit suggest that the MELD score is also predictive of survival in cirrhotics with ascites, and therefore might be a reliable parameter that helps calculate risk in these patients.

Sodium restriction

Because the amount of fluid retained as ascites or oedema depends on the balance between sodium intake and excretion, a reduction in sodium intake helps in achieving a negative sodium balance. Patients with baseline urine sodium greater than 10 mEq/day generally will have a sodium balance ranging from slightly positive to negative when maintained on a sodium-restricted diet (90 mEq/day). Conversely, patients with marked sodium retention (urine sodium excretion less than 10 mEq/day) will have a markedly positive sodium balance even under conditions of sodium restriction of 90 mEq/day. The gain of sodium in these patients will range between 70 and 80 mEq/day (approximately 10 mEq/day of sodium are eliminated in the stools), which is equivalent to an increase of approximately 500 ml/day of extracellular fluid.

Diuretic therapy

The cornerstone of therapy has relied on the administration of diuretics, drugs that increase sodium excretion by reducing the tubular reabsorption of sodium. The diuretic most commonly used in patients with cirrhosis and ascites is spironolactone (50–400 mg/day), a drug that inhibits sodium reabsorption by binding to the mineralocorticoid receptor in the renal collecting tubules, thus blocking the effects of aldosterone[8,25]. Spironolactone may be given alone or in combination with furosemide (20–160 mg/day), which acts by inhibiting the Na^+-K^+-$2Cl^-$ cotransporter in the loop of Henle[25]. The response to diuretic therapy in cirrhotic patients should be evaluated weekly by measuring body weight, urine volume, and sodium excretion. Approximately 10% of patients with ascites either do not respond to diuretic therapy or have diuretic-induced complications that prevent the use of high doses of these drugs. This condition is known as refractory ascites[8,26]. Complications of diuretic therapy in patients with cirrhosis include hepatic encephalopathy, hyponatraemia, renal impairment, potassium disturbances, gynaecomastia, and muscle cramps. Hyponatraemia is a relatively common complication in patients treated with diuretics (see later). It is generally considered that a drop of more than 10 mEq/L to a level lower than 125 mEq/L is an indication to stop diuretic therapy completely. The classic type of impairment in renal function during diuretic therapy is usually intravascular volume depletion, which occurs in patients with a positive response and is rapidly reversible after dose reduction or discontinuation of diuretic therapy. Nevertheless, impairment in renal function may develop in some patients in the absence of a markedly negative sodium balance, and this complication may preclude the use of an effective diuretic dosage[26]. Severe hyperkalaemia (greater than 6 mEq/L) is an uncommon complication of diuretic therapy in patients with ascites without renal failure, but it may develop when patients with renal failure are treated with spironolactone even at low doses. Hypokalaemia may occur only in patients treated with loop diuretics alone. Painful gynaecomastia may complicate chronic spironolactone therapy. In some cases amiloride (10–40 mg/day) can be used as an alternative agent in cases of tender gynaecomastia. However, amiloride is less effective than spironolactone[27]. Tamoxifen (20 mg p. o. b.i.d.) may be useful to decrease pain, but information is limited[28]. Muscle cramps due to diuretic therapy may require a reduction in diuretic dosage. Quinidine (quinidine sulphate, 400 mg/day)[29] and intravenous albumin administration (25 g/week)[30] have been shown to reduce the frequency and intensity of muscle cramps in cirrhotic patients with ascites treated with diuretics.

Therapeutic paracentesis

Therapeutic paracentesis is the treatment of choice in the management of grade 3 ascites[8,9,15,31]. Complete removal of ascites in one tap with intravenous albumin (8 g/L tapped) has been shown to be a fast and effective measure in controlling tense ascites and associated with a lower number of complications than conventional diuretic therapy[31]. After a therapeutic tap a post-paracentesis circulatory dysfunction may develop; this is a circulatory derangement with marked activation of the renin–angiotensin system that occurs a few days after the procedure[32]. This disorder is clinically silent and may be associated with dilutional hyponatraemia, renal impairment, and decreased survival; it may be prevented with administration of plasma expanders[33]. When less than 5 L of ascites are removed, artificial plasma expanders, saline and albumin are equally effective. However, when more than 5 L are removed, albumin is the plasma expander of choice[34,35] (Figure 2). Albumin is administered at a dose of 8 g/L of ascites removed. The recommended doses of dextran 70 and polygeline are 8 g and 150 ml per litre of ascites removed, respectively[34,35]. Patients with ascites and a known history of cirrhosis, and without any associated complications, can be managed as outpatients; however, patients in whom tense ascites is the first manifestation of cirrhosis, or those with associated hepatic encephalopathy, gastrointestinal bleeding, or bacterial infections, require hospitalization. After paracentesis most patients need to be started or continued on diuretics in order to avoid a positive sodium balance and recurrence of ascites.

Other therapeutic methods

Peritoneovenous shunting was commonly used in the past in the treatment of refractory ascites. However, its use has been progressively abandoned due to severe side-effects (shunt occlusion, superior vena cava thrombosis, peritoneal fibrosis), its limited long-term effectiveness, and the introduction of alternative therapies such as therapeutic paracentesis; however, this method is still used in some centres but experience is limited[36]. The most-studied non-phamacological

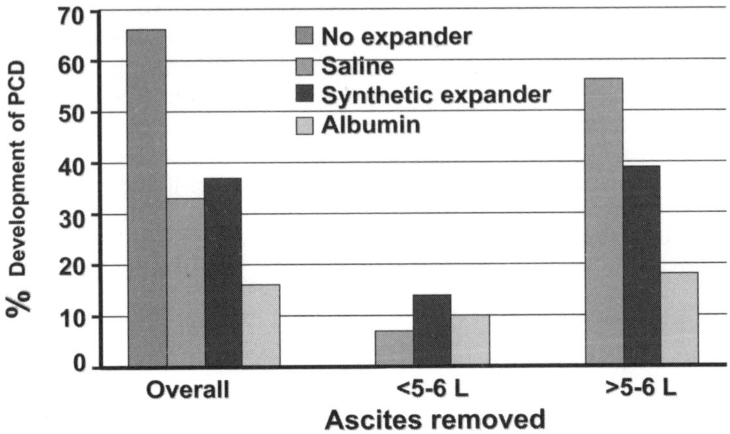

Figure 2 Post-paracentesis circulatory dysfunction (PCD) depends on the type of plasma volume expander and the amount of ascites removed. Two factors have been shown to be independently predictive of the development of PCD: the amount of ascites extracted and the type of plasma volume expander used at the time of large-volume paracentesis (LVP). The incidence of PCD can be as high as 80% if no plasma expander is given at the time of LVP. If less than 5 L of ascites are removed, the incidence of PCD is similar (~10%) using albumin, synthetic expanders or saline solution. However, if more than 5 L are removed, the lowest rate of PCD is observed with albumin (~18%) with significantly higher rates observed with the administration of synthetic plasma expanders (~40%) or saline solution (~55%). Therefore, albumin should be used with LVP when ≥5 L of ascites are removed. (Modified with permission from ref. 34)

intervention for portal hypertension is the transjugular intrahepatic portosystemic shunt (TIPS). TIPS is a non-surgical method of portal decompression that consists of the insertion of an intrahepatic stent between one hepatic vein and the portal vein using a transjugular approach. TIPS was introduced as an alternative treatment to shunt surgery for variceal bleeding in cirrhotic patients. Initial experience demonstrated that patients who received TIPS had resolution of their ascites. This observation, in addition to the fact that ascites formation occurs when portal pressure is ≥12 mmHg, led to the rationale of applying this intervention as therapy for refractory ascites. By reducing portal pressure, TIPS decreases ascites and diuretic requirements in cirrhotic patients. Over the past decade several retrospective, prospective and randomized controlled trials (RCT) evaluating the effects of TIPS in refractory ascites have been published (see later).

Nutritional recommendations

An evaluation by a nutritionist is recommended for appropriate education regarding an appropriate caloric and salt intake. A low-sodium diet with 90 mEq/day is recommended because a more severe restriction is usually unpalatable. Improvement of the nutritional status is of key importance given that cirrhotic patients with advanced disease have decreased intake of nutrients, decreased absorption of nutrients, increased energy expenditure and altered fuel metabolism with an accelerated starvation metabolism[37]. Cirrhotic patients with refractory ascites who undergo a TIPS for therapy of ascites improve their nutritional status significantly as measured by resting energy expenditure, total body nitrogen, body fat and food intake.

Practical management strategies

Grade 1–2 ascites

Patients with grade 1 ascites (only detected by ultrasound) do not require any specific treatment. Patients with grade 2 ascites (moderate amount of fluid with symmetrical distension of the abdomen) are best treated with a low-sodium diet and diuretics[8,15]. The best initial regimen for minimizing ascites is spironolactone (25–100 mg/day). Low doses of furosemide (20–40 mg/day) may be useful at the beginning of therapy (particularly in patients with peripheral oedema or anasarca) because they increase natriuresis; however, higher doses may put the patients at risk for pre-renal failure because these patients usually do not have severe sodium retention. Another therapeutic strategy is to use spironolactone alone in progressively increasing doses up to 400 mg/day and add furosemide only when the highest dose of spironolactone has been achieved if there is no response. If there is no response, compliance with diet and medications should be confirmed, and then diuretics may be increased in a stepwise fashion every 7 days by doubling doses (ratio of 40 mg of furosemide: 100 mg of spironolactone); furosemide up to 160 mg/day and spironolactone up to 400 mg/day. The goal is to achieve an average weight loss of 0.5 kg/day in patients without oedema and 1 kg/day in those with peripheral oedema[8]. Urine sodium should be determined in patients who do not lose weight; those patients with urine sodium >90 mEq/day (i.e. urine sodium greater than sodium intake) are not compliant with sodium restriction. After minimizing ascites, sodium restriction can be maintained while the dose of diuretics may be reduced as needed, or sodium intake can be mildly increased keeping the same dose of diuretics.

Grade 3 ascites

Patients with grade 3 ascites (severe or tense ascites) usually have marked sodium retention (urine sodium excretion less than 10 mEq/day), although large ascites may also develop in some patients with less avid sodium retention. Whereas some patients may have only severe ascites without any other sign of decompensation, other patients have associated conditions such as gastrointestinal bleeding, hepatic encephalopathy, SBP, or hepatocellular carcinoma.

The treatment of choice for patients with grade 3 ascites is total paracentesis plus intravenous albumin[8,9,15,31,33–35]. If albumin is not available, patients may be treated with partial (up to 5 L) paracentesis plus saline, dextran 70 or polygeline[34,35]. Patients with massive peripheral oedema may require a second paracentesis shortly after the first tap because of a rapid shift of fluid from interstitial tissue to the abdominal cavity. Because of the intense sodium retention usually associated with severe ascites, patients should initially receive relatively high doses of diuretics (e.g. spironolactone, 100–200 mg/day, with or without furosemide, 40 mg/day) after paracentesis together with a low-sodium diet[38]. Diuretics should then be adjusted according to individual responses.

Refractory ascites

The definition and diagnostic criteria of refractory ascites are listed in Table 1. The vast majority of patients with refractory ascites have very intense sodium retention and a severely impaired capacity to excrete solute-free water, the latter resulting in dilutional hyponatraemia in a significant proportion of cases[8,26]. Moreover, most patients have a reduction in renal plasma flow and glomerular filtration rate. The difference with non-refractory ascites patients is that in the former group sodium excretion may be increased with the use of diuretics, whereas in the latter group sodium retention cannot be overcome pharmacologically, either because patients do not respond to high doses of diuretics or because side-effects develop and preclude the use of an effective diuretic dosage.

Current treatment strategies include repeated therapeutic paracentesis plus intravenous albumin or TIPS. Therapeutic paracentesis is the most accepted initial therapy for refractory ascites. Patients generally require a tap every 2–4

Table 1 Definition and diagnostic criteria for refractory ascites in cirrhosis

Diuretic-resistant ascites: ascites that cannot be mobilized or the early recurrence of which cannot be prevented because of a lack of response to sodium restriction and diuretic treatment.

Diuretic-intractable ascites: ascites that cannot be mobilized or the early recurrence of which cannot be prevented because of the development of diuretic-induced complications that preclude the use of an effective diuretic dosage.

Requisites:

1. Treatment duration: patients must be on intensive diuretic therapy (spironolactone 400 mg/day and furosemide 160 mg/day) for at least 1 week, and on a salt-restricted diet of less than 90 mmol/day.
2. Lack of response: mean weight loss of <0.8 kg over 4 days and urinary sodium output less than the sodium intake.
3. Early ascites recurrence: reappearance of grade 2 or 3 ascites within 4 weeks of initial mobilization.
4. Diuretic-induced complications: diuretic-induced hepatic encephalopathy is the development of encephalopathy in the absence of any other precipitating factor. Diuretic-induced renal impairment is an increase of serum creatinine by $>100\%$ to a value >2 mg/dl in patients with ascites responding to treatment. Diuretic-induced hyponatraemia is defined as a decrease of serum sodium by >10 mmol/L to a serum sodium of <125 mmol/L. Diuretic induced hypokalaemia or hyperkalaemia is defined as a change in serum potassium to <3 mmol/L or >6 mmol/L despite appropriate measures.

* Modified with permission from ref. 8.

weeks and the majority may be treated as outpatients. This approach is therefore easy to perform and relatively inexpensive[39]. TIPS, although very effective in relieving ascites, is frequently complicated by obstruction of the prosthesis (70% in 1 year)[40]. Nonetheless newer polytetrafluoroethylene-covered prostheses improve TIPS patency and decrease the number of clinical relapses and reinterventions without apparently increasing the risk of encephalopathy[41]. Clinical trials comparing TIPS versus repeated paracentesis show that TIPS is associated with a lower rate of ascites recurrence[39,42–45]; yet hepatic encephalopathy was seen in 30–50% of patients treated with TIPS[39,43,44]. Although two studies showed a survival benefit with TIPS[43,45], two other studies demonstrated no significant difference[39,44]. Finally, the cost of treating patients with refractory ascites with TIPS was higher than the cost of repeated therapeutic paracentesis plus albumin[39]. Three recent meta-analyses of these five RCT similarly concluded that, although TIPS was better at controlling ascites, survival was not improved[46–48]. Therefore large-volume paracentesis is the initial treatment of choice because of its wider applicability, lower cost and fewer side-effects when compared to TIPS[49]. TIPS placement should be evaluated on case-by-case basis and probably reserved for patients with very rapid recurrence of ascites and preserved liver function (bilirubin <3 mg/dl, Child–Pugh score <12, MELD score <14), aged <70, without hepatic encephalopathy or cardiopulmonary disease[8,9,49].

TREATMENT AND PROPHYLAXIS OF SPONTANEOUS BACTERIAL PERITONITIS

SBP is a common and severe complication of cirrhotic patients with ascites characterized by infection of ascitic fluid in the absence of any intra-abdominal source of infection[19,50]. The prevalence of SBP in hospitalized cirrhotic patients ranges between 10% and 30%[50]. Gram-negative bacteria are responsible for nearly 80% of cases, and *Escherichia coli* accounts for most of them. Aerobic Gram-positive bacteria, mostly *Streptococcus viridans, Staphylococcus aureus* and *Enterococcus fecalis* are isolated in approximately 20% of cases[51]. The pathogenesis of SBP involves passage of bacteria from the intestinal lumen to the systemic circulation (bacterial translocation), bacteraemia secondary to the impairment of the reticuloendothelial system (RES) phagocytic activity, and infection due to poor opsonization and defective bactericidal activity of ascitic fluid[52] (Figure 3).

The clinical spectrum of SBP is quite variable, and ranges from a full-blown picture of peritonitis to complete absence of symptoms; therefore the diagnosis relies on a high index of suspicion and prompt examination of the ascitic fluid. Patients with SBP can present with fever, chills, abdominal pain, encephalopathy, and/or rebound abdominal tenderness. Patients with cirrhosis and sepsis-related infections including SBP may develop adrenal insufficiency in 50–70% of cases[53,54]. The development of adrenal insufficiency is associated with an important haemodynamic instability and increased mortality (80% in those with adrenal insufficiency versus 37% without adrenal insufficiency)[53]. Additionally, in cirrhotic patients with adrenal insufficiency,

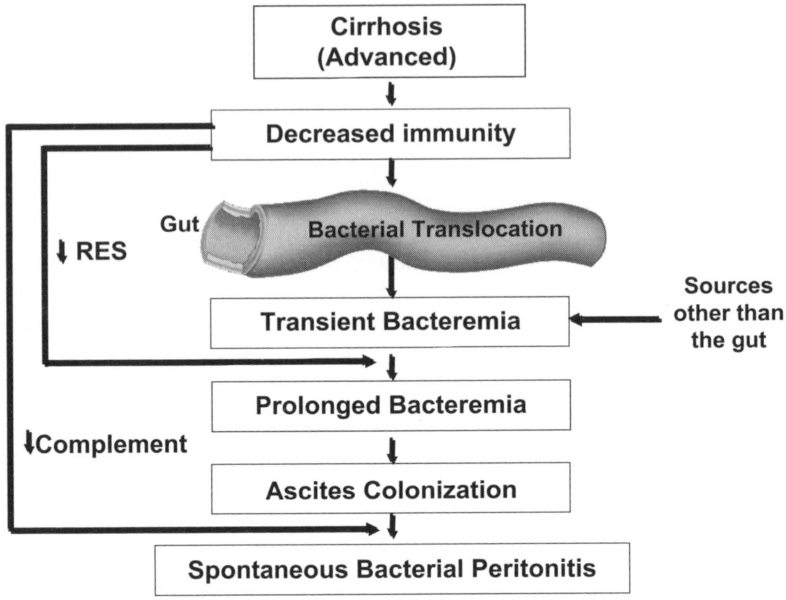

Figure 3 Pathogenesis of spontaneous bacterial peritonitis (SBP) in cirrhosis. In cirrhosis, particularly in the presence of ascites, there is an increase in bacterial translocation from the intestinal lumen to mesenteric lymph nodes (BT). A failure of local and systemic immune defences, more evident with greater severity of liver disease, is a major element in promoting a clinically significant bacterial translocation. Cirrhosis is accompanied by a decrease in bactericidal activity by phagocytic cells and an impaired reticuloendothelial system (RES) activity because of portosystemic shunting that bypasses the liver (thereby escaping the action of the RES) and because of an impaired phagocytic activity of Kupffer cells. Impaired RES allows a transient bacteraemia (bacteria arising from the gut or from other sources) to become prolonged and, in the presence of ascites, bacterial colonization occurs. In patients with reduced local defence mechanisms (e.g. low ascites complement levels), an overt ascites infection will develop (i.e. SBP)

administration of intravenous hydrocortisone injection seems to improve survival in those with advanced Child C cirrhosis[54]. An important clinical feature of SBP is the development of renal failure during the infection as it develops in 30–40% of patients with SBP, and is a major cause of death[55]. The risk may be decreased with infusion of intravenous albumin (see below).

Therapy and prognosis

Antibiotic therapy should be initiated in patients with a polymorphonuclear count in ascitic fluid greater than 250/mm^3 before microbiological results are obtained[19]. Empirical antibiotic therapy with an intravenous third-generation cephalosporin (cefotaxime 2 g every 8–12 h or ceftriaxone 1 g/24 h) for at least 5 days, is required after diagnosis is confirmed[19,50]. Therapy modification depends on results from cultures. Assessment of response to therapy includes frequent clinical evaluation and repeat diagnostic paracentesis 2–3 days after beginning antibiotics. In case of treatment failure (worsening infection or no decrease in PMN count), antibiotic therapy should be revised and appropriately changed. Patients should be treated for at least 5 days until the complete disappearance of all signs of infection and/or decrease of PMN count in ascitic fluid below 250 mm^3 (see ref. 19).

SBP resolves in approximately 90% of cases if treated with the above regimens[19]. However, hospital mortality remains between 10% and 30%, because most of these patients have advanced liver failure and complications such as gastrointestinal bleeding, renal failure and hepatic encephalopathy[50,56,57]. The most important predictor of survival in patients with SBP is the development of renal failure during the infection. Renal failure is triggered by an impairment of circulatory function with activation of the vasoconstrictor systems. The administration of albumin at a dose of 1.5 g/kg at the diagnosis of the infection and 1 g/kg at 48 h later prevents renal failure and improves survival in patients with SBP[56]. Recommendations for the management of SBP are summarized in Table 2.

Prophylaxis

Unfortunately life expectancy after an episode of SBP is short, with a 1-year probability of survival of 30–50% if antibiotic prophylaxis is not given[50,57]. Conditions associated with an increased risk of SBP include: gastrointestinal bleeding, low-protein concentration in ascitic fluid, advanced liver failure (high serum bilirubin and/or markedly prolonged prothrombin time) and past history of SBP[19]. Because most episodes of SBP are caused by Gram-negative bacteria present in the normal intestinal flora, the approach used for the prophylaxis of SBP has been based on the administration of antibiotics, usually quinolones such as norfloxacin or ciprofloxacin. The efficacy of this approach has been demonstrated in patients with gastrointestinal haemorrhage[58–62] and patients who have recovered from the first SBP episode[63], and has been recommended by a panel of experts in an International Consensus Conference on SBP[19].

In patients with gastrointestinal haemorrhage, the short-term administration of norfloxacin or intravenous ceftriaxone markedly reduces the incidence of SBP or bacteraemia as compared with patients not receiving prophylactic antibiotics[59,61,62]. The results of a recent meta-analysis indicate that antibiotic prophylaxis in patients with gastrointestinal bleeding not only prevents infection but also improves survival[60]. Long-term norfloxacin administration is effective in the prevention of SBP recurrence (secondary prophylaxis)[63]. Antibiotic prophylaxis (norfloxacin) also appears to be effective in the prevention of SBP (primary prophylaxis) in patients with low ascitic fluid protein (<15 g/L), who are at high risk of developing the first episode of SBP. A recent study demonstrated that primary prophylaxis with norfloxacin reduced the incidence of SBP, delayed the development of hepatorenal syndrome and improved survival in patients with advanced cirrhosis[64].

Table 2 Recommendations for the management and prevention of spontaneous bacterial peritonitis

---

Therapy

1. After diagnosis of peritonitis has been made (>250 polymorphonuclear cells/mm^3 in ascitic fluid), start with third-generation cephalosporins (i.e. cefotaxime 2 g/8–12 h intravenously or ceftriaxone 1 g/24 h intravenously).
2. Infuse albumin (1.5 g/kg at diagnosis of the infection and 1 g/kg at 48 h later).
3. Maintain antibiotic therapy until disappearance of signs of infection or for at least 5 days. Patients should be evaluated daily to assess signs of infection and a follow-up paracentesis should be performed to determine polymorphonuclear cell count. If there is a reduction of less than 25% of polymorphonuclear cells compared to the basal levels, the presence of a secondary peritonitis and/or bacteria resistant to antibiotic therapy should be suspected and treatment changes made accordingly.
4. After resolution of infection, start long-term norfloxacin 400 mg/day p.o.

Prevention

1. Patients with gastrointestinal haemorrhage:

   Norfloxacin 400 mg/12 h orally or per gastric tube, for 7 days or intravenous ceftriaxone 1 g/day for 7 days.
2. Patients with ascites with a previous episode of SBP:
   (a) Norfloxacin 400 mg/day indefinitely.
   (b) Consider liver transplantation.
3. Patients with ascites and advanced liver disease* without a previous episode of SBP:

   Low ascitic fluid protein concentration (<15 g/L): norfloxacin 400 mg/day indefinitely.

---

*Serum bilirubin >3 mg/dl, Child–Pugh score >10, hyponatraemia and/or renal impairment.

Management practice points (all patients must be evaluated for liver transplantation)

A: Treatment strategy for patients with cirrhosis and grade 2 ascites

Start with a low-sodium diet (90 mmol/day) and spironolactone (25–100 mg/day) to reach goal of weight loss: 500 g/day. If needed, doses to be increased every 5–7 days up to 400 mg/day of spironolactone and 160 mg/day of furosemide.

B: Treatment strategy for patients with cirrhosis and grade 3 ascites and refractory ascites

Total paracentesis plus intravenous albumin (8 g/L of ascites removed) followed by a low-sodium diet (90 mmol/day) and diuretics if patient tolerated them beforehand.

Total paracentesis plus intravenous albumin can be performed when required. Consider use of TIPS in patients with preserved hepatic function, aged <70 years, and no hepatic encephalopathy who require very frequent paracentesis.

References

1. Reuben A. Out came copious water. Hepatology. 2002;36:261–4.
2. Ginès P, Quintero E, Arroyo V et al. Compensated cirrhosis: natural history and prognostic factors. Hepatology. 1987;7:122–8.
3. Guevara M, Cárdenas A, Uriz J, Ginés P. Prognosis of patients with ascites and cirrhosis. In: Ginès P, Arroyo V, Rodes J, Schrier R, editors. Ascites and Renal Dysfunction in Liver Disease, 2nd edn. Oxford: Blackwell Publishing, 2005:260–70.
4. Murray KL, Carithers RL. AASLD practice guidelines: evaluation of the patient for liver transplantation. Hepatology. 2005;41:1407–32.
5. Schrier RW, Arroyo V, Bernardi M, Epstein M, Henriksen JH, Rodés J. Peripheral arterial vasodilation hypothesis: a proposal for the initiation of renal sodium and water retention in cirrhosis. Hepatology. 1988;8:1151–7.
6. Cardenas A, Arroyo V. Mechanisms of water and sodium retention in cirrhosis and the pathogenesis of ascites. Best Pract Res Clin Endocrinol Metab. 2003;17:607–22.
7. Arroyo V, Colmenero J. Ascites and hepatorenal syndrome in cirrhosis: pathophysiological basis of therapy and current management. J Hepatol. 2003;38(Suppl. 1):S69–89.
8. Moore KP, Wong F, Ginès P et al. The management of ascites in cirrhosis: report on the consensus conference of the International Ascites Club. Hepatology. 2003;38:258–66
9. Ginès P, Cardenas, A, Arroyo V, Rodés J. Management of cirrhosis and ascites. N Engl J Med. 2004;350:1646–54.
10. Belghiti J, Durand F. Abdominal wall hernias in the setting of cirrhosis. Semin Liver Dis. 1997;17:219–26.
11. Arroyo V, Bosch J, Gaya J et al. Plasma renin activity and urinary sodium excretion as prognostic indicators in nonazotemic cirrhosis with ascites. Ann Intern Med. 1981;94:198–201.
12. Llach J, Ginès P, Arroyo V et al. Prognostic value of arterial pressure, endogenous vasoactive systems, and renal function in cirrhotic patients admitted to the hospital for the treatment of ascites. Gastroenterology. 1988;94:482–7.
13. Ginès A, Escorsell A, Ginès P et al. Incidence, predictive factors, and prognosis of the hepatorenal syndrome in cirrhosis with ascites. Gastroenterology. 1993;105:229–36.
14. Fernandez-Esparrach G, Sanchez-Fueyo A, Ginès P et al. A prognostic model for predicting survival in cirrhosis with ascites. J Hepatol. 2001;34:46–52.
15. Runyon B. Management of adult patients with ascites due to cirrhosis. Hepatology. 2004;39:841–56
16. Cárdenas A, Bataller R, Arroyo V. Mechanisms of ascites formation. Clin Liver Dis. 2000;4:447–65.
17. Llach J, Rimola A, Navasa M et al. Incidence and predictive factors of first episode of spontaneous bacterial peritonitis in cirrhosis with ascites. Relevance of ascitic fluid protein concentration. Hepatology. 1992;16:742–7.
18. Runyon BA. Low-protein-concentration ascitic fluid is predisposed to spontaneous bacterial peritonitis Gastroenterology. 1986;91:1343–6.
19. Rimola A, Garcia-Tsao G, Navasa M et al. Diagnosis, treatment and prophylaxis of spontaneous bacterial peritonitis: a consensus document. International Ascites Club. J Hepatol. 2000; 32:142–53.
20. Sheer TA, Runyon BA. Spontaneous bacterial peritonitis. Dig Dis. 2005;23:39–46
21. Castellote J, Lopez C, Gornals J et al. Rapid diagnosis of spontaneous bacterial peritonitis by use of reagent strips. Hepatology. 2003;37:893–6.
22. Runyon BA. Monomicrobial nonneutrocytic bacterascites: a variant of spontaneous bacterial peritonitis. Hepatology. 1990;12:710–15.
23. Runyon BA, Hoefs JC. Culture-negative neutrocytic ascites: a variant of spontaneous bacterial peritonitis. Hepatology. 1984;4:1209–11.
24. Kamath P, Wiesner R, Malinchoc M et al. A model to predict survival in patients with end-stage liver disease. Hepatology. 2001;33:464–70.
25. Santos J, Planas R, Pardo A et al. Spironolactone alone or in combination with furosemide in the treatment of moderate ascites in nonazotemic cirrhosis. A randomized comparative study of efficacy and safety. J Hepatol. 2003;39:187–92.
26. Arroyo V, Ginès P, Gerbes A et al. Definition and diagnostic criteria of refractory ascites and hepatorenal syndrome in cirrhosis. Hepatology. 1996;23:164–76

27. Angeli P, Dalla Pria M, De Bei E et al. Randomized clinical study of the efficacy of amiloride and potassium canrenoate in nonazotemic cirrhotic patients with ascites. Hepatology. 1994;19:72–9.
28. Li CP, Lee FY, Hwang SJ et al. Treatment of mastalgia with tamoxifen in male patients with liver cirrhosis: a randomized crossover study. Am J Gastroenterol. 2000;95:1051–5.
29. Lee FY, Lee SD, Tsai YT et al. A randomized controlled trial of quinidine in the treatment of cirrhotic patients with muscle cramps. J Hepatol. 1991;12:236–40.
30. Angeli P, Albino G, Carraro P et al. Cirrhosis and muscle cramps: evidence of a causal relationship. Hepatology. 1996 ;23:264–73.
31. Ginès P, Arroyo V, Quintero E et al. Comparison of paracentesis and diuretics in the treatment of cirrhotics with tense ascites. Results of a randomized study. Gastroenterology. 1987;93:234–41.
32. Ruiz-del-Arbol L, Monescillo A, Jiménez W, Garcia-Plaza A, Arroyo V, Rodés J. Paracentesis-induced circulatory dysfunction: mechanism and effect on hepatic hemodynamics in cirrhosis. Gastroenterology. 1997;113:579–86
33. Ginès P, Tito L, Arroyo V et al. Randomized comparative study of therapeutic paracentesis with and without intravenous albumin in cirrhosis. Gastroenterology. 1988;94:1493–502.
34. Ginès A, Fernández-Esparrach G, Monescillo A et al. Randomized trial comparing albumin, dextran 70, and polygeline in cirrhotic patients with ascites treated by paracentesis. Gastroenterology. 1996;111:1002–110.
35. Moreau R, Valla DC, Durand-Zaleski I et al. Comparison of outcome in patients with cirrhosis and ascites following treatment with albumin or a synthetic colloid: a randomised controlled pilot trail. Liver Int. 2006;26:46–54.
36. Dumortier J, Pianta E, Le Derf Y et al. Peritoneovenous shunt as a bridge to liver transplantation. Am J Transplant. 2005;5:1886–92.
37. Donaghy A. Issues of malnutrition and bone disease in patients with cirrhosis. J Gastroenterol Hepatol. 2002;17:462–6.
38. Fernández-Esparrach G, Guevara M, Sort P et al. Diuretic requirements after therapeutic paracentesis in non-azotemic patients with cirrhosis. A randomized double-blind trial of spironolactone versus placebo. J Hepatol. 1997;26:614–20.
39. Ginès P, Uriz J, Calahorra B et al. Transjugular intrahepatic portosystemic shunting versus paracentesis plus albumin for refractory ascites in cirrhosis. Gastroenterology 2002;123: 1839–47.
40. Boyer TD, Haskal Z. The role of transjugular intrahepatic portosystemic shunt in the management of portal hypertension. Hepatology. 2005;41:386–400.
41. Bureau C, Garcia-Pagan JC, Otal P et al. Improved clinical outcome using polytetrafluoroethylene-coated stents for TIPS: results of a randomized study. Gastroenterology. 2004;126:469–75.
42. Lebrec D, Giuily N, Hadengue A et al. Transjugular intrahepatic portosystemic shunts: comparison with paracentesis in patients with cirrhosis and refractory ascites: a randomized trial. J Hepatol. 1996;25:135–44.
43. Rossle M, Ochs A, Gulberg V et al. A comparison of paracentesis and transjugular intrahepatic portosystemic shunting in patients with ascites. N Engl J Med. 2000;342: 1701–7.
44. Sanyal A, Genning C, Reddy RK et al. The North American Study for Treatment of Refractory Ascites. Gastroenterology. 2003;124:634–41.
45. Salerno F, Merli M, Riggio O. Randomized controlled study of TIPS versus paracentesis plus albumin in cirrhosis with severe ascites. Hepatology. 2004;40:629–35.
46. Deltenre P, Mathurin P, Dharancy S et al. Transjugular intrahepatic portosystemic shunt in refractory ascites: a meta-analysis. Liver Int. 2005;25:349–56
47. Albillos A, Banares R, Gonzales M et al. A meta-analysis of transjugular intrahepatic portosystemic shunt versus paracentesis for refractory ascites. J Hepatol. 2005;43:990–6.
48. Dámico G, Luca A, Morabito A et al. Uncovered transjugular intrahepatic portosystemic shunt for refractory ascites: a meta-analysis. Gastroenterology. 2005;129:1282–93
49. Cardenas A, Gines P. Management of refractory ascites. Clin Gastroenterol Hepatol. 2005; 3:1187–91.
50. Garcia-Tsao G. Bacterial infections in cirrhosis: treatment and prophylaxis. J Hepatol. 2005;42(Suppl. 1):S85–92.

51. Fernández J, Navasa M, G⁴mez J et al. Bacterial infections in cirrhosis: epidemiological changes with invasive procedures and norfloxacin prophylaxis. Hepatology. 2002;35:140–8.
52. Wiest R, Garcia-Tsao R. Bacterial translocation (BT) in cirrhosis. Hepatology. 2005;41: 422–33.
53. Tsai MH, Peng YS, Chen YC et al. Adrenal insufficiency in patients with cirrhosis, severe sepsis and septic shock. Hepatology. 2006;43:673–81.
54. Fernandez J, Escorsell A, Zabalza M et al. Adrenal insufficiency in patients with cirrhosis and septic shock: effect of treatment with hydrocortisone on survival. Hepatology. 2006;44:1288–95.
55. Follo A, Llovet JM, Navasa M et al. Renal impairment after spontaneous bacterial peritonitis in cirrhosis: incidence, clinical course, predictive factors and prognosis. Hepatology. 1994;20:495–501.
56. Sort P, Navasa M, Arroyo V et al. Effect of intravenous albumin on renal impairment and mortality in patients with cirrhosis and spontaneous bacterial peritonitis. N Engl J Med. 1999;341:403–9.
57. Tito L, Rimola A, Gines P, Llach J, Arroyo V, Rodes J. Recurrence of spontaneous bacterial peritonitis in cirrhosis: frequency and predictive factors. Hepatology. 1988;8:2731.
58. Bernard B, Grangé JD, Nguyen K et al. Antibiotics prophylaxis in cirrhotic patients with gastrointestinal bleeding: a meta-analysis. Hepatology. 1999;29:1655–61.
59. Soriano G, Guarner C, Tomás A et al: Norfloxacin prevents bacterial infection in cirrhotics with gastrointestinal hemorrhage. Gastroenterology. 1992;103:1267–72.
60. Soares-Weiser K, Brezis M, Tur-Kaspa R, Leibovici L. Antibiotic prophylaxis for cirrhotic patients with gastrointestinal bleeding. Cochrane Database Syst Rev. 2002;(2):CD002907.
61. Rimola A, Bory F, Terés J et al. Oral, nonabsorbable antibiotics prevent infection in cirrhotics with gastrointestinal hemorrhage. Hepatology. 1995;5:463–7.
62. Fernandez J, Ruiz del Arbol L, Gomez C et al. Norfloxacin versus ceftriaxone in the prophylaxis of infections in patients with advanced cirrhosis and hemorrhage. Gastroenterology. 2006;131:1049–56.
63. Ginès P, Rimola A, Planas R et al. Norfloxacin prevents spontaneous bacterial peritonitis recurrence in cirrhosis: results of a double blind, placebo-controlled trial. Hepatology. 1990;12:716–24.
64. Navasa M, Fernández J, Montoliu S et al. Randomized, double-blind, placebo, controlled trial evaluating norfloxacin in the primary prophylaxis of spontaneous bacterial peritonitis in cirrhotics with renal impairment, hyponatremia or severe liver failure. J Hepatol. 2006:44(Suppl. 2);114A.

Section VII
Sequelae of chronic hepatitis
(Part 2)

Chair: W STREMMEL and R WILLIAMS

18
Advances in the pathogenesis and treatment of type 1 and type 2 hepatorenal syndrome

V. ARROYO and J. FERNÁNDEZ

INTRODUCTION

Hepatorenal syndrome (HRS) is a frequent complication in patients with cirrhosis and ascites. The annual incidence has been estimated as 8%[1–3]. The traditional concept is that HRS is a functional renal failure (low renal perfusion and glomerular filtration rate (GFR) in the absence of significant histological renal lesions) associated with terminal hepatic insufficiency due to intense splanchnic arterial vasodilation and severe impairment in circulatory function (peripheral arterial vasodilation hypothesis)[4–6]. The identification in 1996 of two types of HRS with different courses and prognostic implications was a clear indication that this traditional concept was insufficient to explain the clinical features associated with HRS[7]. During the past decade other evidence has accumulated indicating that these two types of HRS also differ in pathogenesis. Finally, effective treatments of HRS have been developed, demonstrating that it is a reversible condition and not a terminal event in the clinical course of cirrhosis. Some of these studies found significant differences in treatment response between patients with type 1 and type 2 HRS. Taking together all these features, it appears that type 1 and type 2 HRS are not different expressions of an identical disorder but probably two syndromes with different pathogenesis, clinical implications, prognosis and treatment. The aim of this chapter is to review the current knowledge on HRS in cirrhosis, paying particular attention to the new developments in its pathogenesis and treatment.

DIAGNOSIS OF HRS

The first step in the diagnosis of HRS is the demonstration of a reduced GFR, and this is not easy in advanced cirrhosis. The muscle mass, and therefore the release of creatinine, is considerably reduced in these patients and they may present normal serum creatinine concentration in the setting of a very low

GFR. Similarly, urea is synthesized by the liver and may be reduced as a consequence of hepatic insufficiency. Therefore, a false-negative diagnosis of HRS is relatively common[8–10]. There is consensus to establish the diagnosis of HRS when serum creatinine has risen above 1.5 mg/dl or creatinine clearance has decreased to less than 40 ml/min[7]. The second step is the differentiation of HRS from other types of renal failure. The traditional parameters used by nephrologists to differentiate functional renal failure from acute tubular necrosis (oliguria, low urine sodium concentration, urine-to-plasma osmolality ratio greater than unity, normal fresh urine sediment and no proteinuria) are not useful in decompensated cirrhosis with ascites. Acute tubular necrosis in patients with cirrhosis and ascites usually shows with oliguria, low urine sodium concentration and urine osmolality greater than plasma osmolality[11]. On the contrary, a relatively high urinary sodium concentration has been observed in patients with HRS and high serum bilirubin[12].

The diagnosis of HRS is based on the exclusion of other disorders that can cause renal failure in cirrhosis[7]. Acute renal failure of pre-renal origin due to renal (diuretics) or extrarenal fluid losses should be investigated. If renal failure is secondary to volume depletion, renal function improves rapidly after volume expansion, whereas no improvement occurs in HRS. Even if there is no history of fluid losses, renal function should be assessed after diuretic withdrawal and volume replacement to rule out any subtle reduction in plasma volume as the cause of renal failure. The diagnostic criteria of HRS proposed by the International Ascites Club in San Francisco in 2005 consider that volume replacement should be performed with intravenous albumin (1 g/kg body weight up to a maximum of 100 g), rather than with saline. This is based on a randomized study showing that albumin is more effective as plasma expander than a saline solution of hydroxyethyl starch in patients with spontaneous bacterial peritonitis (SBP)[13]. The presence of shock before the onset of renal failure points towards a diagnosis of acute tubular necrosis. On the other hand, cirrhotic patients with infections may develop transient renal failure, which resolves after resolution of the infection. This occurs in approximately one-third of patients[14,15]. Therefore, HRS in cirrhotic patients with bacterial infections should be diagnosed in patients without septic shock, and only if renal failure does not improve following antibiotic administration. Complete resolution of the infection, which was required for the diagnosis of HRS in the initial proposal by the International Ascites Club in 1996[7], is no longer accepted because it may delay the initiation of treatment with vasoconstrictors and albumin (see below). Cirrhotic patients are predisposed to develop renal failure in the setting of treatments with aminoglycosides[16], non-steroidal anti-inflammatory drugs[17] and vasodilators (renin–angiotensin system inhibitors, prazosin, nitrates)[18]. Therefore, treatment with these drugs or other potential nephrotoxic agents in the days preceding the diagnosis of renal failure should be ruled out. Finally, patients with cirrhosis can develop renal failure due to intrinsic renal diseases, particularly glomerulonephritis. These cases can be recognized by the presence of proteinuria, haematuria or both.

CLINICAL FEATURES ASSOCIATED WITH TYPE 1 AND TYPE 2 HRS

Type 1 HRS is characterized by a severe and rapidly progressive renal failure, which has been defined as doubling of serum creatinine reaching a level >2.5 mg/dl in <2 weeks. Although type 1 HRS may arise spontaneously it frequently occurs in close relationship to a precipitating factor, such as severe bacterial infection, mainly SBP, gastrointestinal haemorrhage, major surgical procedure or acute hepatitis superimposed on cirrhosis. The association of HRS and SBP has been carefully investigated[14,19,20]. Type 1 HRS develops in approximately 25% of patients with SBP despite a rapid resolution of the infection with non-nephrotoxic antibiotics. Patients with severe circulatory dysfunction (moderate increase in serum creatinine and high levels of plasma renin activity and norepinephrine concentration) prior to infection, or intense inflammatory response (high concentration of polimorphonuclear leucocytes in ascitic fluid and high cytokine levels in plasma and ascitic fluid) are prone to develop type 1 HRS after the infection. Besides renal failure, patients with type 1 HRS associated with SBP show signs and symptoms of rapid and severe deterioration of liver function (jaundice, coagulopathy and hepatic encephalopathy) and circulatory function (arterial hypotension, very high plasma levels of renin and norepinephrine)[14,19–21]. It is interesting to note that, at variance to SBP, sepsis related to other types of infection in patients with cirrhosis induces type 1 HRS only in the setting of lack of response to antibiotics[15]. In most patients with sepsis unrelated to SBP responding to antibiotics, renal impairment, which is also a frequent event, is reversible. Without treatment, type 1 HRS is the complication of cirrhosis with the poorest prognosis with a median survival time after the onset of renal failure of only 2 weeks[3].

Type 2 HRS is characterized by a moderate (serum creatinine lower than 2.5 mg/dl) and slowly progressive renal failure. Patients with type 2 HRS show signs of liver failure and arterial hypotension but to a lesser degree than patients with type 1 HRS. The dominant clinical feature is severe ascites with poor or no response to diuretics (a condition known as refractory ascites). Patients with type 2 HRS are predisposed to develop type 1 HRS following infections or other precipitating events[14,19,20]. The median survival of patients with type 2 HRS (6 months) is worse than that of patients with non-azotaemic cirrhosis with ascites[22].

PATHOGENESIS OF RENAL DYSFUNCTION AND TYPE 2 HRS: THE PERIPHERAL ARTERIAL VASODILATION HYPOTHESIS

Portal hypertension in cirrhosis is associated with arterial vasodilation in the splanchnic circulation due to the local release of nitric oxide and other vasodilatory substances[23–26]. The peripheral arterial vasodilation hypothesis proposes that renal dysfunction in cirrhosis is related to this feature. Early in the course of the disease, when splanchnic arterial vasodilation is moderate, the decrease in systemic vascular resistance is compensated by the development of a hyperdynamic circulation (increased heart rate and cardiac output)[27–29].

However, as the disease progresses and arterial vasodilation increases, the hyperdynamic circulation is insufficient to correct the changes in circulatory function[30]. Effective arterial hypovolaemia develops, leading to arterial hypotension, activation of high-pressure baroreceptors and reflex stimulation of the renin–angiotensin and sympathetic nervous systems. This increases arterial pressure to normal or near-normal levels but induces sodium and water retention and ascites formation. The stimulation of antidiuretic hormone, which is less sensitive to arterial hypovolaemia than the renin–angiotensin and sympathetic nervous systems, occurs later during the course of the disease. Patients then develop water retention and dilutional hyponatraemia. At this stage of the disease the renin–angiotensin and sympathetic nervous systems are markedly stimulated and arterial pressure is critically dependent on the vascular effect of the sympathetic nervous activity, angiotensin II and antidiuretic hormone. Since the splanchnic circulation is resistant to the effect of angiotensin II, noradrenaline and vasopressin due to the local release of nitric oxide and other vasodilators[31,32], the maintenance of arterial pressure is due to vasoconstriction in extra-splanchnic vascular territories such as the kidneys, muscle, skin and brain[33–36]. Type 2 HRS develops at the latest phase of the disease when there is extreme deterioration in effective arterial blood volume and severe arterial hypotension. The homeostatic stimulation of the renin–angiotensin system, the sympathetic nervous system and antidiuretic hormone is very intense, leading to renal vasoconstriction and marked decrease in renal perfusion and GFR, azotaemia and increased serum creatinine concentration.

CARDIAC DYSFUNCTION IN CIRRHOSIS: THE MODIFIED PERIPHERAL ARTERIAL VASODILATION HYPOTHESIS OF RENAL DYSFUNCTION AND TYPE 2 HRS

Most haemodynamic studies in cirrhosis have been performed in non-azotaemic patients with and without ascites. Based on these investigations it has been asumed that type 2 HRS develops in the setting of a hyperdynamic circulation and increased cardiac output. However, in the few studies assessing cardiovascular function in patients with HRS or refractory ascites (most of them with type 2 HRS), cardiac output was found to be significantly reduced compared to patients without HRS[37,38]. In some cases cardiac output was even lower than in normal subjects, suggesting that circulatory dysfunction associated with HRS is due not only to arterial vasodilation but also to a decrease in cardiac function.

A recent study by Ruiz-del-Arbol et al. supports this feature[39]. This consisted of a longitudinal investigation of 66 non-azotaemic cirrhotic patients with ascites. Forty per cent of patients developed HRS. These patients were studied at inclusion and following the development of HRS. In the initial study those patients who went on to develop HRS had significantly lower mean arterial pressure and cardiac output, and significantly higher plasma renin activity and norepinephrine concentration compared with those who did not develop HRS. Moreover, those who developed type 2 HRS had a further decrease in arterial pressure and cardiac output, and increase in renin and norepinephrine, without changes in peripheral vascular resistance. These findings strongly suggest that progression of circulatory dysfunction in cirrhosis is due to both an increase in arterial vasodilation and a decrease in cardiac function, and that HRS occurs when there is marked splanchnic vasodilation and reduction in cardiac output.

TYPE 1 HRS: A SPECIAL FORM OF MULTIORGAN FAILURE

Type 1 and type 2 HRS have two features in common: they occur in patients with cirrhosis and, on the other hand, renal failure is an important component of both syndromes. However, they show important differences. Type 2 HRS develops imperceptibly in patients with cirrhosis and ascites who are otherwise in a stable clinical condition. These patients do not differ clinically from patients with cirrhosis and ascites without renal failure. They respond poorly to diuretics but this also occurs in a significant number of patients with a serum creatinine concentration below 1.5 mg/dl. Circulatory function, although severely deteriorated, remains steady or progresses slowly over a period of months, as occurs with renal function. Patients had advanced cirrhosis with severe malnourishment but the degree of liver failure is also stable. Hepatic encephalopathy is infrequent. The main clinical problem of patients with type 2 HRS is refractory ascites. In contrast, type 1 HRS is an extremely unstable condition. It frequently develops in the setting of an important clinical event that acts as a precipitating factor. Second, there is a rapid deterioration of circulatory and renal function within days after the onset of the syndrome, leading to severe arterial hypotension and renal failure with intense oliguria. Finally, in addition to the acute and severe impairment in renal function, there is a rapid deterioration of hepatic function, jaundice and encephalopathy. Recent investigations have presented data showing that the deterioration in circulatory function is related to an accentuation of arterial vasodilation and a reduction in cardiac output[21,39,40].

Because this decrease in cardiac output occurs in the setting of a significant decrease in cardiopulmonary pressures, an impairment in cardiac preload is probably an important pathogenetic mechanism. However, the heart of cirrhotic patients also presents intrinsic problems that may also be important pathogenetic factors. For example, despite a marked activation of sympathetic nervous activity during the development of type 1 HRS, there is no increase in heart rate, indicating an impaired cardiac chronotropic function. Recent studies indicates that – in addition to an impairment in the function of the heart, liver, kidneys and brain – patients with type 1 HRS also present a relative adrenal insufficiency which contributes to the poor prognosis of these patients[41–43].

The mechanism of multiorgan failure in type 1 HRS is complex, but probably the rapid deterioration of circulatory function associated with this condition plays an important role. As indicated previously, renal failure in type 1 HRS is due to renal vasoconstriction related to severe impairment in effective arterial blood volume and activation of the renin–angiotensin and sympathetic nervous systems. These vasoconstrictor systems also produce vasoconstriction in the brain and in the liver, and a reduction in cerebral and hepatic blood flow. The reduction in hepatic perfusion could account for the rapid deterioration of liver function and the development of hepatic encephalopathy. Cerebral vasoconstriction could also contribute to encephalopathy. Vasoconstriction of the suprarenal arteries is also an attractive mechanism for the relative adrenal insufficiency in cirrhotic patients with HRS. In patients with severe sepsis, renal failure is frequently associated with adrenal insufficiency[42,43].

TREATMENTS FOR TYPE 1 HRS

Liver transplantation

Liver transplantation is the treatment of choice of HRS[44–47]. Immediately after transplantation a further impairment in GFR may be observed, and many patients require haemodialysis (35% of patients with HRS as compared with 5% of patients without HRS)[44]. After this initial impairment in renal function, GFR starts to improve and reaches an average of 30–40 ml/min by 1–2 months postoperatively. This moderate renal failure persists during follow-up, is more marked than that observed in transplantation patients without HRS, and is probably due to a greater nephrotoxicity of cyclosporin or tacrolimus in patients with renal impairment prior to transplantation. The haemodynamic and neurohormonal abnormalities associated with HRS disappear within the first month after the operation, and patients regain a normal ability to excrete sodium and free water.

Patients with HRS who undergo transplantation have more complications, spend more days in the intensive-care unit, and have a higher in-hospital mortality rate than transplantation patients without HRS. The long-term survival of patients with HRS who undergo liver transplantation, however, is good; with a 3-year probability of survival of 60%. This survival rate is only slightly reduced compared with that of transplantation in patients without HRS (which ranges between 70% and 80%)[45].

The main problem of liver transplantation in type 1 HRS is its applicability. Due to their extremely short survival, most patients die before transplantation can be performed. The introduction of the MELD score, which includes serum creatinine, bilirubin and the international normalized ratio, for listing has partially solved the problem, since patients with HRS are generally allocated the first places of the waiting list. Treatment of HRS with vasoconstrictors and albumin (see below) increases survival in a significant proportion of patients; making possible that a significant proportion of them reach liver transplantation. Therefore, treatment of HRS decreases early morbidity and mortality after transplantation and prolongs long-term survival.

Vasoconstrictors and albumin

The intravenous administration of vasoconstrictor agents (vasopressin, ornipressin, terlipressin, noradrenaline) or the combination of oral midodrine (an alpha-agonistic agent) and intravenous or subcutaneous octreotide over a period of 1–3 weeks is an effective treatment of type 1 HRS. Eleven pilot studies including 154 patients with HRS (132 with type 1 HRS) have so far been published on this topic[48–58]. In most patients intravenous albumin was also given. The overall rate of positive response was 61.6% (95 patients). In eight of these studies (including 128 patients) a positive response was considered when there was reversal of HRS as defined by a decrease of serum creatinine below 1.5 mg/dl. This feature was observed in 79 patients (61.7%). A second important observation was that type 1 HRS does not recur after discontinuation of treatment in most patients. Five studies including 65 patients have reported data on this feature. Forty-four patients responded to therapy and HRS recurred in only nine. These findings contrast sharply with those from seven studies in patients with type 1 HRS not receiving specific treatment, or treated with plasma volume expansion alone or associated with vasodilators (dopamine) or octreotride or with peritoneovenous shunting[14,20,50,57,59–61]. Reversal of HRS was observed in only four of the 137 patients (2.9%) included in these studies. Survival data were recorded in 13 studies (eight using vasoconstrictors and five using other treatments). Forty (41.6%) and 29 (30%) of the 96 patients with type 1 HRS treated with vasoconstrictors were alive at 1 and 3 months after treatment. The corresponding figures in 65 patients receiving other treatments were two (3%) and none (0%), respectively.

A retrospective survey in 99 patients with type 1 HRS admitted to 22 hospitals in France and treated with terlipressin (all cases) and albumin (70% of cases) showed a rate of improvement in renal function of 58%[62]. The probability of survival was 40% at 1 month and 22% at 3 months. Improvement of survival was related to reversal of HRS. This study, which can be taken as an indication reflecting what occurs in regular clinical practice, therefore confirms the results obtained in the pilot studies.

Two randomized controlled studies have recently been completed comparing albumin versus albumin plus terlipressin in patients with type 1 HRS (Sanyal et al. and Guevara et al., unpublished results) and they confirm the results obtained in the pilot studies. Reversal of HRS was significantly more frequent in patients treated with terlipressin and albumin. On the other hand, survival of patients responding to treatment was significantly longer than that of patients not responding to treatment, or treated with albumin alone.

These studies clearly indicate that vasoconstrictors associated with intravenous albumin should be recommended for the management of patients with type 1 HRS since they normalize serum creatinine in a high proportion of patients and may improve survival. Terlipressin has been the most widely used vasoconstrictor agent in type 1 HRS; it is very effective and is associated with a low incidence of side-effects. The efficacy of the association of oral midodrine and intravenous or subcutaneous octreotide is probably due exclusively to the vasoconstrictor effect of midodrine[63]. Noradrenaline has also been shown to be effective and safe; however, whereas there is much experience with terlipressin, noradrenaline and midodrine have been used in only a few studies. Based on these considerations terlipressin should be the drug of choice for the treatment of type 1 HRS.

Reversal of type 1 HRS in two pilot studies in which terlipressin was given alone (seven of 28 patients, 25%)[53,55] was lower than that in the studies in which vasoconstrictors were associated with intravenous albumin, suggesting that the albumin is an important component in the pharmacological treatment of type 1 HRS. Two recent studies[13,64] suggest that the beneficial effect of albumin on circulatory and renal function in patients with type 1 HRS is related not only to the expansion of plasma volume but also to a direct vasoconstrictor effect on the peripheral arterial circulation.

Terlipressin dosage should be progressive, starting with 0.5 mg/4 h. If serum creatinine does not decrease by more than 30% in 3 days the dose should be doubled. The maximal dose of terlipressin has not been defined, although there was consensus that patients not responding to 12 mg/day will not respond to higher doses. Albumin should be given starting with a priming dose of 1 g/kg body weight followed by 20–40 g/day. It is advisable to monitor central venous pressure. In patients responding to therapy, treatment should be maintained until normalization of serum creatinine (<1.5 mg/dl).

Transjugular intrahepatic portacaval shunt (TIPS)

Three pilot studies have evaluated TIPS in type 1 HRS[58,65,66]. In the first study[65], 14 patients with type 1 HRS (12 with alcoholic cirrhosis, nine with active alcoholism) and 17 with refractory ascites (some of them with type 2 HRS) not suitable for liver transplantation were treated. Patients with bilirubin >15 mg/dl, Child–Pugh score >12 or hepatic encephalopathy were excluded. Eleven of the 31 patients developed *de-novo* hepatic encephalopathy or deterioration of previous hepatic encephalopathy. The 3, 6 and 12 month survival rates in patients with type 1 HRS were 64%, 50% and 20%, respectively. The second study[66] was performed in seven patients (four alcoholics) with type 1 HRS and a Child–Pugh score <12. Marked decrease in serum creatinine was observed in six patients and reversal of HRS in four. Five patients developed episodes of hepatic encephalopathy after TIPS but they responded satisfactorily to medical treatment. Five patients were alive after 1 month of TIPS but only two after 3 months. The third study[58] was performed in 14 patients (13 with alcoholic cirrhosis) with type 1 HRS treated initially with vasoconstrictors (midodrine and octreotide) plus albumin. Reversal of HRS was obtained in 10 patients. TIPS was subsequently inserted in five of these patients who had bilirubin <5 mg/dl, INR <2, Child–Pugh score <12. Normalization of GFR was obtained in all cases and all were alive between 6 and 30 months after TIPS. TIPS, therefore, is effective at normalizing serum creatinine in a significant proportion of patients with cirrhosis and severe azotaemia, and is an alternative treatment for type 1 HRS.

Extracorporeal albumin dialysis (MARS)

Three pilot studies including 29 patients (26 with type 1 HRS and 21 with alcoholic cirrhosis and/or severe acute alcoholic hepatitis) aimed at assessing MARS in patients with type 1 HRS have been reported[67–69]. Since MARS incorporates a standard dialysis machine or a continuous veno-venous haemofiltration monitor, and GFR was not measured, it is not possible to establish the effect of this treatment on renal function. The decrease in serum creatinine observed in most patients could be related to the dialysis process. However, clear beneficial effects on systemic haemodynamics and on hepatic encephalopathy, were observed. The survival rate at 1 and 3 months after treatment was 41% (12 patients) and 34% (10 patients), respectively. A recent randomized controlled trial in a large series of cirrhotic patients with hepatic encephalopathy[70], many of whom had HRS, has demonstrated a clear beneficial effect of MARS on the rate and time of recovery of encephalopathy. Since the endpoint of this trial was encephalopathy, no conclusion could be drawn in relation to survival.

TREATMENTS FOR TYPE 2 HRS

Transjugular intrahepatic portosystemic shunt

Five trials comparing TIPS versus paracentesis in patients with refractory and/or recidivant ascites have so far been published[71–75]. In total 172 patients were treated by TIPS in these studies. Unfortunately, very few of these patients had HRS. Patients with serum creatinine >3 mg/dl were excluded from three studies. Only two studies gave the number of patients with HRS included (six out of 53). Finally, mean serum creatinine was below 1.5 mg/dl in the groups treated by TIPS in these five studies. Therefore, data from these five trials are not valid for the assessment of TIPS in the management of patients with type 2 HRS.

There are only two pilot studies specifically assessing TIPS in type 2 HRS[65,76]. In one study[76] a significant reduction of serum creatinine (from 2.1 ± 0.3 to 1.4 ± 0.3 at 1 month after TIPS) was observed in eight out of nine patients. This was associated with significant improvement in the control of ascites. Four of these patients died, two within the first month and two at 12 and 14 months after the procedure. The remaining five patients had longer survival. No data were given on the type or rate of complications associated with TIPS. A second study included 14 patients with HRS and 17 with type 2 HRS treated by TIPS[65]. Mean baseline serum creatinine concentration in patients with type 2 HRS was of only 1.44 ± 0.3 mg/dl, but mean creatinine clearance was 28 ± 14 ml/min. A significant improvement in serum creatinine and creatinine clearance was observed in the whole group of 31 patients, as well as an improvement in the control of ascites in 24 cases. Six patients developed TIPS dysfunction and 11 developed hepatic encephalopathy during follow-up. The 1-year probability of survival in the 17 patients with type 2 HRS treated by TIPS was 70%. TIPS is therefore effective in reversing type 2 HRS, although more data on complication rate and survival are needed before advocating widespread use of this procedure. The introduction of covered stents should be a stimulus to re-evaluate the role of TIPS in the management of refractory ascites and type 2 HRS.

Vasoconstrictors and albumin

Three pilot studies provided data on the effect of terlipressin plus albumin in 26 patients with type 2 HRS[51,55,76]. Reversal of HRS was obtained in most cases (21 cases, 80%). In one of these studies[76] in 11 patients the course of renal function after stopping treatment was assessed, and HRS recurred in all cases. There were no data on survival. The current state of knowledge on vasoconstrictor therapy in type 2 HRS is therefore very poor; it appears to be less effective than in type 1 HRS due to the high rate of HRS recurrence.

PREVENTION OF HRS

Three randomized controlled studies in large series of patients have shown that HRS can be prevented in specific clinical settings. In the first study[61], the administration of albumin (1.5 g/kg intravenously at infection diagnosis and 1 g/kg intravenously 48 h later) to patients with cirrhosis and SBP markedly reduced the incidence of circulatory dysfunction and type 1 HRS (10% incidence of type 1 HRS in patients receiving albumin vs 33% in the control group). Hospital mortality rate (10% vs 29%) and 3-month mortality rate (22% vs 41%) were lower in patients receiving albumin. The second study was performed in cirrhotic patients with a high risk of developing SBP and type 1 HRS (Fernandez J et al., unpublished data). Primary prophylaxis of SBP using long-term oral norfloxacin was given to 35 patients with low protein ascites (<15 g/L) and impaired renal function (serum creatinine ⩾1.2 mg/dl, BUN ⩾25 mg/dl or serum sodium ⩽130 mEq/L) or severe liver failure (Child–Pugh score ⩾9 points with serum bilirubin ⩾3 mg/dl), 33 patients received placebo. Norfloxacin administration was associated with a significant decrease in 1-year probability of development of SBP (7% vs 61%) and HRS (28% vs 41%) and a significant increase in the 3-month and 1-year probability of survival (94% vs 62% and 60% vs 48%, respectively). In the third study[77], the administration of the tumour necrosis factor inhibitor pentoxyfilline (400 mg three times a day) to patients with severe acute alcoholic hepatitis reduced the occurrence of HRS (8% in the pentoxyfilline group vs 35% in the placebo group) and hospital mortality (24% vs 46%, respectively). Because bacterial infections and acute alcoholic hepatitis are important precipitating factors of type 1 HRS, these prophylactic measures may decrease the incidence of this complication.

Acknowledgements

This study was supported in part by grants from the Instituto de Salud Carlos III (C03/2).

References

1. Gines P, Rodes J. Clinical disorders of renal function in cirrhosis with ascites. In: Arrovo V, Gines P, Rodes J, Schrier RW, editors. Ascites and Renal Dysfunction in Liver Disease: Pathogenesis, Diagnosis, and Treatment. Malden: Blackwell Science, 1999:36–62.

2. Hecker R, Sherlock S. Electrolyte and circulatory changes in terminal liver failure. Lancet. 1956;271:1121–5.
3. Gines A, Escorsell A, Gines P et al. Incidence, predictive factors, and prognosis of the hepatorenal syndrome in cirrhosis with ascites. Gastroenterology. 1993;105:229–36.
4. Koppel MH, Coburn JW, Mims MM, Goldstein H, Boyle JD, Rubini ME. Transplantation of cadaveric kidneys from patients with hepatorenal syndrome – evidence for functional nature of renal failure in advanced liver disease. N Engl J Med. 1969;280:1367–8.
5. Iwatsuki S, Popovtze MM, Corman JL et al. Recovery from hepatorenal syndrome after orthotopic liver transplantation. N Engl J Med. 1973;289:1156–9.
6. Gines P, Guevara M, Arroyo V, Rodes J. Hepatorenal syndrome. Lancet. 2003;362:1819–27.
7. Arroyo V, Gines P, Gerbes AL et al. Definition and diagnostic criteria of refractory ascites and hepatorenal syndrome in cirrhosis. Hepatology. 1996;23:164–76.
8. Papadakis MA, Arieff AI. Unpredictability of clinical evaluation of renal function in cirrhosis – prospective study. Am J Med. 1987;82:945–52.
9. Caregaro L, Menon F, Angeli P et al. Limitations of serum creatine level and creatine clearance as filtration markers in cirrhosis. Arch Int Med. 1994;154:201–5.
10. Sherman DS, Fish DN, Teitelbaum I. Assessing renal function in cirrhotic patients: problems and pitfalls. Am J Kidney Dis. 2003;41:269–78.
11. Cabrera J, Arroyo V, Ballesta AM et al. Aminoglycoside nephrotoxicity in cirrhosis – value of urinary beta-2-microglobulin to discriminate functional renal failure from acute tubular damage. Gastroenterology. 1982;82:97–105.
12. Dudley FJ, Kanel GC, Wood LJ, Reynolds TB. Hepatorenal syndrome without avid sodium retention. Hepatology. 1986;6:248–51.
13. Fernandez J, Monteagudo J, Bargallo X et al. A randomized unblinded pilot study comparing albumin versus hydroxyethyl starch in spontaneous bacterial peritonitis. Hepatology. 2005;42:627–34.
14. Follo A, Llovet JM, Navasa M et al. Renal impairment after spontaneous bacterial peritonitis in cirrhosis – incidence, clinical course, predictive factors and prognosis. Hepatology. 1994;20:1495–501.
15. Terra C, Guevara M, Torre A et al. Renal failure in patients with cirrhosis and sepsis unrelated to spontaneous bacterial peritonitis: value of MELD score. Gastroenterology. 2005;129:1944–53.
16. Hampel H, Bynum GD, Zamora E, El Serag HB. Risk factors for the development of renal dysfunction in hospitalized patients with cirrhosis. Am J Gastroenterol. 2001;96:2206–10.
17. Brater DC, Anderson SA, Browncartwright D, Toto RD. Effects of nonsteroidal antiinflammatory drugs on renal function in patients with renal insufficiency and in cirrhotics. Am J Kidney Dis. 1986;8:351–5.
18. Westphal JF, Brogard JM. Drug administration in chronic liver disease. Drug Safety. 1997; 17:47–73.
19. Toledo C, Salmeron JM, Rimola A et al. Spontaneous bacterial peritonitis in cirrhosis – predictive factors of infection resolution and survival in patients treated with cefotaxime. Hepatology. 1993;17:251–7.
20. Navasa M, Follo A, Filella X et al. Tumor necrosis factor and interleukin-6 in spontaneous bacterial peritonitis in cirrhosis: relationship with the development of renal impairment and mortality. Hepatology. 1998;27:1227–32.
21. Ruiz-del-Arbol L, Urman J, Fernandez J et al. Systemic, renal, and hepatic hemodynamic derangement in cirrhotic patients with spontaneous bacterial peritonitis. Hepatology. 2003; 38:1210–18.
22. Rodes J, Arroyo V, Bosch J. Clinical types and drug therapy of renal impairment in cirrhosis. Postgrad Med J. 1975;51:492–7.
23. Goyal RK, Hirano I. Mechanisms of disease – the enteric nervous system. N Engl J Med. 1996;334:1106–15.
24. Gupta S, Morgan TR, Gordan GS. Calcitonin gene-related peptide in hepatorenal syndrome – a possible mediator of peripheral vasodilation. J Clin Gastroenterol. 1992;14: 122–6.
25. Bendtsen F, Schifter S, Henriksen JH. Increased circulating calcitonin gene-related peptide (Cgrp) in cirrhosis. J Hepatol. 1991;12:118–23.

26. McNicol PL, Liu G, Shulkes A, Hardy KJ, Jones RM. Vasoactive intestinal peptide and calcitonin gene-related peptide levels and hemodynamics during human liver transplantation. Transplant Proc. 1993;25:1830–1.
27. Benoit JN, Granger DN. Splanchnic hemodynamics in chronic portal hypertension. Semin Liver Dis. 1986;6:287–98.
28. Vorobioff J, Bredfeldt JE, Groszmann RJ. Increased blood flow through the portal system in cirrhotic rats. Gastroenterology. 1984;87:1120–6.
29. Vorobioff J, Bredfeldt JE, Groszmann RJ. Hyperdynamic circulation in portal hypertensive rat model – a primary factor for maintenance of chronic portal hypertension. Am J Physiol. 1983;244:G52–7.
30. Schrier RW, Arroyo V, Bernardi M, Epstein M, Henriksen JH, Rodes J. Peripheral arterial vasodilation hypothesis – a proposal for the initiation of renal sodium and water retention in cirrhosis. Hepatology. 1988;8:1151–7.
31. Lee FY, Albillos A, Colombato LA, Groszmann RJ. The role of nitric-oxide in the vascular hyporesponsiveness to methoxamine in portal hypertensive rats. Hepatology. 1992;16: 1043–8.
32. Sieber CC, Lopeztalavera JC, Groszmann RJ. Role of nitric oxide in the *in vitro* splanchnic vascular hyporeactivity in ascitic cirrhotic rats. Gastroenterology 1993;104:1750–4.
33. Maroto A, Gines P, Arroyo V et al. Brachial and femoral artery blood flow in cirrhosis – relationship to kidney dysfunction. Hepatology. 1993;17:788–93.
34. Maroto A, Gines A, Salo J et al. Diagnosis of functional kidney failure of cirrhosis with doppler sonography – prognostic value of resistive index. Hepatology. 1994;20:839–44.
35. Fernandez-Seara J, Prieto J, Quiroga J et al. Systemic and regional hemodynamics in patients with liver cirrhosis and ascites with and without functional renal failure. Gastroenterology 1989;97:1304–312.
36. Guevara M, Bru C, Gines P et al. Increased cerebrovascular resistance in cirrhotic patients with ascites. Hepatology. 1998;28:39–44.
37. Tristani FE, Cohn JN. Systemic and renal hemodynamics in oliguric hepatic failure – effect of volume expansion. J Clin Invest. 1967;46:1894–??
38. Lebrec D, Kotelanski B, Cohn JN. Splanchnic hemodynamic factors in cirrhosis with refractory ascites. J Lab Clin Med. 1979;93:301–9.
39. Ruiz-del-Arbol L, Monescillo A, Arocena C et al. Circulatory function and hepatorenal syndrome in cirrhosis. Hepatology. 2005;42:439–47.
40. Ma ZH, Lee SS. Cirrhotic cardiomyopathy: getting to the heart of the matter. Hepatology. 1996;24:451–9.
41. Harry R, Auzinger G, Wendon J. The clinical importance of adrenal insufficiency in acute hepatic dysfunction. Hepatology. 2002;36:395–402.
42. Tsai MH, Peng YS, Chen YC et al Adrenal insufficiency in patients with cirrhosis, severe sepsis and septic shock. Hepatology. 2006;43:673–81.
43. Fernandez J, Escorsell A, Zabalza M et al. Adrenal insufficiency in patients with cirrhosis and septic shock: effect of treatment with hydrocortisone on survival. Hepatology. 2006;44: 1288–95.
44. Gonwa TA, Morris CA, Goldstein RM, Husberg BS, Klintmalm GB. Long-term survival and renal function following liver transplantation in patients with and without hepatorenal syndrome – experience in 300 patients. Transplantation. 1991;51:428–30.
45. Gonwa TA, Klintmalm GB, Levy M, Jennings LS, Goldstein RM, Husberg BS. Impact of pretransplant renal function on survival after liver transplantation. Transplantation. 1995;59:361–5.
46. Seu P, Wilkinson AH, Shaked A, Busuttil RW. The hepatorenal syndrome in liver transplant recipients. Am Surgeon. 1991;57:806–9.
47. Rimola A, Gavaler JS, Schade RR, Ellankany S, Starzl TE, Vanthiel DH. Effects of renal impairment on liver transplantation. Gastroenterology. 1987;93:148–56.
48. Guevara M, Gines P, Fernandez-Esparrach G et al. Reversibility of hepatorenal syndrome by prolonged administration of ornipressin and plasma volume expansion. Hepatology. 1998;27:35–41.
49. Gulberg V, Bilzer M, Gerbes AL. Long-term therapy and retreatment of hepatorenal syndrome type 1 with ornipressin and dopamine. Hepatology. 1999;30:870–5.
50. Angeli P, Volpin R, Gerunda G et al. Reversal of type 1 hepatorenal syndrome with the administration of midodrine and octreotide. Hepatology. 1999;29:1690–7.

51. Uriz J, Gines P, Cardenas A et al. Terlipressin plus albumin infusion: an effective and safe therapy of hepatorenal syndrome. J Hepatol. 2000;33:43–8.
52. Mulkay JP, Louis H, Donckier V et al. Long term terlipressin administration improves renal function in cirrhotic patients with type 1 hepatorenal syndrome: a pilot study. Acta Gastro-Enterol Belg. 2001;64:15–19.
53. Colle I, Durand F, Pessione F et al. Clinical course, predictive factors and prognosis in patients with cirrhosis and type 1 hepatorenal syndrome treated with Terlipressin: a retrospective analysis. J Gastroenterol Hepatol. 2002;17:882–8.
54. Halimi C, Bonnard P, Bernard B et al. Effect of terlipressin (Glypressin®) on hepatorenal syndrome in cirrhotic patients: results of a multicentre pilot study. Eur J Gastroenterol Hepatol. 2002;14:153–8.
55. Ortega R, Gines P, Uriz J et al. Terlipressin therapy with and without albumin for patients with hepatorenal syndrome: results of a prospective, nonrandomized study. Hepatology. 2002;36:941–8.
56. Duvoux C, Zanditenas D, Hezode C et al. Effects of noradrenalin and albumin in patients with type I hepatorenal syndrome: a pilot study. Hepatology. 2002;36:374–80.
57. Solanki P, Chawla A, Garg R, Gupta R, Jain M, Sarin SK. Beneficial effects of terlipressin in hepatorenal syndrome: a prospective, randomized placebo-controlled clinical trial. J Gastroenterol Hepatol. 2003;18:152–6.
58. Wong F, Pantea L, Sniderman K. Midodrine, octreotide, albumin, and TIPS in selected patients with cirrhosis and type 1 hepatorenal syndrome. Hepatology. 2004;40:55–64.
59. Linas SL, Schaefer JW, Moore EE, Good JT, Giansiracusa R. Peritoneovenous shunt in the management of the hepatorenal syndrome. Kidney Int. 1986;30:736–40.
60. Pomier-Layrargues G, Paquin SC, Hassoun Z, Lafortune M, Tran A. Octreotide in hepatorenal syndrome: a randomized, double-blind, placebo-controlled, crossover study. Hepatology. 2003;38:238–43.
61. Sort P, Navasa M, Arroyo V et al. Effect of intravenous albumin on renal impairment and mortality in patients with cirrhosis and spontaneous bacterial peritonitis. N Engl J Med. 1999;341:403–9.
62. Moreau R, Durand F, Poynard T et al. Terlipressin in patients with cirrhosis and type 1 hepatorenal syndrome: a retrospective multicenter study. Gastroenterology. 2002;122:923–30.
63. Pomier-Layrargues G, Paquin SC, Hassoun Z, Lafortune M, Tran A. Octreotide in hepatorenal syndrome: a randomized, double-blind, placebo-controlled, crossover study. Hepatology. 2003;38:238–43.
64. Fernandez J, Navasa M, Garcia-Pagan JC et al. Effect of intravenous albumin on systemic and hepatic hemodynamics and vasoactive neurohormonal systems in patients with cirrhosis and spontaneous bacterial peritonitis. J Hepatol. 2004;41:384–90.
65. Brensing KA, Textor J, Perz J et al. Long term outcome after transjugular intrahepatic portosystemic stent–shunt in non-transplant cirrhotics with hepatorenal syndrome: a phase II study. Gut. 2000;47:288–95.
66. Guevara M, Gines P, Bandi JC et al. Transjugular intrahepatic portosystemic shunt in hepatorenal syndrome: effects on renal function and vasoactive systems. Hepatology. 1998;28:416–22.
67. Mitzner SR, Stange J, Klammt S et al. Improvement of hepatorenal syndrome with extracorporeal albumin dialysis MARS: results of a prospective, randomized, controlled clinical trial. Liver Transplant. 2000;6:277–86.
68. Catalina MV, Barrio J, Anaya F et al. Hepatic and systemic haemodynamic changes after MARS in patients with acute on chronic liver failure. Liver Int. 2003;23:39–43.
69. Jalan R, Sen S, Steiner C, Kapoor D, Alisa A, Williams R. Extracorporeal liver support with molecular adsorbents recirculating system in patients with severe acute alcoholic hepatitis. J Hepatol. 2003;38:24–31.
70. Stange I, Hassanein TI, Mehta R et al. Short-term survival of patients with severe intractable hepatic encephalopathy: the role of albumin dialysis. Hepatology. 2005;42:286A.
71. Gines P, Uriz J, Calahorra B et al. Transjugular intrahepatic portosystemic shunting versus paracentesis plus albumin for refractory ascites in cirrhosis. Gastroenterology. 2002;123: 1839–47.

72. Lebrec D, Giuily N, Hadengue A et al. Transjugular intrahepatic portosystemic shunts: comparison with paracentesis in patients with cirrhosis and refractory ascites: a randomized trial. J Hepatol. 1996;25:135–44.
73. Rossle M, Ochs A, Gulberg V et al. A comparison of paracentesis and transjugular intrahepatic portosystemic shunting in patients with ascites. N Engl J Med. 2000;342: 1701–7.
74. Sanyal AJ, Genning C, Reddy KR et al. The North American study for the treatment of refractory ascites. Gastroenterology. 2003;124:634–41.
75. Salerno F, Merli M, Riggio O et al. Randomized controlled study of TIPS versus paracentesis plus albumin in cirrhosis with severe ascites. Hepatology. 2004;40:629–35.
76. Alessandria C, Venon WD, Marzano A, Barletti C, Fadda M, Rizzetto M. Renal failure in cirrhotic patients: role of terlipressin in clinical approach to hepatorenal syndrome type 2. Eur J Gastroenterol Hepatol. 2002;14:1363–8.
77. Akriviadis E, Botla R, Briggs W, Han S, Reynolds T, Shakil O. Pentoxifylline improves short-term survival in severe acute alcoholic hepatitis: a double-blind, placebo-controlled trial. Gastroenterology. 2000;119:1637–48.

19
Hepatopulmonary syndrome

M. J. KROWKA

The classic triad of chronic liver disease, arterial hypoxaemia on room air, and intrapulmonary vascular dilation characterizes hepatopulmonary syndrome (HPS)[1,2]. The clinical presentation of HPS includes dyspnoea and fatigue (manifestations of hypoxaemia at rest, with exertion and during sleep). Physical examination findings associated with HPS are nailbed and lip cyanosis, as well as skin spider angiomas. Screening for hypoxaemia can be accomplished by finger pulse oximetry or arterial blood gas analysis. Oximetry saturations $<90\%$ or PaO_2 <80 mmHg merit testing for pulmonary vascular dilations.

Profound hypoxemia (PaO_2 <50 mmHg) is not uncommon. Spontaneous resolution of the syndrome with normalization of arterial oxygenation can occur, but is rare. Although hypoxaemia is common (40–50%) in the setting of chronic liver disease, the prevalence of HPS is uncommon. Approximately 20–30% of HPS patients may have other comorbidities that cause significant hypoxaemia (i.e. ascites, hydrothorax, pneumonia, pulmonary fibrosis and COPD). Approximately 5% of patients referred to the Mayo Clinic liver transplant programme have HPS.

Non-invasive methods used to determine the existence of pulmonary vascular dilations are based upon the concept of either microbubbles or technetium-labelled macroaggregated albumen (^{99m}Tc MAA) traversing dilated pulmonary vessels. The echocardiography test is qualitative and more sensitive in detecting pulmonary vascular dilation than the lung perfusion test. The lung perfusion–brain uptake study gives a quantitative measure of the degree of pulmonary vascular dilation. Pulmonary angiography as a means to document (or further characterize) pulmonary vascular dilation is not necessary in most cases.

The pathophysiology of HPS may be related to circulating mediators that result in excess local effects of nitric oxide and bacterial translocation, up-regulated pulmonary endothelium endothelin B receptors causing vasodilation, and pulmonary capillary proliferation via angiogenesis. Yet no proven medical treatments exist for HPS, other than nasal cannula O_2 supplementation. There is no proven role for TIPS. As the definitive treatment for HPS, orthotopic liver transplantation (OLT; cadaveric or living donor) is strongly advised, assuming no other major co-morbidities complicate advanced liver disease.

HPS alone has prognostic significance in patients with cirrhosis[3]. In addition, a 30–38% mortality within 12 months of OLT and 16% transplant hospitalization mortality have been reported. PaO_2 <50 mmHg and ^{99m}Tc MAA lung perfusion–brain uptake >20% is associated with higher post-OLT mortality[2]. However, complete resolution of HPS can occur even in the setting of severe hypoxaemia prior to OLT, and the time to resolution (usually many months) is related to the severity of hypoxaemia. The 5-year survivals with and without liver transplantation are 76% versus 23%, respectively. PaO_2 <50 mmHg with or without OLT is associated with worse survival compared to those with PaO_2 >50 mmHg[4].

Current questions of significance:

1. What is the most recent thought about pathophysiology?
2. Should endothelin antagonists be considered as experimental treatment in HPS?
3. Should HPS patients receive increased priority for liver transplantation?
4. Do the lungs of patients with HPS really normalize after liver transplantation?

References

1. Krowka MJ. Leading article: Hepatopulmonary syndromes. Gut. 2000;40:1–4.
2. Rodrigeuz-Roisin R, Krowka MJ, Fallon MB, Herve P. Pulmonary–hepatic vascular disorders: report of a task force. Eur Respir J. 2004;24:861–80.
3. Schenk P, Schoniger-Hekele M, Fuhrmann V, Madl C, Silberhumer G, Muller C. Prognostic significance of the hepatopulmonary syndrome in patients with cirrhosis. Gastroenterology. 2003;125:1042–52.
4. Swanson KL, Wiesner RH, Krowka MJ. Natural history of hepatopulmonary syndrome: impact of liver transplantation. Hepatology. 2005;41:1122–9.

20
Hepatocellular carcinoma

A. FORNER, C. RODRÍGUEZ DE LOPE, M. VARELA and J. BRUIX

INTRODUCTION

Hepatocellular carcinoma (HCC) has experienced a rising worldwide incidence in the past two decades, due not only to an augmented capacity to diagnose the disease, but also to the impact of the long-term chronic hepatitis C virus (HCV) infection, particularly in developed countries such as the United States[1]. This fact, linked with the improvement in the treatment of the complications associated to cirrhosis, has converted HCC into the main cause of death in cirrhotic patients[2]. In addition, HCC has become the fifth most common cancer in the world and the third cause of cancer-related mortality[3].

HCC is a neoplasm with well-defined risk factors. Cirrhosis underlies HCC in the majority of cases in developed areas, so any agent that causes cirrhosis should be considered as a potential risk factor for HCC. The current potential curative treatments (surgical resection, percutaneous ablation and liver transplantation) are applicable only in a relatively small proportion of asymptomatic early-stage cases. For these reasons, despite there being no unequivocal evidence that surveillance improves patient survival, screening for HCC in cirrhotics is recommended, and relies on biannual abdominal ultrasound in patients affected with cirrhosis who would benefit from effective treatment if HCC were diagnosed[4].

DIAGNOSIS

The proper diagnosis of liver nodules detected by screening ultrasound is a major clinical problem. Biopsy using fine needles has until recently been the only way to achieve a conclusive HCC diagnosis. However, its accuracy is far from optimal, particularly in nodules smaller than 2 cm, in which a high rate of false-negatives due to sampling error or misdiagnosis between high- or low-grade dysplasia and HCC has been described. In addition, fine-needle biopsy is not exempt from adverse events, particularly bleeding related to clotting impairment and a potential risk of needle track seeding. During recent years the progressive improvement in imaging techniques has allowed detection of a specific vascular pattern of HCC, characterized by an intense contrast uptake during the arterial phase followed by washout in the delayed venous phase (Figure 1). Taking into account these two premises, the AASLD has recently

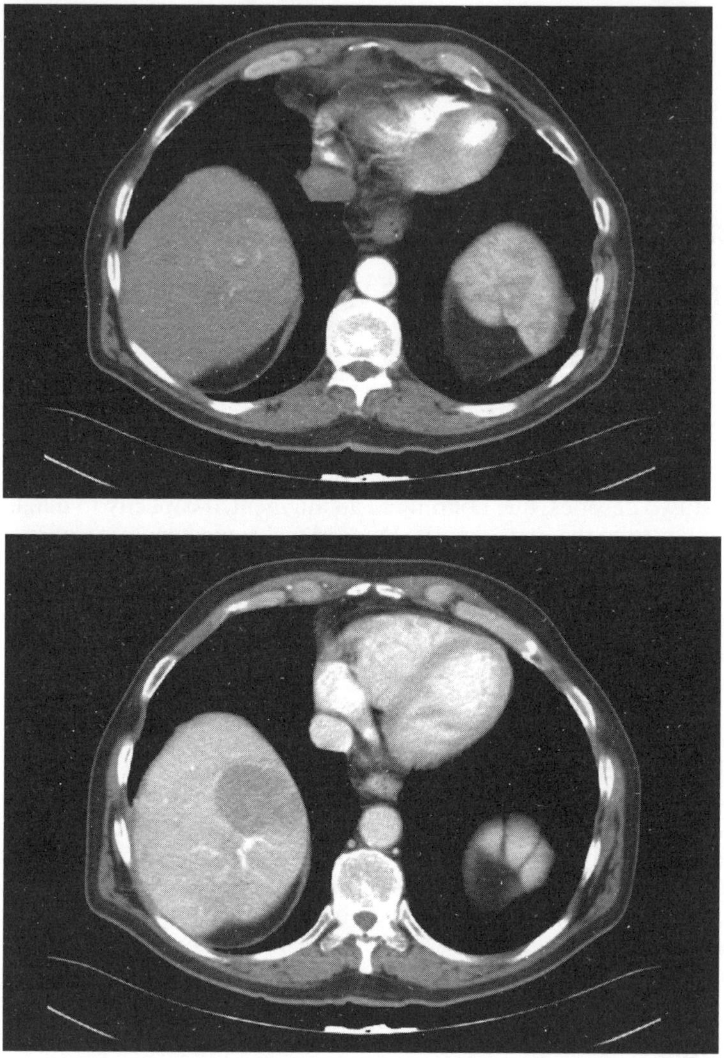

a

b

Figure 1 Abdominal CT scan of a patient affected with liver cirrhosis and HCC. Figure 1a shows an intense vascular intake during the arterial phase after contrast administration; Figure 1b illustrates the specific contrast washout during the delayed/venous phase

Figure 2 (Opposite) Diagnostic strategy to apply to a hepatic nodule detected during screening ultrasound in patients with cirrhosis. Reproduced with permission from ref. 4

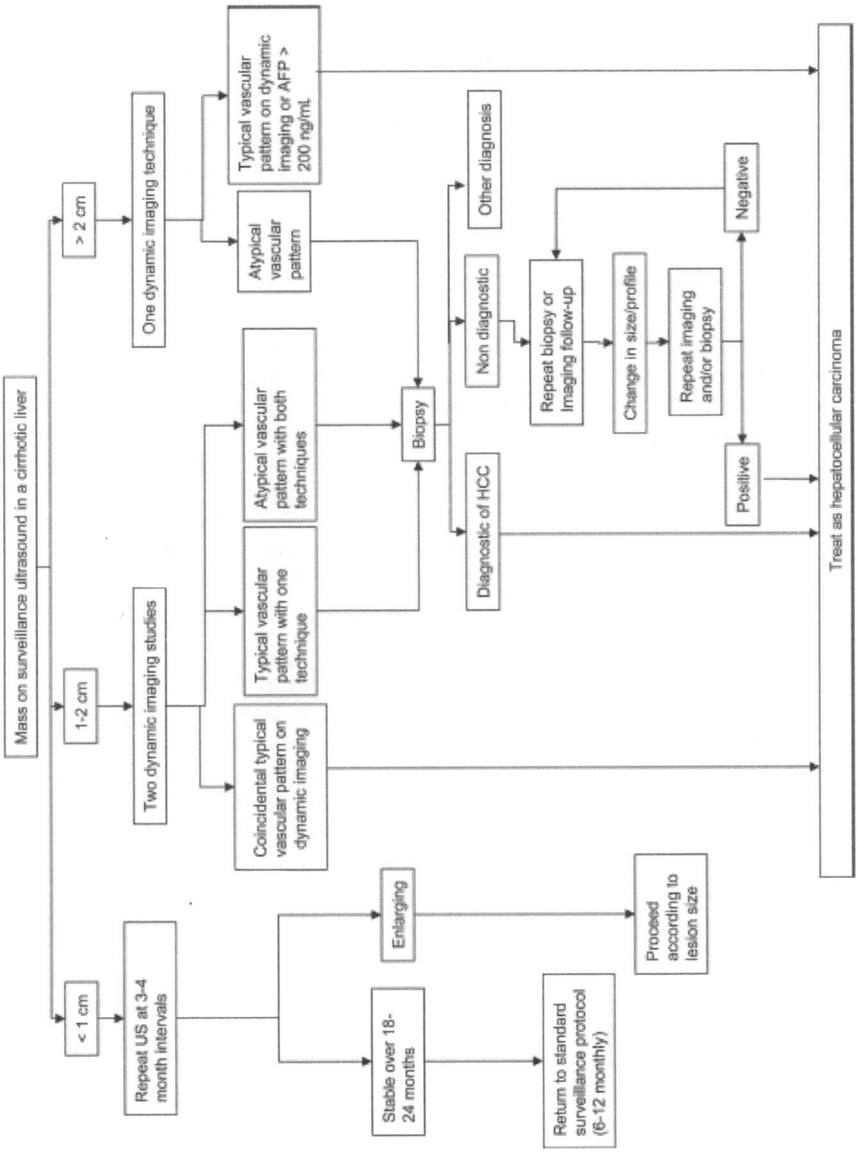
Mass on surveillance ultrasound in a cirrhotic liver
< 1 cm
1-2 cm
> 2 cm
Repeat US at 3-4 month intervals
Two dynamic imaging studies
One dynamic imaging technique
Coincidental typical vascular pattern on dynamic imaging
Typical vascular pattern with one technique
Atypical vascular pattern with both techniques
Atypical vascular pattern
Typical vascular pattern on dynamic imaging or AFP > 200 ng/mL
Biopsy
Diagnostic of HCC
Non diagnostic
Other diagnosis
Repeat biopsy or Imaging follow-up
Change in size/profile
Repeat imaging and/or biopsy
Positive
Negative
Treat as hepatocellular carcinoma
Enlarging
Stable over 18-24 months
Proceed according to lesion size
Return to standard surveillance protocol (6-12 monthly)

published a diagnostic strategy for diagnosis of nodules detected by screening ultrasound in cirrhotic patients[4], which is summarized in Figure 2. A nodule smaller than 1 cm has low probability of malignancy, and its accurate diagnosis is difficult, so a closer follow-up is recommended; if growth is detected one should act according to the new size. In nodules greater than 1 cm a non-invasive diagnosis is possible if the specific vascular pattern described previously is detected on dynamic imaging techniques (contrast-enhanced ultrasound, triphasic CT or MRI with gadolinium). In nodules between 1 and 2 cm this specific vascular pattern should be demonstrated coincidentally in two imaging techniques to enhance the specificity of the findings. If imaging is not conclusive a biopsy is recommended, but taking into account its low diagnostic accuracy; therefore a negative result does not rule out HCC diagnosis. These recommendations have been recently validated (Forner et al., AASLD, 2006).

PROGNOSIS PREDICTION

Prognosis prediction is key to deciding the best therapeutic option and to giving correct information to patients and their relatives. Taking into account that cirrhosis underlies HCC in the majority of patients, prognosis depends not only on tumour stage, but also on the degree of liver functional reserve. In recent years several prognosis prediction systems have been proposed[5]. The prognosis system recommended by the AASLD and the EASL is the BCLC system[6]. This has been externally validated and links prognostic predictors to treatment indication. Briefly, the BCLC system considers four prognostic groups (Figure 3):

1. Early stage: patients with well-preserved liver function and an asymptomatic single HCC with a maximum diameter of 5 cm or up to 3 nodules each less than 3 cm. Such patients will benefit from curative therapies, including resection, liver transplantation (LT), and percutaneous ablation. Patients with a single nodule less than 2 cm in the setting of very well preserved liver function and absence of relevant portal hypertension fit into the so-called very early stage. This tumour may be considered as a carcinoma *in situ* and 5-year survival after surgical resection may exceed 90%.

2. Intermediate stage: patients exceeding these limits, but who have preserved liver function and are free of cancer-related symptoms and vascular invasion or extrahepatic spread. Such patients may benefit from palliation with chemoembolization.

3. Advanced stage: patients with mild cancer-related symptoms and/or vascular invasion or extrahepatic spread, but with maintained liver function. In this stage there is no effective therapy, and these patients may profit from new therapies in the setting of randomized controlled trials (RCT).

Figure 3 (Opposite) The BCLC staging and treatment strategy. Reproduced with permission from ref. 7

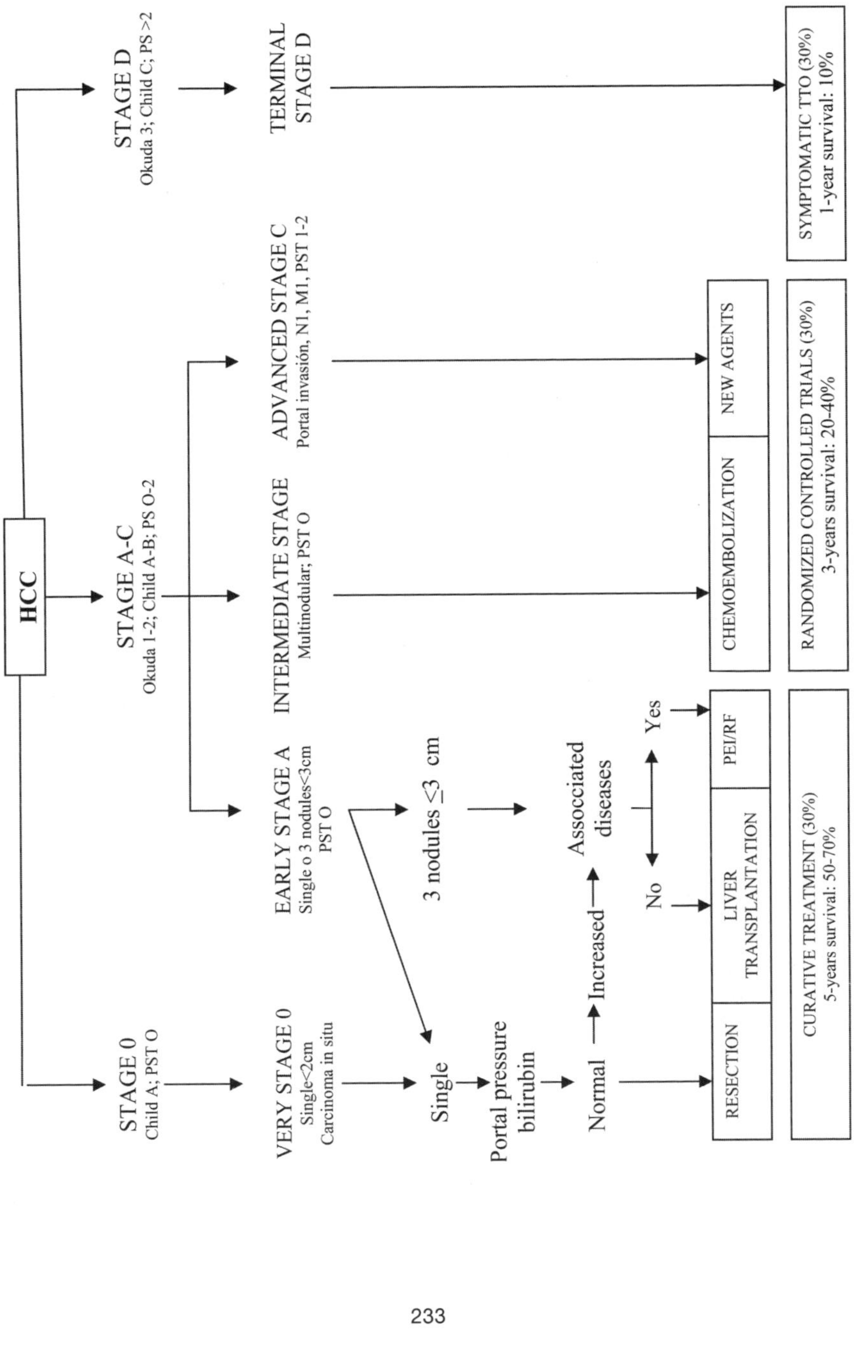

HCC
STAGE 0
Child A; PST O
STAGE A-C
Okuda 1-2; Child A-B; PS O-2
STAGE D
Okuda 3; Child C; PS >2
VERY STAGE 0
Single<2cm
Carcinoma in situ
EARLY STAGE A
Single o 3 nodules<3cm
PST O
INTERMEDIATE STAGE
Multinodular; PST O
ADVANCED STAGE C
Portal invasión, N1, M1, PST 1-2
TERMINAL
STAGE D
Single
3 nodules ≤3 cm
Portal pressure
bilirubin
Normal
Increased
Associated
diseases
No
Yes
RESECTION
LIVER
TRANSPLANTATION
PEI/RF
CHEMOEMBOLIZATION
NEW AGENTS
CURATIVE TREATMENT (30%)
5-years survival: 50-70%
RANDOMIZED CONTROLLED TRIALS (30%)
3-years survival: 20-40%
SYMPTOMATIC TTO (30%)
1-year survival: 10%

4. Terminal stage: patients with severe cancer-related symptoms or great tumour burden or Child–Pugh class C patients who are no candidate for liver transplantation. Such patients can only benefit from symptomatic treatment.

TREATMENT

Nowadays, only three approaches have proven curative capacities: surgical resection, percutaneous ablation and liver transplantation. Unfortunately these are applicable only in patients in whom an early-stage HCC is diagnosed. In well-selected patients who carry a tumour exceeding the criteria to indicate radical treatments, the only treatment that has shown benefit in terms of survival is transarterial chemoembolization (TACE). Disappointingly, despite the wide use of screening programmes in developed countries, almost 50% of patients cannot benefit from these treatments, and only best supportive treatment can be offered. Emerging therapies, mainly acting through molecular pathways or the immune system, are under evaluation.

CURATIVE TREATMENTS

Surgical resection

Surgical resection has for decades been the only curative treatment available until the development of percutaneous ablation and liver transplantation. It is still the treatment of choice in non-cirrhotic patients, in whom an extended hepatectomy is feasible. Contrarily, surgical resection is formally contraindicated in decompensated cirrhotic patients due to its unacceptably high rate of postoperative complications. Nowadays, surgical resection is reserved for patients presenting a solitary HCC with compensated cirrhosis, normal bilirubin levels and absence of clinically significant portal hypertension (platelet count $<100\,000/\text{mm}^2$, splenomegaly, oesophageal varices, and hepatic vein pressure gradient >10 mmHg)[4]. With the improvements in surgical techniques and anaesthesiology management, in these well-selected patients the perioperative mortality should be $<3\%$, the transfusion rate $<10\%$, and the 5-year survival rate $>50\%$[7]. Recognition of the molecular mechanisms that control liver regeneration or those that determine postoperative liver failure could allow better selection of surgical candidates and could potentially expand current proposals. The main drawback of resection is the high rate of recurrence during follow-up, which may reach 70% at 5 years[8]. From genomic investigations it has been determined that 60–70% of recurrences are intrahepatic metastases undetected at the time of resection, whereas 30–40% are *de-novo* tumours, which occur later, usually more than 2 years after resection. Some pathological characteristics, such as vascular invasion, satellites and poor differentiation degree, can predict recurrence, and we have proposed enlisting patients for liver transplantation before appearance of recurrence when analysis of the resected specimen reveals vascular invasion and/or additional satellites[9]. Several strategies to diminish the recurrence rate, such as immunotherapy with activated lymphocytes, chemoembolization, internal radiation, retinoids, and interferon administration, have been evaluated but, despite promising results with all of them, data lack enough strength to introduce them in routine clinical practice and their preventive value and survival impact need further investigation.

Liver transplantation

This option offers the theoretical advantage of being able to both resect the entire potentially tumour-bearing liver and eliminate cirrhosis. During the pioneering years the poor outcome obtained due to recruitment of candidates with far-advanced disease brought into debate the adequacy of orthotopic liver transplantation (OLT) as a treatment for HCC. The finding that patients with incidental asymptomatic tumours had the same outcome as patients with non-malignant disease encouraged reconsideration of liver transplantation as an efficient treatment in well-selected patients. In this way Mazzaferro and colleagues[10] reported in 1996 a 5-year survival rate of 70% and a recurrence rate below 15% when OLT was performed in patients with a single HCC $\leqslant 5$ cm or up to three nodules each <3 cm. Since then HCC has become a well-established transplantation indication. However, the scenario has changed radically. The shortage of donors has led to an increase in waiting times, with dropout rates of 20% and reduced survival according to the intention-to-treat principle[8]. Several strategies to improve survival after liver transplantation according to the intention-to-treat principle have been proposed. First, increasing the pool of donors by the use of living donors, domino/split liver transplantation, or high-risk donors. Living donor liver transplantation (LDLT) has become the most feasible alternative to cadaveric liver transplantation and a cost-effectiveness study has shown that it is beneficial when the waiting time is expected to exceed 7 months[11]. However, its low applicability (less than 20% of the patients evaluated for LDLT are finally transplanted) and concerns about the donors' risk and the postoperative complications, mainly of biliary origin, temper the potential benefit of this approach. Second, prioritizing those patients with high probability of dropout; however, there are still no good proposals for this issue. The MELD system, widely applied in the USA and Europe, has failed to properly prioritize HCC patients with preserved liver function. Finally, adjuvant therapies (such as percutaneous ablation and chemoembolization), aiming to reduce progression while waiting, are frequently used, although there is no robust scientific evidence of a survival benefit. A cost-effectiveness analysis using a Markov model suggested a survival improvement for neoadjuvant treatments when waiting times are expected to exceed 6 months[12]. Future approaches could be based on the development of new agents that, while being able to be safely applied in all patients, would prevent, or at least delay, tumour progression. The relationship between immunosuppressive regimens and cancer promotion is controversial, but some preliminary studies suggest that rapamycin, a m-TOR inhibitor, may reduce the probability of recurrence during follow-up[13].

Percutaneous ablation

This minimally invasive approach has demonstrated efficacy with low cost and low rate of adverse events, competing with surgical resection. Currently it has become the first-line treatment for patients affected with early HCC who are not candidates for surgical resection or OLT. The most widely accepted ablative therapies are percutaneous ethanol injection (PEI) and radiofrequency (RF)[14]; other less frequently employed modalities are cryotherapy, microwave therapy, and laser therapy.

PEI obtains its ablative capacity with the injection of absolute ethanol directly into the tumours through a fine needle under ultrasound guidance. It has low cost and is a well-tolerated procedure with few side-effects. This procedure can achieve complete necrosis in 70–80% of single tumours $\leqslant$3 cm, and in Child–Pugh A patients obtains a 5-year survival rate of 50%[14]. Worse results are obtained in tumours between 3 and 5 cm, due to the presence of intra-tumoral septa that prevent the spread of the injected ethanol, in which complete necrosis is achieved in less than 50%. RF produces thermal injury in tissue via an alternating electric current in the radiofrequency range (460–500 kHz) through a device (single or multiple cooled-tip electrodes or single electrodes associated with J-hook needles) placed by ultrasound guidance. This achieves complete necrosis in significantly fewer sessions and provides a better rate of responses in tumours larger than 3 cm[15]. The disadvantages with respect to PEI are the high cost of the equipment, and the higher rate and severity of adverse effects, with a reported complication rate of up to 12% and mortality of 0.5%[14]. In a recent RCT, RF was directly compared with PEI in 238 Child–Pugh A–B patients with up to three nodules $<$3 cm each. Four-year survival was significantly higher in RFA than in PEI (74% vs 57%) with similar incidences of adverse events[16]. Recurrence after successful percutaneous treatment is as frequent as after surgical resection, exceeding 70% at 5 years. There are no RCT that have properly compared surgical resection versus percutaneous ablation in early-stage HCC; however, cohort studies have not detected significant differences in survival between these two therapeutic modalities[17].

PALLIATIVE TREATMENTS

Patients with compensated cirrhosis and HCC exceeding the criteria of early stage, who have not yet presented cancer-related symptoms or vascular invasion or extrahepatic spread, correspond to the intermediate stage. Several therapeutic modalities in these patients have been evaluated through RCT; from all these studies only transarterial chemoembolization has been proved to significantly improve survival in well-selected candidates. Other therapeutic modalities such as internal radiation offer promising results, but impact on survival has not yet been proven.

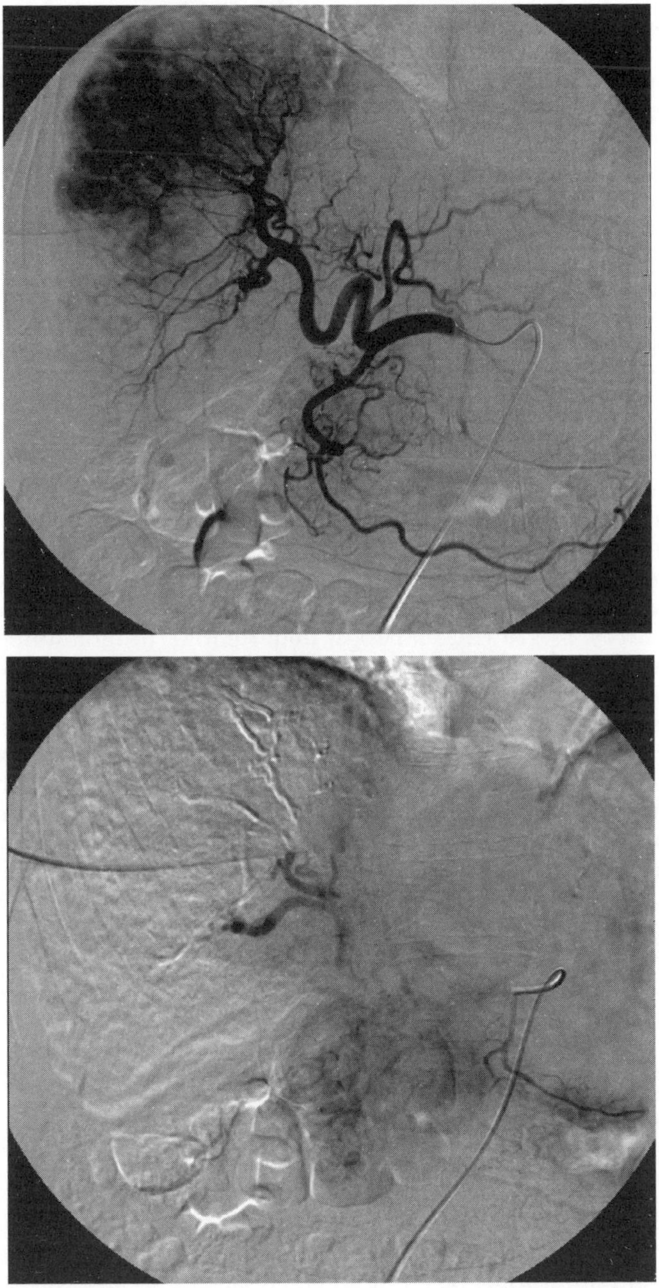

a

b

Figure 4 An example of transarterial chemoembolization. Figure 4a demonstrates the typical arterial hypervascularization of HCC on arteriography. Figure 4b illustrates the result after selective embolization with gelfoam

Transarterial chemoembolization (TACE)

The predominant arterial vascularization of HCC is the pathological basis for the effectiveness of TACE. The procedure consists in intra-arterial infusion of a chemotherapeutic agent (doxorubicin or cisplatin) suspended in a carrier substance, usually lipiodol (an oily contrast agent used for lymphographic studies that is selectively retained within the tumour), followed by a blockage of feeding arterial branches with an embolizing agent, commonly gelatin sponge prepared as 0.5–1 mm cubes (Figure 4). To avoid injury to the surrounding non-tumoral liver the embolization procedure should be as selective as possible. This procedure is not innocuous. The majority of patients suffer a postembolization syndrome, consisting of transient abdominal pain, ileus, and fever. Serious complications related to the procedure are infrequent (less than 10%) but may be severe, such as ischaemic cholecystitis, hepatic abscess and biliary strictures. Death may occur in 4% of cases. TACE should be reserved for patients with correct liver and renal function, preserved portal vein flow, normal clotting parameters and no extrahepatic spread. The efficacy of TACE in terms of inducing objective responses is clear, achieving partial or complete tumour necrosis in approximately 35–43% of cases[18]. However, this initial response is lost during follow-up, due to the tumour capacity to recover its vascularization through an intensive collateral formation, which is accompanied by re-growing capacity and tumour progression. In fact although this procedure has been used for decades, its efficacy in terms of survival improvement was not demonstrated until the recent publication of two RCT[18,19] and a cumulative meta-analysis[20].

Several topics are under investigation to increase the benefits of chemoembolization. Efforts are directed to finding the best embolization agent and chemotherapeutic drug, and discovering how to increase intratumoral concentrations of the drug. Encouraging results have been obtained in a recent phase II trial using calibrated drug-eluting beads, an embolizing device that slowly releases chemotherapy. This diminishes the passage of chemotherapy to systemic circulation (and thus, the adverse events), while the arterial blood flow is effectively obstructed[21]. The real impact in terms of survival benefit versus that of the usual gelfoam-based procedure should be evaluated in a larger RCT. Another issue is clarifying the best treatment scheme for repeat treatment, whether it should be undertaken at a fixed time or on disease progression after initial response. Finally, it is worth evaluating the use of adjuvant therapies to avoid, or at least delay, tumour progression after TACE. In this way, being aware of the key role of angiogenesis in the arterial collateral formation after the procedure, which permits tumour persistence and progression, angiogenesis blockade should increase the duration of treatment response and hopefully translate into survival benefit.

Intra-arterial chemotherapy

Intra-arterial chemotherapy consists in intra-arterial infusion of chemotherapeutic agents, with no occlusion of hepatic artery flow. It has been proposed for patients in whom portal thrombosis precludes hepatic artery flow occlusion. At present there is no robust scientific evidence of survival benefit, and its use is not recommended[22].

Internal radiation

Internal radiation relies on the delivery of high radiation doses to arterial-fed tumours through a transarterial approach, thereby sparing the normal liver parenchyma from damaging radiation. Promising results have been obtained in preliminary studies using yttrium-90 microspheres[23], and a large, multicentre RCT is awaited to evaluate the real survival benefit of internal radiation in unresectable HCC.

Systemic treatments

Several cytotoxic drugs and chemotherapeutic schemes have been evaluated in advanced HCC, but HCC has demonstrated to be chemoresistant, and their use is usually associated with an inadmissible rate of severe adverse effects. Therefore, systemic chemotherapy should be currently discouraged in cirrhotic patients with advanced HCC. Anti-oestrogenic drugs, such as tamoxifen, have been widely evaluated in advanced HCC. Despite promising initial results, tamoxifen has definitively shown no survival benefit[20]. Other substances, such as octreotide, antiandrogens, megestrol, and interferon associated with cytotoxic drugs and seocalcitol, have demonstrated no efficacy.

EMERGING TREATMENTS

Due to the absence of an effective treatment in advanced HCC, those patients affected with advanced HCC, and who are still in good physical condition regardless of the extensive tumour burden, are candidates to participate in RCT evaluating new therapies.

In recent years we have witnessed significant advances in knowledge of the molecular pathogenesis of HCC[24] and of the immune system response against HCC. This has led to the development of new strategies that act by activation of the immune system against the tumour (immunotherapy) or through some disrupted molecular pathways (biological therapies).

Immunotherapy

Advances in knowledge of the immunological response against cancer and its potential usefulness is supported by a Japanese study showing a significant recurrence reduction after surgery with adoptive immunotherapy[25]. Currently, active research in several fields, such as cytokines, anticancer antibodies and the key role of dendritic cells and $CD4^+$ and $CD25^+$ regulatory T cells, is ongoing. However, the majority of results are still preliminary[26].

Biological therapies

Several approaches are currently under evaluation[27].

1. Tyrosine kinase pathways: several agents act through the inhibition of tyrosine kinase pathways. For instance, Sorafenib (Bay 43-9006) blocks both raf kinase and VEGF, and has shown delay of progression in a preliminary phase 2 study, and the results of a large, multicentre, double-blinded, randomized controlled trial have been reported to be positive. If confirmed, this agent will become the standard of care for patients with advanced HCC.

2. Ubiquitin-proteasome pathway: this system is key in protein degradation of several proteins involved in cancer biology such as p53, p27, EGFR, pRb and TGF-β. Bortezomib, a potent, selective, and reversible inhibitor of the 26S proteasome used in multiple myeloma, has shown positive results in preliminary studies.

3. PI3Kinase/Akt/mTOR pathway: interleukin 2, insulin and insulin-like growth factors act through this pathway, leading to cell cycle progression. Overexpression of mTOR has been observed in about 15% of human HCC. Rapamycin, a mTOR kinase inhibitor used as an immunosuppressive agent, could have some anticancer activity[13], and its capability to reduce recurrence after liver transplantation is under evaluation.

4. Angiogenesis pathway: HCC growth depends particularly on the development of an intense net of arterial vascularization. Therefore, the blockade of factors with powerful angiogenic capacity, such as VEGF and FGF, is a major hope.

5. Apoptotic pathway: cancer cells develop several mechanisms to prevent cell death due to apoptosis. Inhibition of these mechanisms to preserve natural apoptosis could be a potential therapeutic target. In this way inhibitors of telomerase activity and survivin are good examples, and results of ongoing phase II trials are expected.

PREVENTION

Liver cancer is a neoplasm with well-defined risk factors. Most of the patients diagnosed with hepatocellular carcinoma present an underlying liver disease related to chronic viral infection (due to HCV or HBV infection) that in some areas is associated with aflatoxin intake, excessive alcohol consumption or with metabolic conditions such as haemochromatosis. This offers the possibility of being successful in primary prevention by avoiding the acquisition of the risk factors that will induce chronic liver damage and, ultimately, liver cancer. Vaccination against HBV has shown its efficacy in preventing the dissemination of this agent, and also in reducing the incidence of HCC. Hence

HCC is the first neoplasm that has been successfully prevented by vaccination[28]. There is no vaccination for HCV; thus dissemination of HCV infection is prevented solely by avoiding procedures involving potentially contaminated blood (uncontrolled transfusion, needle sharing, piercing, risky sex practices). Aflatoxin intake can be diminished by storing food in proper conditions, and there are attempts to increase its body clearance through pharmacological intervention, but as yet there is no proof of efficacy. Screening for haemochromatosis should detect disease at a pre-cirrhotic stage, and thus avoid progression to cirrhosis and HCC. In patients with chronic viral infection it is evident that successful antiviral therapy will prevent chronic inflammation with fibrosis deposition, and thus progression to cirrhosis. In this setting antiviral therapy should be considered effective in preventing HCC, or at least in diminishing the risk. However, if cirrhosis is already present, the impact of antiviral therapy or just chronic interferon administration is not well established despite several promising studies. Ongoing investigations (EPIC and HALT trials) should unequivocally establish whether such intervention is effective. However, in some patients antiviral treatment may not be feasible or effective and, in such patients, the sole option to diminish deaths related to cancer is to achieve early detection and to apply radical therapy.

SUMMARY

Despite major advances in the past decade, and the implementation of screening programmes to diagnose HCC at an early stage, only 30–40% of HCC patients will benefit from effective treatments: surgical resection, liver transplantation, percutaneous ablation and transarterial chemoembolization. Further progresses should be achieved in diagnosis, prognosis and treatment. In particular, prevention of the risk factors associated with cirrhosis and HCC, and development of new treatments, are mandatory. The ability to diagnose HCC at a very early stage, when it has not yet acquired a fully overt malignant phenotype, the improvement of tumour staging and the detection of new prognostic parameters that properly define the natural history of tumours should be the main goals in the immediate future. Accomplishment of all these expectations will come in part from genomic and proteomic research that is currently being conducted. Hopefully, in the next decade we will be able to change the nihilistic situation that currently characterizes the management of hepatocellular carcinoma, particular at an advanced stage.

References

1. El-Serag HB, Mason AC. Rising incidence of hepatocellular carcinoma in the United States. N Engl J Med. 1999;340:745–50.
2. Sangiovanni A, Prati GM, Fasani P et al. The natural history of compensated cirrhosis due to hepatitis C virus: a 17-year cohort study of 214 patients. Hepatology. 2006;43:1303–10.
3. Parkin DM, Bray F, Ferlay J, Pisani P. Global cancer statistics, 2002. CA Cancer J Clin. 2005;55:74–108.
4. Bruix J, Sherman M. Management of hepatocellular carcinoma. Hepatology. 2005;42: 1208–36.

5. Sala M, Forner A, Varela M, Bruix J. Prognostic prediction in patients with hepatocellular carcinoma. Semin Liver Dis. 2005;25:171–80.
6. Llovet JM, Bru C, Bruix J. Prognosis of hepatocellular carcinoma: the BCLC staging classification. Semin Liver Dis. 1999;19:329–38.
7. Llovet JM, Burroughs A, Bruix J. Hepatocellular carcinoma. Lancet. 2003;362:1907–17.
8. Llovet JM, Fuster J, Bruix J. Intention-to-treat analysis of surgical treatment for early hepatocellular carcinoma: resection versus transplantation. Hepatology. 1999;30:1434–40.
9. Sala M, Fuster J, Llovet JM et al. High pathological risk of recurrence after surgical resection for hepatocellular carcinoma: an indication for salvage liver transplantation. Liver Transplant. 2004;10:1294–300.
10. Mazzaferro V, Regalia E, Doci R et al. Liver transplantation for the treatment of small hepatocellular carcinomas in patients with cirrhosis. N Engl J Med. 1996;334:693–9.
11. Sarasin FP, Majno PE, Llovet JM, Bruix J, Mentha G, Hadengue A. Living donor liver transplantation for early hepatocellular carcinoma: a life-expectancy and cost-effectiveness perspective. Hepatology. 2001;33:1073–9.
12. Llovet JM, Mas X, Aponte JJ et al. Cost effectiveness of adjuvant therapy for hepatocellular carcinoma during the waiting list for liver transplantation. Gut. 2002;50: 123–8.
13. Sahin F, Kannangai R, Adegbola O, Wang J, Su G, Torbenson M. mTOR and P70 S6 kinase expression in primary liver neoplasms. Clin Cancer Res. 2004;10:8421–5.
14. Beaugrand M, N'Kontchou G, Seror O, Ganne N, Trinchet JC. Local/regional and systemic treatments of hepatocellular carcinoma. Semin Liver Dis. 2005;25:201–11.
15. Lencioni R, Cioni D, Crocetti L et al. Early-stage hepatocellular carcinoma in patients with cirrhosis: long-term results of percutaneous image-guided radiofrequency ablation. Radiology. 2005;234:961–7.
16. Shiina S, Teratani T, Obi S et al. A randomized controlled trial of radiofrequency ablation with ethanol injection for small hepatocellular carcinoma. Gastroenterology. 2005;129:122–30.
17. Castells A, Bruix J, Bru C et al. Treatment of small hepatocellular carcinoma in cirrhotic patients: a cohort study comparing surgical resection and percutaneous ethanol injection. Hepatology. 1993;18:1121–6.
18. Llovet JM, Real MI, Montana X, Planas R, Coll S, Aponte J et al. Arterial embolisation or chemoembolisation versus symptomatic treatment in patients with unresectable hepatocellular carcinoma: a randomised controlled trial. Lancet. 2002;359:1734-9.
19. Lo CM, Ngan H, Tso WK et al. Randomized controlled trial of transarterial lipiodol chemoembolization for unresectable hepatocellular carcinoma. Hepatology. 2002;35:1164–71.
20. Llovet JM, Bruix J. Systematic review of randomized trials for unresectable hepatocellular carcinoma: chemoembolization improves survival. Hepatology. 2003;37:429–42.
21. Varela M, Real M, Burrel M et al. Chemoembolization of hepatocellular carcinoma with drug eluting beads. Efficacy and doxorubicin pharmacokinetics. J Hepatol. 2007 (In press).
22. Chung YH, Song IH, Song BC et al. Combined therapy consisting of intraarterial cisplatin infusion and systemic interferon-alpha for hepatocellular carcinoma patients with major portal vein thrombosis or distant metastasis. Cancer. 2000;88:1986–91.
23. Salem R, Lewandowski RJ, Atassi B et al. Treatment of unresectable hepatocellular carcinoma with use of 90Y microspheres (TheraSphere): safety, tumor response, and survival. J Vasc Interv Radiol. 2005;16:1627–39.
24. Thorgeirsson SS, Grisham JW. Molecular pathogenesis of human hepatocellular carcinoma. Nat Genet. 2002;31:339–46.
25. Takayama T, Sekine T, Makuuchi M et al. Adoptive immunotherapy to lower postsurgical recurrence rates of hepatocellular carcinoma: a randomised trial. Lancet. 2000;356:802–7.
26. Greten TF, Manns MP, Korangy F. Immunotherapy of hepatocellular carcinoma. J Hepatol. 2006;45:868–78.
27. Roberts LR, Gores GJ. Hepatocellular carcinoma: molecular pathways and new therapeutic targets. Semin Liver Dis. 2005;25:212–25.
28. Chang MH, Chen CJ, Lai MS et al. Universal hepatitis B vaccination in Taiwan and the incidence of hepatocellular carcinoma in children. Taiwan Childhood Hepatoma Study Group. N Engl J Med. 1997;336:1855–9.

Index

Falk Symposium Series

82B. Paumgartner G, Beuers U, eds.: *Bile Acids in Liver Diseases.* International Falk Workshop. 1995 ISBN 0-7923-8891-7

83. Dobrilla G, Felder M, de Pretis G, eds.: *Advances in Hepatobiliary and Pancreatic Diseases: Special Clinical Topics.* Falk Symposium 83. 1995. ISBN 0-7923-8892-5

84. Fromm H, Leuschner U, eds.: *Bile Acids – Cholestasis – Gallstones: Advances in Basic and Clinical Bile Acid Research.* Falk Symposium 84. 1995 ISBN 0-7923-8893-3

85. Tytgat GNJ, Bartelsman JFWM, van Deventer SJH, eds.: *Inflammatory Bowel Diseases.* Falk Symposium 85. 1995 ISBN 0-7923-8894-1

86. Berg PA, Leuschner U, eds.: *Bile Acids and Immunology.* Falk Symposium 86. 1996 ISBN 0-7923-8700-7

87. Schmid R, Bianchi L, Blum HE, Gerok W, Maier KP, Stalder GA, eds.: *Acute and Chronic Liver Diseases: Molecular Biology and Clinics.* Falk Symposium 87. 1996 ISBN 0-7923-8701-5

88. Blum HE, Wu GY, Wu CH, eds.: *Molecular Diagnosis and Gene Therapy.* Falk Symposium 88. 1996 ISBN 0-7923-8702-3

88B. Poupon RE, Reichen J, eds.: *Surrogate Markers to Assess Efficacy of TReatment in Chronic Liver Diseases.* International Falk Workshop. 1996 ISBN 0-7923-8705-8

89. Reyes HB, Leuschner U, Arias IM, eds.: *Pregnancy, Sex Hormones and the Liver.* Falk Symposium 89. 1996 ISBN 0-7923-8704-X

89B. Broelsch CE, Burdelski M, Rogiers X, eds.: *Cholestatic Liver Diseases in Children and Adults.* International Falk Workshop. 1996 ISBN 0-7923-8710-4

90. Lam S-K, Paumgartner P, Wang B, eds.: *Update on Hepatobiliary Diseases 1996.* Falk Symposium 90. 1996 ISBN 0-7923-8715-5

91. Hadziselimovic F, Herzog B, eds.: *Inflammatory Bowel Diseases and Chronic Recurrent Abdominal Pain.* Falk Symposium 91. 1996 ISBN 0-7923-8722-8

91B. Alvaro D, Benedetti A, Strazzabosco M, eds.: *Vanishing Bile Duct Syndrome – Pathophysiology and Treatment.* International Falk Workshop. 1996 ISBN 0-7923-8721-X

92. Gerok W, Loginov AS, Pokrowskij VI, eds.: *New Trends in Hepatology 1996.* Falk Symposium 92. 1997 ISBN 0-7923-8723-6

93. Paumgartner G, Stiehl A, Gerok W, eds.: *Bile Acids in Hepatobiliary Diseases – Basic Research and Clinical Application.* Falk Symposium 93. 1997 ISBN 0-7923-8725-2

94. Halter F, Winton D, Wright NA, eds.: *The Gut as a Model in Cell and Molecular Biology.* Falk Symposium 94. 1997 ISBN 0-7923-8726-0

94B. Kruse-Jarres JD, Schölmerich J, eds.: *Zinc and Diseases of the Digestive Tract.* International Falk Workshop. 1997 ISBN 0-7923-8724-4

95. Ewe K, Eckardt VF, Enck P, eds.: *Constipation and Anorectal Insufficiency.* Falk Symposium 95. 1997 ISBN 0-7923-8727-9

96. Andus T, Goebell H, Layer P, Schölmerich J, eds.: *Inflammatory Bowel Disease – from Bench to Bedside.* Falk Symposium 96. 1997 ISBN 0-7923-8728-7

97. Campieri M, Bianchi-Porro G, Fiocchi C, Schölmerich J, eds. *Clinical Challenges in Inflammatory Bowel Diseases: Diagnosis, Prognosis and Treatment.* Falk Symposium 97. 1998 ISBN 0-7923-8733-3

98. Lembcke B, Kruis W, Sartor RB, eds. *Systemic Manifestations of IBD: The Pending Challenge for Subtle Diagnosis and Treatment.* Falk Symposium 98. 1998 ISBN 0-7923-8734-1

99. Goebell H, Holtmann G, Talley NJ, eds. *Functional Dyspepsia and Irritable Bowel Syndrome: Concepts and Controversies.* Falk Symposium 99. 1998 ISBN 0-7923-8735-X

Falk Symposium Series

100. Blum HE, Bode Ch, Bode JCh, Sartor RB, eds. *Gut and the Liver.* Falk Symposium 100. 1998 ISBN 0-7923-8736-8
101. Rachmilewitz D, ed. *V International Symposium on Inflammatory Bowel Diseases.* Falk Symposium 101. 1998 ISBN 0-7923-8743-0
102. Manns MP, Boyer JL, Jansen PLM, Reichen J, eds. *Cholestatic Liver Diseases.* Falk Symposium 102. 1998 ISBN 0-7923-8746-5
102B. Manns MP, Chapman RW, Stiehl A, Wiesner R, eds. *Primary Sclerosing Cholangitis.* International Falk Workshop. 1998. ISBN 0-7923-8745-7
103. Häussinger D, Jungermann K, eds. *Liver and Nervous System.* Falk Symposium 102. 1998 ISBN 0-7924-8742-2
103B. Häussinger D, Heinrich PC, eds. *Signalling in the Liver.* International Falk Workshop. 1998 ISBN 0-7923-8744-9
103C. Fleig W, ed. *Normal and Malignant Liver Cell Growth.* International Falk Workshop. 1998 ISBN 0-7923-8748-1
104. Stallmach A, Zeitz M, Strober W, MacDonald TT, Lochs H, eds. *Induction and Modulation of Gastrointestinal Inflammation.* Falk Symposium 104. 1998 ISBN 0-7923-8747-3
105. Emmrich J, Liebe S, Stange EF, eds. *Innovative Concepts in Inflammatory Bowel Diseases.* Falk Symposium 105. 1999 ISBN 0-7923-8749-X
106. Rutgeerts P, Colombel J-F, Hanauer SB, Schölmerich J, Tytgat GNJ, van Gossum A, eds. *Advances in Inflammatory Bowel Diseases.* Falk Symposium 106. 1999 ISBN 0-7923-8750-3
107. Špičák J, Boyer J, Gilat T, Kotrlik K, Marećek Z, Paumgartner G, eds. *Diseases of the Liver and the Bile Ducts – New Aspects and Clinical Implications.* Falk Symposium 107. 1999 ISBN 0-7923-8751-1
108. Paumgartner G, Stiehl A, Gerok W, Keppler D, Leuschner U, eds. *Bile Acids and Cholestasis.* Falk Symposium 108. 1999 ISBN 0-7923-8752-X
109. Schmiegel W, Schölmerich J, eds. *Colorectal Cancer – Molecular Mechanisms, Premalignant State and its Prevention.* Falk Symposium 109. 1999 ISBN 0-7923-8753-8
110. Domschke W, Stoll R, Brasitus TA, Kagnoff MF, eds. *Intestinal Mucosa and its Diseases – Pathophysiology and Clinics.* Falk Symposium 110. 1999 ISBN 0-7923-8754-6
110B. Northfield TC, Ahmed HA, Jazwari RP, Zentler-Munro PL, eds. *Bile Acids in Hepatobiliary Disease.* Falk Workshop. 2000 ISBN 0-7923-8755-4
111. Rogler G, Kullmann F, Rutgeerts P, Sartor RB, Schölmerich J, eds. *IBD at the End of its First Century.* Falk Symposium 111. 2000 ISBN 0-7923-8756-2
112. Krammer HJ, Singer MV, eds. *Neurogastroenterology: From the Basics to the Clinics.* Falk Symposium 112. 2000 ISBN 0-7923-8757-0
113. Andus T, Rogler G, Schlottmann K, Frick E, Adler G, Schmiegel W, Zeitz M, Schölmerich J, eds. *Cytokines and Cell Homeostasis in the Gastrointestinal Tract.* Falk Symposium 113. 2000 ISBN 0-7923-8758-9
114. Manns MP, Paumgartner G, Leuschner U, eds. *Immunology and Liver.* Falk Symposium 114. 2000 ISBN 0-7923-8759-7
115. Boyer JL, Blum HE, Maier K-P, Sauerbruch T, Stalder GA, eds. *Liver Cirrhosis and its Development.* Falk Symposium 115. 2000 ISBN 0-7923-8760-0
116. Riemann JF, Neuhaus H, eds. *Interventional Endoscopy in Hepatology.* Falk Symposium 116. 2000 ISBN 0-7923-8761-9

Falk Symposium Series

116A. Dienes HP, Schirmacher P, Brechot C, Okuda K, eds. *Chronic Hepatitis: New Concepts of Pathogenesis, Diagnosis and Treatment*. Falk Workshop. 2000 ISBN 0-7923-8763-5

117. Gerbes AL, Beuers U, Jüngst D, Pape GR, Sackmann M, Sauerbruch T, eds. *Hepatology 2000 – Symposium in Honour of Gustav Paumgartner*. Falk Symposium 117. 2000 ISBN 0-7923-8765-1

117A. Acalovschi M, Paumgartner G, eds. *Hepatobiliary Diseases: Cholestasis and Gallstones*. Falk Workshop. 2000 ISBN 0-7923-8770-8

118. Frühmorgen P, Bruch H-P, eds. *Non-Neoplastic Diseases of the Anorectum*. Falk Symposium 118. 2001 ISBN 0-7923-8766-X

119. Fellermann K, Jewell DP, Sandborn WJ, Schölmerich J, Stange EF, eds. *Immunosuppression in Inflammatory Bowel Diseases – Standards, New Developments, Future Trends*. Falk Symposium 119. 2001 ISBN 0-7923-8767-8

120. van Berge Henegouwen GP, Keppler D, Leuschner U, Paumgartner G, Stiehl A, eds. *Biology of Bile Acids in Health and Disease*. Falk Symposium 120. 2001 ISBN 0-7923-8768-6

121. Leuschner U, James OFW, Dancygier H, eds. *Steatohepatitis (NASH and ASH)*. Falk Symposium 121. 2001 ISBN 0-7923-8769-4

121A. Matern S, Boyer JL, Keppler D, Meier-Abt PJ, eds. *Hepatobiliary Transport: From Bench to Bedside*. Falk Workshop. 2001 ISBN 0-7923-8771-6

122. Campieri M, Fiocchi C, Hanauer SB, Jewell DP, Rachmilewitz R, Schölmerich J, eds. *Inflammatory Bowel Disease – A Clinical Case Approach to Pathophysiology, Diagnosis, and Treatment*. Falk Symposium 122. 2002 ISBN 0-7923-8772-4

123. Rachmilewitz D, Modigliani R, Podolsky DK, Sachar DB, Tozun N, eds. *VI International Symposium on Inflammatory Bowel Diseases*. Falk Symposium 123. 2002 ISBN 0-7923-8773-2

124. Hagenmüller F, Manns MP, Musmann H-G, Riemann JF, eds. *Medical Imaging in Gastroenterology and Hepatology*. Falk Symposium 124. 2002 ISBN 0-7923-8774-0

125. Gressner AM, Heinrich PC, Matern S, eds. *Cytokines in Liver Injury and Repair*. Falk Symposium 125. 2002 ISBN 0-7923-8775-9

126. Gupta S, Jansen PLM, Klempnauer J, Manns MP, eds. *Hepatocyte Transplantation*. Falk Symposium 126. 2002 ISBN 0-7923-8776-7

127. Hadziselimovic F, ed. *Autoimmune Diseases in Paediatric Gastroenterology*. Falk Symposium 127. 2002 ISBN 0-7923-8778-3

127A. Berr F, Bruix J, Hauss J, Wands J, Wittekind Ch, eds. *Malignant Liver Tumours: Basic Concepts and Clinical Management*. Falk Workshop. 2002 ISBN 0-7923-8779-1

128. Scheppach W, Scheurlen M, eds. *Exogenous Factors in Colonic Carcinogenesis*. Falk Symposium 128. 2002 ISBN 0-7923-8780-5

129. Paumgartner G, Keppler D, Leuschner U, Stiehl A, eds. *Bile Acids: From Genomics to Disease and Therapy*. Falk Symposium 129. 2002 ISBN 0-7923-8781-3

129A. Leuschner U, Berg PA, Holtmeier J, eds. *Bile Acids and Pregnancy*. Falk Workshop. 2002 ISBN 0-7923-8782-1

130. Holtmann G, Talley NJ, eds. *Gastrointestinal Inflammation and Disturbed Gut Function: The Challenge of New Concepts*. Falk Symposium 130. 2003 ISBN 0-7923-8783-X

131. Herfarth H, Feagan BJ, Folsch UR, Schölmerich J, Vatn MH, Zeitz M, eds. *Targets of Treatment in Chronic Inflammatory Bowel Diseases*. Falk Symposium 131. 2003 ISBN 0-7923-8784-8

Falk Symposium Series

132. Galle PR, Gerken G, Schmidt WE, Wiedenmann B, eds. *Disease Progression and Carcinogenesis in the Gastrointestinal Tract*. Falk Symposium 132. 2003 ISBN 0-7923-8785-6
132A. Staritz M, Adler G, Knuth A, Schmiegel W, Schmoll H-J, eds. *Side-effects of Chemotherapy on the Gastrointestinal Tract*. Falk Workshop. 2003 ISBN 0-7923-8791-0
132B. Reutter W, Schuppan D, Tauber R, Zeitz M, eds. *Cell Adhesion Molecules in Health and Disease.* Falk Workshop. 2003 ISBN 0-7923-8786-4
133. Duchmann R, Blumberg R, Neurath M, Schölmerich J, Strober W, Zeitz M. *Mechanisms of Intestinal Inflammation: Implications for Therapeutic Intervention in IBD.* Falk Symposium 133. 2004 ISBN 0-7923-8787-2
134. Dignass A, Lochs H, Stange E. *Trends and Controversies in IBD – Evidence-Based Approach or Individual Management?* Falk Symposium 134. 2004 ISBN 0-7923-8788-0
134A. Dignass A, Gross HJ, Buhr V, James OFW. *Topical Steroids in Gastroenterology and Hepatology.* Falk Workshop. 2004 ISBN 0-7923-8789-9
135. Lukáš M, Manns MP, Špičák J, Stange EF, eds. *Immunological Diseases of Liver and Gut.* Falk Symposium 135. 2004 ISBN 0-7923-8792-9
136. Leuschner U, Broomé U, Stiehl A, eds. *Cholestatic Liver Diseases: Therapeutic Options and Perspectives.* Falk Symposium 136. 2004 ISBN 0-7923-8793-7
137. Blum HE, Maier KP, Rodés J, Sauerbruch T, eds. *Liver Diseases: Advances in Treatment and Prevention.* Falk Symposium 137. 2004 ISBN 0-7923-8794-5
138. Blum HE, Manns MP, eds. *State of the Art of Hepatology: Molecular and Cell Biology.* Falk Symposium 138. 2004 ISBN 0-7923-8795-3
138A. Hayashi N, Manns MP, eds. *Prevention of Progression in Chronic Liver Disease: An Update on SNMC (Stronger Neo-Minophagen C).* Falk Workshop. 2004 ISBN 0-7923-8796-1
139. Adler G, Blum HE, Fuchs M, Stange EF, eds. *Gallstones: Pathogenesis and Treatment.* Falk Symposium 139. 2004 ISBN 0-7923-8798-8
140. Colombel J-F, Gasché C, Schölmerich J, Vucelic C, eds. *Inflammatory Bowel Disease: Translation from Basic Research to Clinical Practice.* Falk Symposium 140. 2005. ISBN 1-4020-2847-4
141. Paumgartner G, Keppler D, Leuschner U, Stiehl A, eds. *Bile Acid Biology and its Therapeutic Implications.* Falk Symposium 141. 2005 ISBN 1-4020-2893-8
142. Dienes H-P, Leuschner U, Lohse AW, Manns MP, eds. *Autoimmune Liver Disease.* Falk Symposium 142. 2005 ISBN 1-4020-2894-6
143. Ammann RW, Büchler MW, Adler G, DiMagno EP, Sarner M, eds. *Pancreatitis: Advances in Pathobiology, Diagnosis and Treatment.* Falk Symposium 143. 2005 ISBN 1-4020-2895-4
144. Adler G, Blum AL, Blum HE, Leuschner U, Manns MP, Mössner J, Sartor RB, Schölmerich J, eds. *Gastroenterology Yesterday – Today – Tomorrow: A Review and Preview.* Falk Symposium 144. 2005 ISBN 1-4020-2896-2
145. Henne-Bruns D, Buttenschön K, Fuchs M, Lohse AW, eds. *Artificial Liver Support.* Falk Symposium 145. 2005 ISBN 1-4020-3239-0
146. Blumberg RS, Gangl A, Manns MP, Tilg H, Zeitz M, eds. *Gut–Liver Interactions: Basic and Clinical Concepts.* Falk Symposium 146. 2005 ISBN 1-4020-4143-8
147. Jewell DP, Colombel JF, Peña AS, Tromm A, Warren BS, eds. *Colitis: Diagnosis and Therapeutic Strategies.* Falk Symposium 147. 2006 ISBN 1-4020-4315-5

Falk Symposium Series

148. Kruis W, Forbes A, Jauch K-W, Kreis ME, Wexner SD, eds. *Diverticular Disease: Emerging Evidence in a Common Condition.* Falk Symposium 148. 2006 ISBN 1-4020- 4317-1
149. van Cutsem E, Rustgi AK, Schmiegel W, Zeitz M, eds. *Highlights in Gastrointestinal Oncology.* Falk Symposium 149. 2006. ISBN 1-4020-5108-5
150. Galle PR, Gerken G, Schmidt WE, Wiedenmann B, eds. *Disease Progression and Disease Prevention in Hepatology and Gastroenterology.* Falk Symposium 150. 2006 ISBN 1-4020-5109-3
151. Fraser A, Gibson PR, Hibi T, Qian J-M, Schölmerich, eds. *Emerging Issues in Inflammatory Bowel Disease.* Falk Symposium 151. 2006 ISBN-13 978-1-4020-5701-4
152. Not published.
153. Dignass A, Rachmilewitz D, Stange E-F, Weinstock JV, eds. *Immunoregulation in Inflammatory Bowel Diseases – Current Understanding and Innovation.* Falk Symposium 153. 2007 ISBN-13 978-1-4020-5888-2
154. Adler G, Fiocchi C, Lazebnik LB, Vorobiev GI, eds. *Inflammatory Bowel Disease – Diagnostic and Therapeutic Strategies.* Falk Symposium 154. 2007 ISBN-13 978-1-4020-6115-8
155. Keppler D, Beuers U, Leuschner U, Stiehl A, Trauner M, Paumgartner G, eds. *Bile Acids: Biological Actions and Clinical Relevance.* Falk Symposium 155. 2007 ISBN 978-1-4020-6251-3
156. Blum HE, Cox DW, Häussinger D, Jansen PLM, Kullak-Ublick GA, eds. *Genetics in Liver Diseases.* Falk Symposium 156. 2007 ISBN 978-14020-6393-0
157. Diehl AM, Hayashi N, Manns MP, Sauerbruch T, eds. *Chronic Hepatitis: Metabolic, Cholestatic, Viral and Autoimmune.* Falk Symposium 157. 2007 ISBN 978-1-4020-6522-4